Camel Meat and Meat Products

Camel Meat and Meat Products

Edited by

I.T. Kadim

Department of Animal & Veterinary Sciences, College of Agricultural and Marine Sciences, Sultan Qaboos University, Sultanate of Oman

O. Mahgoub

Department of Animal & Veterinary Sciences, College of Agricultural and Marine Sciences, Sultan Qaboos University, Sultanate of Oman

B. Faye

FAO Consultant, Camel and Range Research Centre, Al-Jouf-Sakaka, Saudi Arabia

M.M. Farouk

Food Assurance and Meat Quality, Food and Biobased Products, AgResearch Limited, Ruakura Research Centre, Hamilton, New Zealand

www.cabi.org

CABI is a trading name of CAB International

CABI
Nosworthy Way
Wallingford
Oxfordshire OX10 8DE
UK

Tel: +44 (0)1491 832111
Fax: +44 (0)1491 833508
E-mail: info@cabi.org
Website: www.cabi.org

CABI
38 Chauncey Street
Suite 1002
Bostan, MA 02111
USA

T: +1 800 552 3083 (toll free)
T: +1 (0)617 395 4051
E-mail: cabi-nao@cabi.org

A catalogue record for this book is available from the British Library, London, UK.

Library of Congress Cataloging-in-Publication Data

Camel meat and meat products / edited by I.T. Kadim ... [et al.].
p. cm.
Includes bibliographical references and index.
ISBN 978-1-78064-101-0 (alk. paper)
1. Camels. 2. Camel meat. I. Kadim, I. T. (Isam T.)

TX556.C35C35 2012
636.2'95--dc23

2012013382

ISBN: 978 1 78064 101 0

Commissioning editor: Sarah Hulbert
Editorial assistant: Alexandra Lainsbury
Production editor: Tracy Head

Typeset by SPi, Pondicherry, India.
Printed and bound in the UK by MPG Books Group.

Contents

Contributors

R. Al Jassim, School of Agriculture and Food Sciences, Faculty of Science, The University of Queensland, Gatton Campus, Gatton, QLD 4343, Australia. E-mail: r.aljassim@uq.edu.au

S. Al-Mugheiry, Muscat Municipality, PO Box 361, Madinat Qaboos 115, Sultanate of Oman. E-mail: salem_20@hotmail.co.uk

S. Al-Zadjali, Muscat Municipality, PO Box 361, Madinat Qaboos 115, Sultanate of Oman. E-mail: alzadjali@hotmail.com

A.E. Bekhit, Department of Food Science, Division of Sciences, University of Otago, PO Box 56, Dunedin, New Zealand. E-mail: aladin.bekhit@otago.ac.nz

A. Carne, Department of Biochemistry, Otago School of Medical Sciences, University of Otago, PO Box 56, Dunedin, New Zealand. E-mail: alan.carne@otago.ac.nz

M.M. Farouk, Food Assurance and Meat Quality, Food and Biobased Products, AgResearch Limited, Ruakura Research Centre, East Street, Private Bag 3123, Hamilton, New Zealand. E-mail: mustafa.farouk@agresearch.co.nz

B. Faye, FAO Consultant, Camel and Range Research Centre, PO Box 322, Al-Jouf-Sakaka, Saudi Arabia. E-mail: bernard.faye@cirad.fr

J. Hogan, 81 Boscombe Road, Brookfield, QLD 4069, Australia. E-mail: jim_hogan81@hotmail.com

D.L. Hopkins, NSW Department of Primary Industries, Centre for Red Meat and Sheep Development, PO Box 129, Cowra, NSW 2794, Australia. E-mail: david.hopkins@dpi.nsw.gov.au

I.T. Kadim, Department of Animal and Veterinary Sciences, College of Agricultural and Marine Sciences, Sultan Qaboos University, PO Box 34, Al-Khod 123, Muscat, Sultanate of Oman. E-mail: isam@squ.edu.om

O. Mahgoub, Department of Animal and Veterinary Sciences, College of Agricultural and Marine Sciences, Sultan Qaboos University, PO Box 34, Al-Khod 123, Muscat, Sultanate of Oman. E-mail: osmahgob@squ.edu.om

M. Mbaga, Department of Natural Resource Economics, College of Agricultural and Marine Sciences, Sultan Qaboos University, PO Box 34 Al-Khod 123, Muscat, Sultanate of Oman. E-mail: msafiri@squ.edu.om

H.J. Swatland, Ontario Agricultural College, Department of Animal and Poultry Science, University of Guelph, Canada. E-mail: swatland@uoguelph.ca

M.H. Tageldin, Department of Animal and Veterinary Sciences, College of Agricultural and Marine Sciences, Sultan Qaboos University, PO Box 34, Al-Khod 123, Muscat, Sultanate of Oman. E-mail: tage@squ.edu.om

Preface

This *Camel Meat and Meat Products* book is intended as a foundation text with an emphasis on general principles and practical applications, as well as more advanced aspects of camel meat sciences to a wide range of scientists and professionals who are concerned with camel meat research and education. It is also designed to serve as a reference for extension personnel working with camel herders, food scientists and biology, animal science and veterinary technology students. The descriptions and illustrations of the 14 chapters are based on material prepared by several researchers between 1950 and 2012.

Camels were domesticated and developed approximately 5000 years ago and throughout those years have played an integral role in the daily life of the camel owner. They are distributed in African and Asian arid and semi-arid areas, where other livestock farming systems cannot be easily implemented. According to the FAO statistics, there are more than 25 million camels worldwide today, with the dromedary accounting for 95%, and 80% of them are located in Africa. Somalia and the Sudan together have more than 10 million camels. With its remarkable anatomy and physiology, the camel has several advantages over other domesticated animals in terms of adaptation to a harsh environment. The dromedary camel's morphological features make them well adapted to survive in the difficult and drought-stricken arid, semi-arid and mountainous regions. Although camels have been neglected as an economically productive animal or as an animal with the potential for food production, camel products are becoming increasingly available in many countries for economical and health reasons. Moreover, the camel is envisaged to become an important source of food both in drought areas of the world, where famine occurs periodically, as well as in the growing cities of the arid countries where the urban demand for camel products is increasing.

Although the camel population is growing, lack of effort to improve camel productivity is still the main constraint for developing marketable products for different parts of the world. With an increasing human population and decreasing animal products, there is an urgent need to optimize the utilization of semi-arid and arid rangelands through appropriate camel production. The camel, in general, is considered a less than conventional source of meat compared with other domestic animals, but its large body mass and high quality of lean meat gives it an advantage as a meat producer. Interest in camel research has been growing in the past two decades, with most scientific publications covering veterinary aspects such as anatomy, physiology and diseases, but carcass and meat quality characteristics received little attention.

Camels have an important role as meat producers because of the versatile role they play as a symbol of social prestige that has recently declined. Approximately 250,000 camels are slaughtered annually in different countries. About 50% of the camels slaughtered are young males aged around 4 years, but their market has not yet developed well locally in spite of the existing regional camel market. Camel meat is mainly exported from the Horn of Africa to the Arabian Peninsula and a second market extends from western Africa to northern countries of Africa. Camel meat valorization through local or regional markets would be a strong opportunity for the integration of pastoralists into the market.

Camel meat is a good source of high-quality protein with less fat, less cholesterol and a relatively higher amount of polyunsaturated fatty acids compared with other meat animals. Camel meat is also used as a remedy for diseases such as hyperacidity, hypertension, pneumonia and respiratory disease according to certain beliefs. Furthermore, camel meat is considered as having significant health benefits and has been endorsed by the Australian National Heart Foundation.

With the growing demand for meat in developing countries, there is a need for more research to highlight the potential and contribution of the camel in sustainable food production. Despite centuries of interaction with the camel, there is still a scarcity of available information pertaining to many aspects of the productivity of this unique animal. In the present health-conscious era, the consumption of camel meat seems to have strong points in its favour. However, despite all the benefits associated with camel production, camels still face challenges in their natural environment such as diseases, drought and predation, which expose the pastoralist to risks of losing their sources of livelihood.

This book consists of 14 chapters with the objective of providing basic information of what will be a continuing process of scientific work to understand the meat production of camels. Studying the basic live weight, growth and development, carcass and meat quality characteristics are necessary elements to understand camel meat production. This information could be utilized by government bodies and international and national development agencies interested in developing markets for camel meat and professionals concerned with camel meat research and education. We have great pleasure in acknowledging the hard work done by many researchers whose published information is used in this book. Their efforts have been amply acknowledged in the text, tables and figures. Our sincere wish is that this book will be of use to all meat scientists, students and those in the camel meat industry who are intrigued by the legendary qualities and modern possibilities and usefulness of this remarkable creature.

The editors would like to thank CABI for publishing the book and sincerely acknowledge the contributions of the authors and their collaborators. We would also like to thank everybody who has supported research regarding camel meat production and quality including camel owners, technical and research staff and students worldwide.

I.T. Kadim
O. Mahgoub
B. Faye
M.M. Farouk

1 Classification, History and Distribution of the Camel

Bernard Faye
FAO Consultant, Camel and Range Research Centre, Al-Jouf-Sakaka, Saudi Arabia

1.1 Taxonomy

Camelidae is a family belonging to the order Artiodactyla and to the sub-order Tylopoda. Artiodactyla comprises three sub-orders: the suiforms (notably Suidae family), the ruminantia (notably Bovidae family) and the tylopodes, which have a padded foot. Camelidae is the only family in this sub-order. Therefore, despite being ruminating animals, the camelids are not in the ruminant order. The family Camelidae comprises two main types (large and small camelids) distributed into three genera: *Camelus*, *Lama* and *Vicugna* (Fig. 1.1).

The small camelids originated from the Andes Mountains of South America and include two domestic species (llama and alpaca) and two wild species (guanaco in genus *Lama*, and vicuna in genus *Vicugna*). The large camelids are represented by two domesticated species, the one-humped camel (dromedary) and the two-humped camel (Bactrian). The dromedary lives in the hot arid lands of northern Africa and eastern Asia, and the Bactrian in the cold steppes and deserts in Central Asia. A new large camelid has been described a few times. It is a wild species living in very remote areas between Mongolia and China, and is called the Tartary camel (*Camelus bactrianus ferus*); it has been distinguished from the domestic double-humped camel (Ji *et al.*, 2009).

1.2 Succinct History

The Camelidae family had probably appeared in North America by the Oligocene period, 35 million years ago (Epstein, 1971). The first representative of this family was the Poebrotherium. The direct ancestor of today's camel migrated to Asia through the Bering Strait 3 or 4 million years ago. It then rapidly occupied the dry zone of the Northern Hemisphere. A direct ancestor known as *Camelus thomasi* was present over much of Europe and Asia. *Camelus dromedarius* separated from the northern branch and spread across Arabia and moved into Africa. During the late Pleistocene, *C. dromedarius* ranged from the Atlantic to northern India, but it had become extinct in the African continent (Fig. 1.2). It was reintroduced into Africa after being domesticated. For a long time *C. thomasi* was considered to be the direct ancestor of the dromedary, but no skeleton of this ancestor was discovered in the Arabian Peninsula, despite domestication probably occurring in this part of the world. According to some authors, there was probably another species that existed

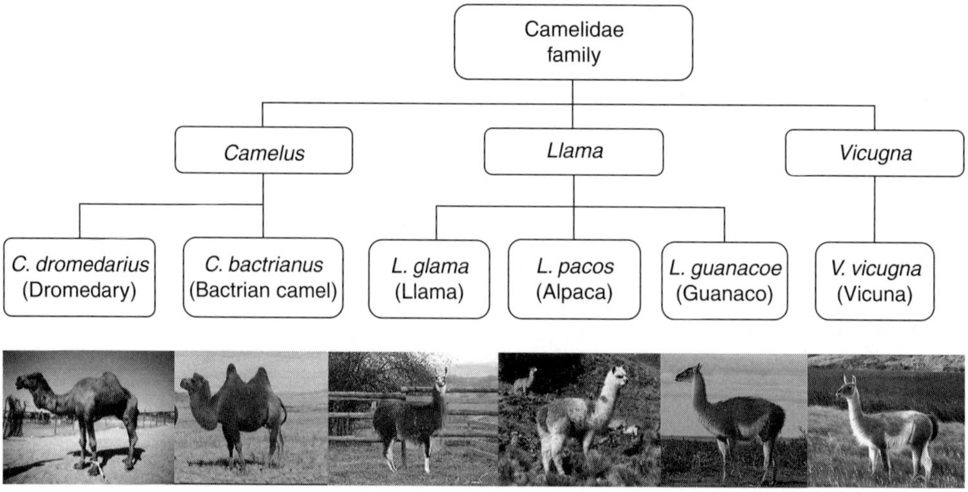

Fig. 1.1. Taxonomy of the Camelidae family.

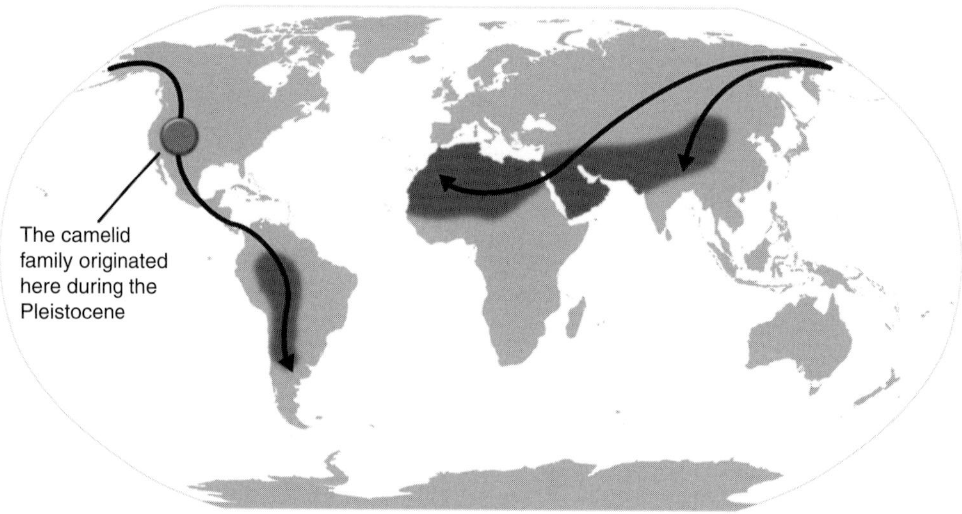

Fig. 1.2. Historical migrations of the Camelidae family (source: Wikipedia).

after *C. thomasi* and before *C. dromedarius*. According to Uerpman and Uerpman (2002), a wild form of dromedary was hunted in 5000 BC in the Arab Emirates, which means that domestication was not achieved at that time. For long time it was considered that the dromedary was domesticated 4000 years BP in South Arabia from wild population living in arid valleys located in Hadramaout (Wilson, 1984). Recent studies have shown that the domestication, which is a long and slow process, was achieved in the Iron Age between 2000 and 1000 years BC. The first historic sources describing domestic camels were written 1100 years BC during the battle between the northern Arab tribes and those on the Mediterranean coasts.

The first domestic camels were probably used mainly as ride and pack animals in connection with the trade of spices, incense and possibly salt. The story of Makeda (the Queen of Sheba) visiting King

Solomon (950 BC) describes the imposing caravan of camels carrying goods of the royal suite. Indeed, the caravan trade was flourishing between Arabia and the Near East and some prestigious civilizations took on this activity, such as the Nabateans who founded Petra. Some authors (Lhote, 1953) proposed that there were other foci for domestication in northern Africa but this hypothesis is difficult to support with today's archaeological evidence.

From this focus of domestication, the spread of the dromedary camel was associated with the dryness of the climate. From 3500 BP onwards, the camel was present in Socotra Island, from where it occupied the Horn of Africa and then migrated through Sahelian zones. At around 3000 BP, the camel was used in the whole Near East and Middle East for pack and milk production. The spread of the domestic camel was, however, also linked to military uses (troops and materials transport, and warfare/charges), as attested by various documents, in Palestine, Mesopotamia, Assyria and in India during the invasion by Alexander the Great at about 2300 BP.

The timing of the introduction of the camel into Egypt has been debated (Epstein, 1971) because the camel was not depicted on any temple or palace wall in ancient Egypt. It is accepted now that most of the camels entered Egypt around 2200–2100 BP through the Sinai Peninsula or by way of the Red Sea, in connection with the incense trade with Upper Egypt. The spread of the dromedary camel in northern Africa was simultaneous with the Roman invasion: the first mentioning of camels in this area dates back to 46 BC when Caesar reported that he captured 22 of them from the Numidia army (in Upper Egypt). It is probably the Romans who included camels in the rural economy and used the animals not only for pack and riding but also for ploughing and wheeled transport, as attested by illustrations on roman tombs.

At the time of the Muslim Arab conquest (after 639), the camel distribution zone widened: camels were present in Spain in 1020 and Sicily in 1059. In 1405, the camel was introduced into the Canary Islands by a French landowner. Other introductions that occurred later include: the USA in 1856 by President Jefferson Davis for the US army; the Kalahari Desert (southern Africa) at the beginning of the 20th century; arid zones of South America (Brazil, Peru); and some parts of European countries (France, Italy, Spain and Cyprus). It was, however, in Australia where camel introduction was really perennial: six camels were imported in 1849; 121 in 1866; and in 1895, the number of dromedary camels was 6000. They were widely used for riding, ploughing and also for the exploration of arid areas in central Australia. In 1920, farmers, postmen and policemen used 12,000 domestic camels. The number of domestic animals decreased, however, with motorization and only 2000 were accounted for in the 1960s. At the same time, many camels returned to the wild condition ('feralization'). In 1960, the population of wild camels in Australia was estimated at 90,000 heads. In 1996, estimates of the size of the feral population varied between 100,000 and 500,000 heads (Gee, 1996). The population in 2010 was more than 1 million feral camels, which started to cause some environmental problems because of the high density of camels. A Camel Destruction Act was published to limit the pressure of the wild population on the Australian desert environment.

1.3 The Camel Population in the World

It is difficult to determine exactly the number of camels in the world. First, this is because it is mainly an animal of nomadic people and pastoralists who are moving frequently. Secondly, it is because camels are not usually subjected to obligatory vaccination. So, an exhaustive census for the camels is quite difficult. Officially, the total number of camels in the world was around 25 million heads (FAOstat, 2009). This number is probably underestimated. In the Sahelian countries (Mauritania, Mali, Niger, Chad, Sudan and Ethiopia) particularly, when the number of camel heads was adjusted after an appropriate census, the recorded number was greater than the former estimation of the population. For instance, in Chad, the camel population was re-adjusted from 800,000 to more than 1.3 million heads after an appropriate census by the Ministry of Animal Resources. Thus, by considering both the wild Australian camel population and the different national estimations, the camel world population is probably around 30 million heads. As a whole, this population represents less than 1% of the total herbivorous domestic population in the world, however (Fig. 1.3).

More than 80% of the world population lives in Africa, with 60% in the Horn of Africa. The most important countries for the camel economy with a camel population of more than 1 million are, in order: Somalia, Sudan, Ethiopia, Niger, Mauritania, Chad, Kenya, Mali and Pakistan (Fig. 1.4).

The world camel population is increasing regularly with a yearly growth of 3.4%. Since 1961 (the date of the first FAO statistics), the world camel population has more than doubled (Fig. 1.5).

The growth rate was not similar, however, for all the countries. We can distinguish five types of trends:

1. Countries with high recent growth (Algeria, Chad, Mali, Mauritania, Oman, Qatar, Syria, UAE, Yemen, Ethiopia and Eritrea).
2. Countries with regular growth (Bahrain, Burkina Faso, Djibouti, Egypt, Iran, Kenya, Niger, Nigeria, Pakistan, Saudi Arabia, Somalia, Sudan, Tunisia and Western Sahara).
3. Countries with stable population (Lebanon, Libya and Senegal).
4. Countries with declining population (Afghanistan, China, India, Israel, Jordan, Mongolia and former Soviet republics from Central Asia).
5. Countries with a high rate of decline (Iraq, Morocco and Turkey).

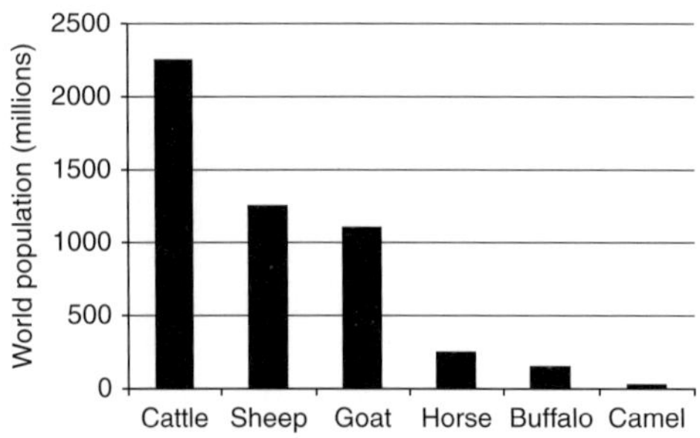

Fig. 1.3. The size of the camel population in relation to other herbivorous populations of the world (in millions of heads).

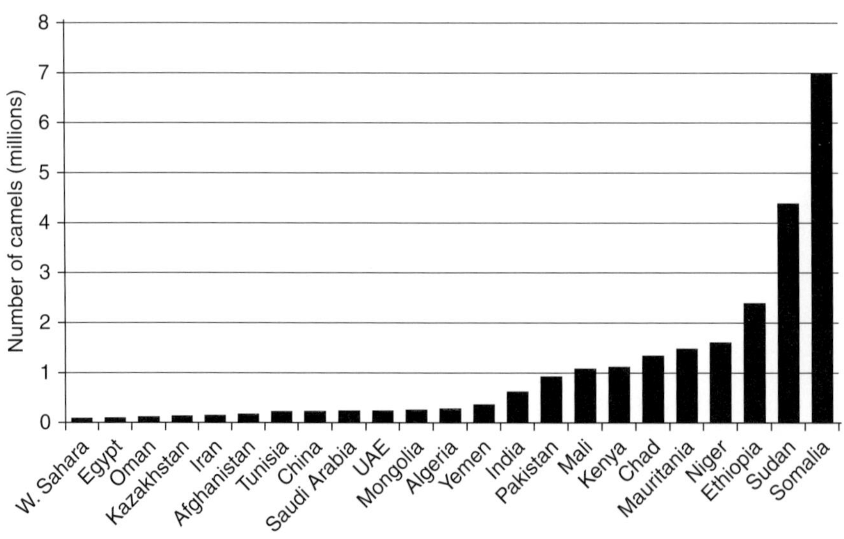

Fig. 1.4. Number of camels in countries that have more than 0.1 million camels.

The dromedary is obviously linked to arid countries and, in sociological aspects, mainly (but not exclusively) to Muslim countries (Fig. 1.6).

In countries of desert nature (e.g. Mauritania, Saudi Arabia and Gulf countries), the camel farming systems are found all over the country, but only a small space is devoted to camel rearing in sub-arid countries. For instance in India, only the north-western area (Rajasthan, Gujarat states) is favourable for camel farming. In Ethiopia, only the lowlands (below 1500 m altitude) are regularly occupied by camels. Similar patterns are

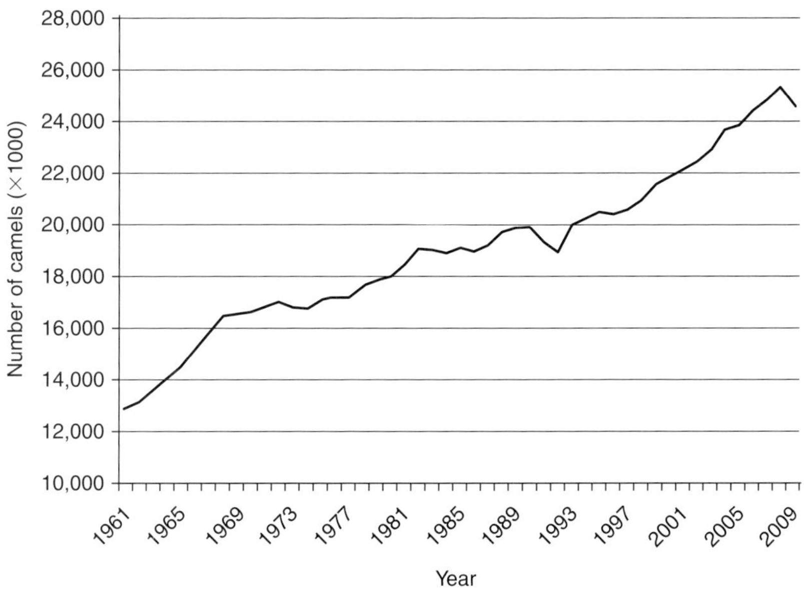

Fig. 1.5. Camel world population growth between 1961 and 2009.

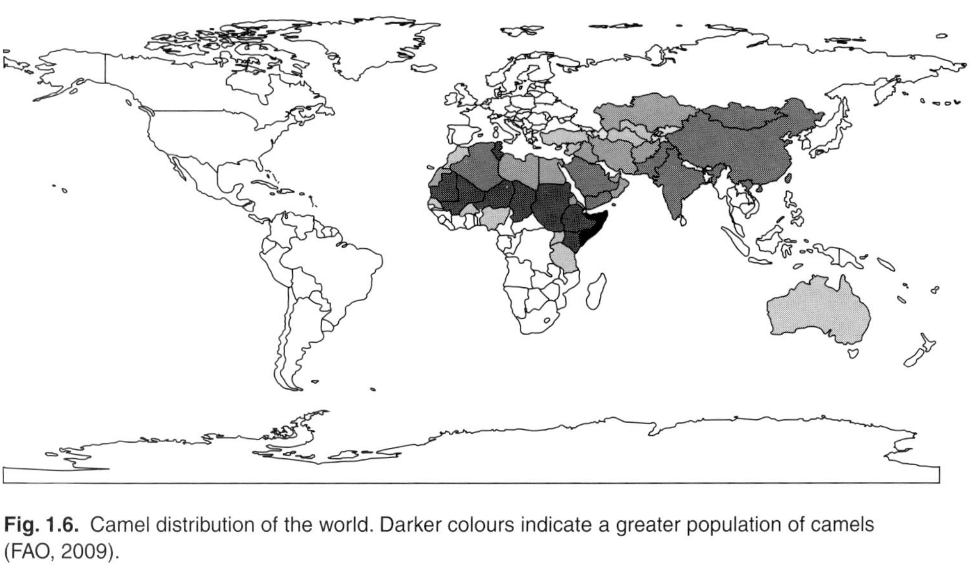

Fig. 1.6. Camel distribution of the world. Darker colours indicate a greater population of camels (FAO, 2009).

observed for the Bactrian camel. For instance, in China and Kazakhstan, the Bactrian camel is present only in the arid part of the country (the Gobi Desert in China and Moyoum-Koum Desert in Kazakhstan).

The camel was introduced in other countries, either for leisure as in the circus or zoological gardens, or for rearing in multipurpose activities such as a tourist attraction, for walking in remote places and beaches, and sometimes for milk production. Some camel farms in Western Europe (Faye *et al.*, 1995) or in North America were established, but their significance remains quite marginal. Even the dromedary introduction to the South-African desert (Kalahari) was poorly developed. The main success of camel introduction out of its original home countries was in Australia but the major part of the herd is now feral (Faye *et al.*, 2002). With a wild camel estimated population of approximately 1 million heads, the camel in Australia is regarded mainly as a big environmental problem in the central desert area of the country rather than as a potential source of meat (Saalfeld and Edward, 2010).

1.4 Conclusion

The camel has accompanied humans for thousands of years and provided many facilities in desert areas. Even if the camel population is marginal at the world level, the role of the camel in the countries where desert is predominant is quite essential. The camel population is regularly increasing, in spite of modernization and growing urbanization. It is likely that the place of the camel in our future is assured: the camel is compatible with the modern way of life.

References

Epstein, H. (1971) History and origin of the African camel. *The Origin of the Domestic Animals in Africa.* African Publishing Corporation, New York, pp. 558–564.

Faye, B., Grech, S. and Korchani, T. (2002) Le dromadaire, entre féralisation et intensification. *Anthropozoologica* 39, 7–13.

Faye, B., Jouany, J.P., Chacornac, J.P. and Ratovonanahary, M. (1995) L'élevage des grands camélidés. Analyse des initiatives réalisées en France. *INRA Productions Animales* 8, 3–17.

FAO (2009) Food and Agricultural Organization, United Nations. Available at: http://faostat.fao.org/site/339/default.aspx (accessed 7 July 2011).

Gee, P. (1996) The camel in Australia. *Journal of Camel Practice and Research* 3, 139–140.

Ji, R., Cui, P., Ding, F., Geng, J., Gao, H., Zhang, H., Yu, J., Hu, S. and Meng H. (2009) Monophyletic origin of domestic Bactrian camel (*Camelus bactrianus*) and its evolutionary relationship with the extant wild camel (*Camelus bactrianus ferus*). *Animal Genetics* 40, 377–382.

Lhote, H. (1953) Le cheval et le chameau dans les peintures et gravures rupestres du Sahara. *Bulletin de IFAN (Institut Français d'Afrique Noire)* 15, 1212–1225.

Saalfeld, W.K. and Edward, G.P. (2010) Distribution and abundance of the feral camel (*Camelus dromedarius*) in Australia. *The Rangeland Journal* 32, 1–9.

Uerpman, H.P. and Uerpman, M. (2002) The appearance of the domestic camel in SE-Arabia. *Journal of Oman Studies* 12, 235–260.

Wilson, R.T. (1984) *The Camel.* Longman Publishers, London, UK.

2 Camel Meat in the World

Bernard Faye
FAO Consultant, Camel and Range Research Centre, Al-Jouf-Sakaka, Saudi Arabia

2.1 Introduction

The camel is a multipurpose animal. It can be used for milk, meat, wool, transport, races, tourism, agricultural work and beauty contests. No other domestic animal is able to provide as many variable services to humans. Meat production is the only one of these purposes that requires the camel to be slaughtered. Consequently, meat production is linked to proper herd management in terms of the selection of animals to be slaughtered, such as young males that are not kept for reproduction or other activities and culled female or males, and to market organization at a local and regional level.

The available statistics on camel meat production in the world are limited to the number of slaughtered animals and the mean carcass weight upon which meat production is estimated. There are, however, no available statistics on the type of camels slaughtered or meat processing. A significant number of camels are slaughtered out of official channels so they are not included in the statistics, suggesting that camel meat production is probably underestimated.

2.2 Slaughtering Rate

The percentage of slaughtered camels has regularly increased since 1960, in the range of 5–7% (Fig. 2.1). This increase could be explained by a better organization of the camel meat commodity channels and a decrease in the unofficial slaughtering, although unofficial slaughtering in camels is less important than for small ruminants or even for cattle. Indeed, the heavy weight of the camel does not usually encourage killing one animal for few guests, unlike the goat or sheep.

The slaughtering rate is obviously higher in male than in female camels. Only local statistics are available for camel meat production. For example, in a slaughterhouse in Laâyoune (south Morocco/western Sahara), monitoring the age pyramid of slaughtered animals for 5 months (B. Faye, unpublished results) has shown that 44% of the slaughtered males were less than 1 year old compared with 14% of the females. The culled adult females represented 28% of the slaughtered females versus only 7.7% for adult culled males (Fig. 2.2).

The slaughtering rate is rather variable in different countries. It is lower in Africa (5.7%) than in Asia (7.6%) and Europe (11%).[1] On a regional level, the highest rates were in western Asia (31.2%), eastern Asia (20.8%), northern Africa (8.3%) and western Africa (7.6%). In other regions, the slaughtering rate appears below the world level: eastern Africa (4.3%), central Asia

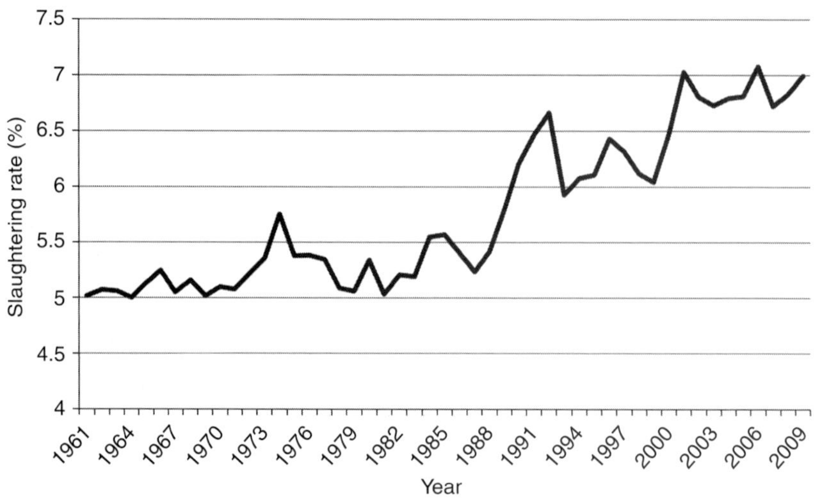

Fig. 2.1. Change in the percentage of slaughtered animals since 1961 (source: FAO, 2011).

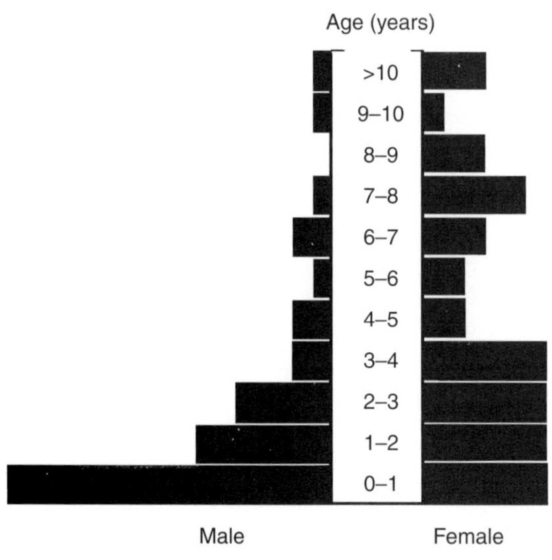

Fig. 2.2. Age pyramid of the slaughtered camels at the Laâyoune abattoir (Morocco).

(2.5%) and southern Asia (1.5%). The low rate observed in this last region is linked to religion because Hindus are mainly vegetarian. In eastern Africa, the main purpose of camel farming is milk production. The high slaughtering rate in some parts of the world is linked to a decline of the camel population (eastern Asia) and this is probably also related to countries where camel importation for meat production is important (western Asia).

2.3 Contribution to Meat Production

From 1961 to 2009, camel meat production increased at a rate of 2.8 from 123,000 to 356,000 t. The more important camel meat producers are Sudan, Egypt, Saudi Arabia (KSA) and Somalia (Fig. 2.3), but some of these countries are mainly exporting (Sudan and Somalia), whereas others are importing (KSA and Egypt).

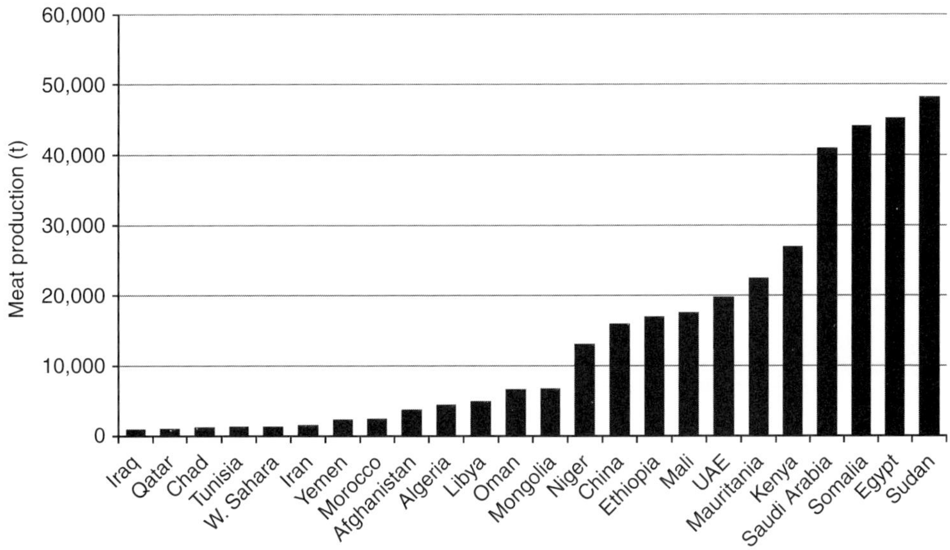

Fig. 2.3. Camel meat production in the countries producing more than 1000 t.

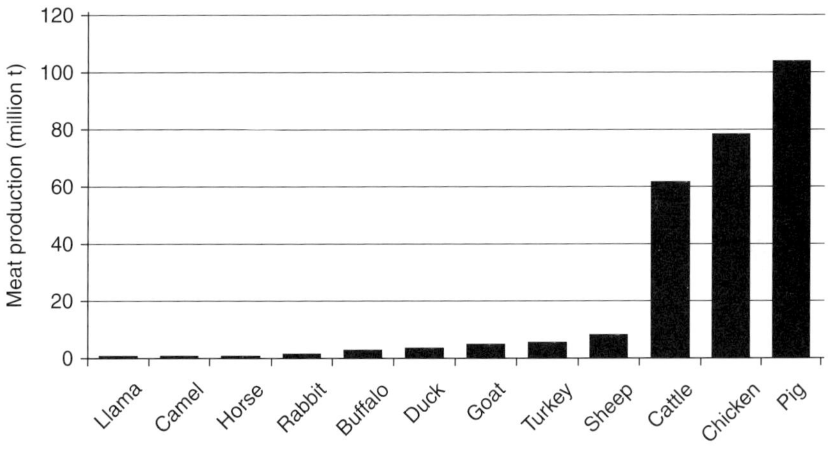

Fig. 2.4. Meat production in the world according to different species.

The contribution of camel meat to world meat production is rather marginal owing to the minor significance of camels among the herbivores. Compared with all meat-producing types (except fish), camel meat represents 0.13% of the total meat produced in the world and 0.45% of red meat from herbivores (Fig. 2.4).

However, because the camel is confined to arid areas, its contribution must be assessed in the countries where it is found. Camel meat is produced mainly in Africa (67.7%) and in Asia (27.6%). In South America, the meat comes from small camelids (llama and alpaca), with Peru being the major producer (Table 2.1). The most important regions for camelid meat production are northern Africa (29%), eastern Africa (24%), western Asia (19.7%), western Africa (14.3%), eastern Asia (6.1%), South America (4.7%) and southern Asia (1.5%).

The contribution of camel meat to total red meat production varies, however, according to the regions. Camel meat production

Table 2.1. The contribution of camels to world meat production compared with other animal groups (t) in 2009 (source: FAOstat, 2011).

Region	Camelid	Cattle	Buffalo	Goat	Sheep	Equines	Pig	Rabbit	Poultry
Eastern Africa	89,594	1,599,148	0	302,226	195,915	19	388,385	7,939	427,008
Middle Africa	1,296	380,812	0	84,168	38,404	673	88,667	1,956	69,312
North Africa	10,8427	1,045,000	270,000	266,010	581,668	2,853	3,083	76,840	1,722,111
South Africa	0	879,720	0	50,429	112,593	480	304,320	900	1,002,907
West Africa	53,511	1,009,828	0	463,893	320,342	30,115	343,679	6,750	515,092
Africa	252,829	4,914,509	270,000	1,166,728	1,248,924	34,140	1,128,135	94,385	3,736,431
North America	0	13,126,861	0	22,601	97,878	45,000	12,540,115	0	21,109,562
Central America	0	2,102,544	0	43,843	52,691	84,436	1,334,106	4,250	3,336,158
Caribbean	0	230,770	0	12,420	10,616	6,780	330,526	278	631,688
South America	17,500	15,022,353	0	72,591	245,286	106,476	4,742,366	262,024	15,782,531
Americas	17,500	30,482,529	0	151,456	406,471	242,692	18,947,115	266,552	40,859,941
Central Asia	1,046	1,206,545	0	34,396	366,856	86,816	247,471	2,081	108,820
East Asia	22,860	6,682,923	306,437	1,891,555	2,051,404	480,187	49,675,528	753,850	15,531,205
South Asia	5,460	2,332,950	2,291,413	1,143,833	874,453	0	497,580	0	3,189,893
South-east Asia	0	1,255,367	374,338	158,025	52,788	4,748	6,427,418	0	5,769,693
West Asia	73,697	957,043	3,032	185,616	771,520	2,050	98,012	1,365	3,085,038
Asia	103,063	12,434,829	2,975,221	3,413,426	4,117,022	573,801	56,946,011	757,296	27,684,649
East Europe	172	3,222,874	100	36,664	235,628	71,143	6,374,974	73,237	4,904,014
North Europe	0	1,912,124	0	630	427,166	8,033	3,429,571	373	2,134,098
South Europe	0	2,142,851	2,517	74,394	376,326	38,119	6,350,720	317,786	2,828,162
West Europe	0	3,749,693	0	9,392	144,646	11,211	10,287,646	86,965	4,296,288
Europe	172	11,027,542	2,617	121,080	1,183,766	128,506	26,442,911	478,361	14,162,562
Oceania	0	2,810,107	0	18,595	1,292,002	26,178	518,787	0	1,022,275
World	373,565	61,669,517	3,247,839	4,871,286	8,248,186	1,005,318	103,982,960	1,596,594	87,465,861

represents 3% of the total meat market and 4.1% of the red meat market in eastern Africa, 2.7 and 4.8%, respectively, in northern Africa, 2.0 and 2.9% in western Africa, and 1.4 and 3.7% in western Asia. In other regions, the contribution of camel meat is less than 1% of the red meat produced in these areas. The main contribution is in Africa (2% of the total meat and 3.2% of the red meat), i.e. more than Asia (0.1 and 0.45%, respectively).

In spite of the low contribution of camel to the world meat production, it is noticeable that the growth is higher than for cattle, sheep and horse meat. Using the index 100 in 1961,[2] the index of meat production in 2009 was 448 for goat, 309 for buffalo, 285 for camel, 223 for cattle, 165 for sheep and only 136 for horse (Fig. 2.5).

From 1961 to 2009, the mean camel carcass weight at the world level increased from 180 kg to 200 kg indicating either a slight increase in the meat productivity or an increase in the mean age at slaughtering. As a consequence, camel meat production increased at the world level because of both the higher slaughtering rate and the mean weight of the carcasses.

2.4 The Camel Meat Market

The 2009 slaughtering rates per country (Fig. 2.6) might partly help to explain the camel meat market:

- Countries such as Egypt, Saudi Arabia, Bahrain, the United Arab Emirates, Libya, Kuwait, Oman and Qatar, as well as Morocco, have a slaughtering rate of more than 20%, which is not compatible with the simultaneous growing camel population in these countries. However, these countries also import live camels for their local meat market. For instance, with a slaughtering rate of 121%, Egypt is slaughtering the equivalent of more than its own camel population.
- In China, the high slaughtering rate (30.4%) could explain the strong decline in its camel population because no camel importation is taking place.
- On the contrary, in the Horn of Africa, the slaughtering rate is lower than the world mean value: in Somalia (3.7%), Ethiopia (4.2%), Sudan (5%) and even Djibouti (6.4%) or Kenya (6.7%). These

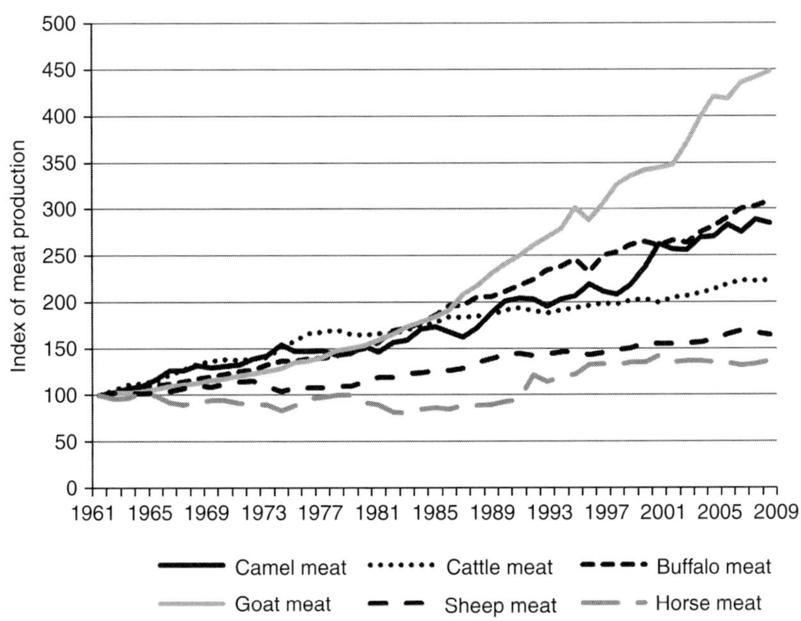

Fig. 2.5. Growth of red meat production in the world since 1961 (index 100).

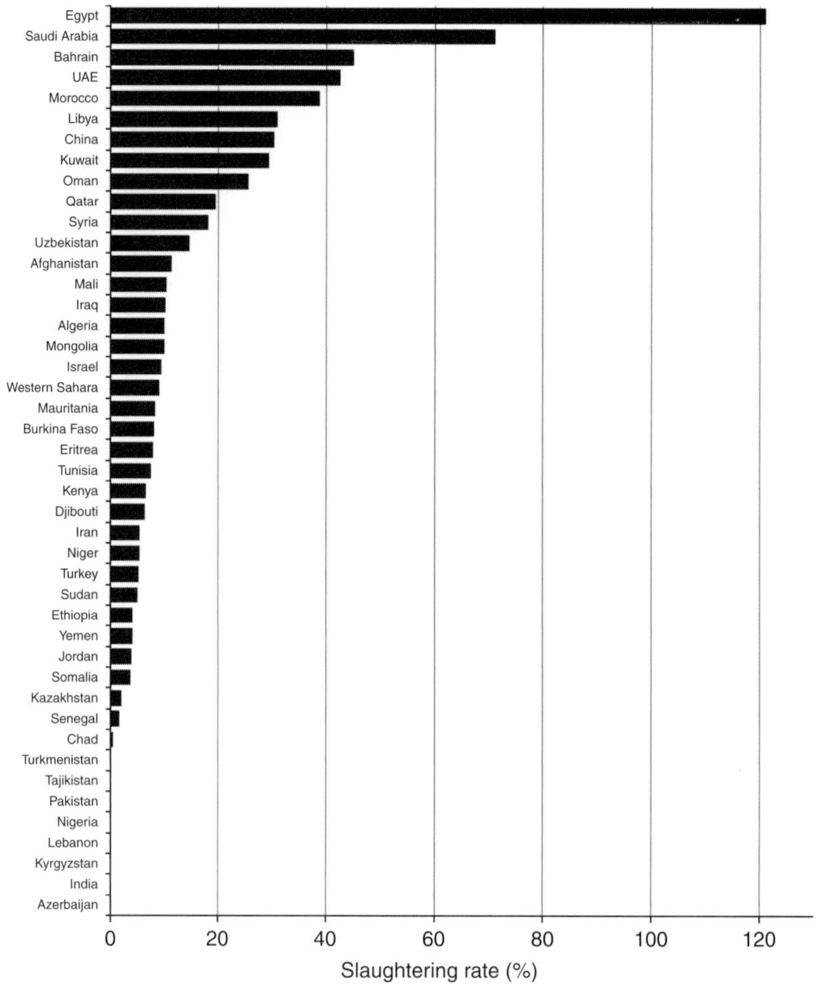

Fig. 2.6. Classification of countries with camels according to their slaughtering rate (source: FAO, 2011).

are countries where the export of live camels is quite important.

- A very low slaughtering rate could be explained by the lack of data (unofficial slaughtering such as in central Asia), the exportation of live camels through unofficial channels (Chad and Niger) or a human diet that generally lacks meat (India).

In contrast to camel milk, which is limited to the local market, camel meat is included in regional markets as the above statistics suggest. The export is, however, mainly of live animals rather than of carcasses. The main exporting countries are in the Horn of Africa and in Sahel, whereas the importing countries are mainly the Gulf States and in northern Africa.

2.5 The Camel Meat Market in the Horn of Africa

In eastern Africa, the camel stock amounts to around 11.8 million heads, which is 62% of the world camel stock in 2009. Less than 4.3% of this stock is slaughtered for local consumption. The official annual exportation doesn't exceed 41,000 heads but, as shown in Table 2.2, the number of camels slaughtered is below the expected level. Thus, there is a gap between the available live capital and its economic development through the living animal market for slaughtering (Alary and Faye, 2012); for example, the official market was around 5030 heads from Djibouti, Ethiopia and Somalia.

The official data of exportation from the Berbera and Bossasso Ports register 7636 heads in 2004. But according to estimations of the capacities of holding areas in Ethiopia, around 57,000 camels could be exported. In a survey conducted in 2007 in the Somali region of Ethiopia, the exportations are estimated at around 37,000 heads with a profit margin ranging from US$22 to US$33 per head. This is higher than profit from cattle, sheep and goats (Alary and Faye, 2012).

At the regional level, the official exportations would represent 10% only of the expected export potential. These gaps between the data and the apparent reality raise a number of questions related to camel economic development. The lack of reliable data on the true activity of the camel market might explain that camels often remain for policy makers rather an emblem or myth of the pastoral area rather than an efficient source of income for the local economy.

At the end of the 19th century, the 'Somali' pastoral area that covered the eastern part of Ethiopia, northern part of Kenya and central and northern part of current Somalia established a well-functioning marketing chain to supply the British garrison established at Aden in 1839. The international trade extended to the Arabian Gulf region and the Indian subcontinent. This period experienced the development of a very dynamic network of pastoral traders and brokers (based mainly on strong parental ties or lineages) that registered an increase of activity until the 1970s. This was mainly due to the explosion of the demand in Saudi Arabia (due to the sacrifice for pilgrimage to Mecca and the petroleum boom). The livestock market involved mainly sheep, but camels were also a significant component of this trade and one of the most beneficial for the traders and local economy.

The various wars in the 1970s and 1980s (Somalia – Ethiopia in 1977; the civil war in the northern part of Somalia in 1988) have perturbed the organization of the legal market mainly based on family networks as well as seeing the emergence of a smuggling market. This period also witnessed the emergence of new competitors in the region

Table 2.2. Estimation of exportations of live animals from Somali, Harari, East and West Harerghe and Dire Dawa regions from the declarations of exporters (number of heads).

		Estimation from declarations of exportations in 2006		Estimation from the capacity of each holding area	
Type of exporters	Species	Volume per exporter (heads)	Total	Capacity per month (heads)	Total
Medium exporter	Cattle			300	1,800
	Shoat[a]	11,000	22,000	1,500	24,000
	Camel	300	600	250	500
Large exporter	Cattle	18,000[b]	24,000[b]	10,000	60,000
	Shoat	30,000–100,000[b]	60,000–200,000[b]	25,000	400,000
	Camel	5,000[b]	10,000[b]	8,000	16,000
Total for the region	Cattle		24,000[b]		61,800
	Shoat		222,000[b]		424,000
	Camel		10,600[b]		16,500

[a]Sheep and goat.
[b]Estimation for all the holding area of the large exporter.

such as Australia, New Zealand and Egypt for sheep or Sudan and Ethiopia for camels and goats. Since 1991, the civil war in Somalia has induced the disorganization of official services such as the veterinary services, the customs and bank services, mainly in ports for exporting livestock, and consequently favoured the official position of Djibouti and Port Sudan in the international market of live animals for the region. This has led to the development of a veritable network of smugglers in the sub-region of Djibouti–Ethiopia–Somalia that export animals via Yemeni traders who re-export animals to Saudi Arabia. Although the Somaliland area in the north of Somalia has been relatively peaceful, the capacities of negotiations of the traditional traders' networks in the area have been relatively weakened by the global political context.

The reinforcement of the smuggling livestock trade activity resulting from the insecurity affecting the region since the mid-1970s also benefited from the strong family relationships. These include kinship-, ethnic- and clan-based affiliations in the pastoral area that covers eastern Ethiopia (Ogaden), Somaliland (Somalia) and the northern part of Kenya (Little *et al.*, 1998). The Somali or Boran traders ensure the transfer of livestock between the three countries and their trekking to the Ports of Somalia (Berbera, Mogadishu and Kismayo or even Djibouti). This smuggling activity is also permitted because of camels that ensure the transport of merchandises from the Ports to the remote areas.

Moreover, during the recent ban on livestock export imposed by Saudi Arabia (1998–2000 and 2001–2004) because of outbreaks of Rift Valley Fever in Ethiopia and Somalia, and insufficient veterinary control (Faye, 2003), the pastoralists suffered from an economic crisis (Pratt *et al.*, 2005); the camel economic activity allowed an export activity to be maintained in the area either by the exportation of live animals or by the increase in the illegal market. Only camel exportation increased from 50,000 in 1995 to 61,400 in 2004.

Camels are imported by the Gulf States, primarily for racing but some are slaughtered. Camels for slaughter are mainly marketed in Egypt and Libya. But the exports of camels from Djibouti have dropped off, mainly because of the new facilities provided by the port of Djibouti and the quarantine at Nagad (Faye, 2004).

2.6 The Camel Meat Market in Central and Western Africa

In sub-Saharan Africa, interestingly the main exporting flow of live camels is from south to north (Sahel to northern Africa; Morocco to Egypt), whereas the live cattle export is from north to south (from Sahel to coastal countries such as Nigeria, Cameroon, Côte d'Ivoire, Ghana, etc.). The gap between potential production and official export in this region is more important than in the Horn of Africa with a significant lack of data on this market. In Chad, a part of the camel stock is exported to Egypt through Sudan. In Niger, the main export flow is to Libya and Algeria. In Mali and Mauritania, the exportation is organized for supplying the markets of Morocco and Algeria. Contrary to the Horn of Africa, the export is mainly by road (by trucks but also, in a large part, on foot). The traders organize their exportation roads on the basis of ethnic or family relationships where the Islamic agreements are quite central in the convention, especially in unofficial trade.

In general, however, there is a need for economic data to be collected to quantify the export flow (mainly for the meat market, contrary to the export to the Gulf countries where camels are used for racing and reproduction) and assess the economic contribution of camel in the regional meat market.

2.7 The Camel Meat Market in Asia

There is no official camel meat market in India where the meat consumption for most of the Indian communities is a taboo, and the farmers are not willing to sell animals for butchery. Most of the camels sold from farms in Rajasthan are young males sold presumably for carting and ploughing (Benard *et al.*, 2008); however, their final use once out of the hands of the farmer within Rajasthan or out of the state or out of India is difficult to know (Faye *et al.*, 2010). In spite of the declining camel population in India (Köhler-Rollefson, 2004), the local market is not based, contrary to some suppositions, on the marketing of females because of the reservations of the camel owners for cultural and economic reasons.

In central Asia, since the Soviet Union collapse, the centralized agricultural markets are deeply disturbed and the restructuring through the private sector is still not yet completely achieved. The camel meat market remains local with a high proportion of self-consumption. The demand is, however, increasing with the development of collective restaurants in industrial complexes. Yet, the camel meat chain is still short, directly from the producers to the consumers. The farming systems are based on the Bactrian camel: its meat is more appreciated by consumers than that of the dromedary, which was introduced from Turkmenistan to get hybrids that have a higher milk potential than pure Bactrian (Faye *et al.*, 2008). In that sense, camel meat is also a by-product of the milk production. In Turkmenistan, the dromedary meat (Arvana breed) is highly appreciated by the local population. Young camels lately weaned are prepared especially for slaughtering and a selection on growth was achieved in collective farms (Saparov and Annageldiyev, 2005). Heavy adult animals for slaughtering (up to 1300 kg) are reported in some local reports. Contrary to the African continent and the Near East, the camel meat market in Asia remains local.

2.8 The Farming Systems for Meat Production

Globally, the farming systems are still not very well developed or organized to ensure good camel meat supply. There is, however, a traditional (and efficient) farming system for fattening camels in pastoral areas of eastern Africa (Somalia, Ethiopia, Djibouti and Sudan), but intensive feed-lots are not yet well developed, except in some parts of northern Africa. For example, in Tunisia fattening camel calves is now encouraged to reach a body weight of 250 kg at slaughtering to get a better contribution of camel meat for satisfying the population demand (Khorchani *et al.*, 2005). In many cases, however, camel meat may be considered as a by-product of milk production. The pastoral system is still the most important way for camel meat production. The regional market is frequently disturbed by health constraints such as the ban by Saudi Arabia after the Rift Valley Fever outbreak in Ethiopia and Somalia. However, the modernization of camel production for supplying more meat, especially for the growing urbanized market, is on the way, and the export of camel stock to satisfy this market at a regional level is a strong opportunity for developing pastoral camel farming systems in the sub-Saharan countries.

2.9 Conclusion: Future Trends

Because of its expected dietetic quality (Kadim *et al.*, 2008), camel meat could have an advantage for human consumers. Moreover, with climatic changes and the desertification process in some countries, camel meat production could increase its distribution area. And finally, the extensive farming system could guarantee the production of environmentally friendly meat. Thus, the conditions for increasing the contribution of camel meat to the world meat supply could be met. This progress will, however, be possible only with an improvement in meat productivity, which is low in this species compared with other domestic animals, with an efficient market, with a better control of veterinary services and with better communication on the dietetic and nutritive aspects of camel meat.

Notes

[1] Camels in Europe are limited to the Canary Islands (Spain).

[2] The index is calculated by the formula $I = (X_n/X_i)*100$ where X_n is the value at the year n and X_i, the value at the year 1961.

References

Alary, V. and Faye, B. (2012) Overview of the camel chains in East of Africa - Importance of gaps between the data and the apparent reality. *Journal of Camelid Science*, in press.

Benard, C., Faye, B., Moulin, C.H. and Kohler-Rollefson I. (2008) A typology of the camel keepers in the Jaisalmer district, Rajasthan, India. *Journal of Camel Practice and Research* 15, 231–238.

Faye, B. (2003) Surveillance and control procedures for camel diseases. *Proceedings of the workshop on the surveillance and control of camels and wildlife diseases in the Middle East*, Regional workshop OIE, Sanaa, Yemen, 10–12 March 2003. OIE Regional Office, Beirut, Lebanon, pp. 3–10.

Faye, B. (2004) L'activité d'élevage dans la Corne de l'Afrique: un enjeu pour le développement durable. *Proceedings of the colloquium Les nouvelles d'Addis/ Les verts: Grande Corne de l'Afrique/Mer Rouge: Enjeux stratégiques de la région dans la mondialisation*, Assemblée Nationale, Paris, 2 February 2004. Les Nouvelles d'Addis, Paris, France, pp. 4–5.

Faye, B., Konuspayeva, G., Messad, S. and Loiseau, G. (2008) Discriminant milk components of Bactrian camel (*Camelus bactrianus*), dromedary (*Camelus dromedarius*) and hybrids. *Dairy Sciences and Technology* 88, 607–617.

Faye, B., Benard, C., Köhler-Rollefson, I. and Boutonnet, J.P. (2010) Camel market channels in the Jaisalmer district of Rajasthan, India. *Journal of Camel Practice and Research* 17, 201–205.

FAO (2011) FAOSTAT database. Food and Agricultural Organization, United Nations. Available at: http://faostat.fao.org/site/339/default.aspx (accessed 29 March 2011).

Kadim, I.T., Mahgoub, O. and Purchas, R.W. (2008) A review of the growth and of the carcass and meat quality characteristics of the one-humped camel (*Camelus dromedaries*). *Meat Science* 80, 555–569.

Khorchani, T., Hammadi, M. and Moslah, M. (2005) Artificial nursing of camel calves: an effective technique for calves safeguard and improving herd productivity. In: Faye, B. and Esenov, P. (eds) *Proceedings of the International Workshop, Desertification Combat and Food Safety: The Added Value of Camel Producers*, Ashkabad, Turkmenistan, 19–22 April 2004. In: vol. 362 NATO Sciences Series, Life and Behavioural Sciences. IOS Press, Amsterdam, The Netherlands, pp. 177–182.

Kölher-Rollefson, I. (2004) The camel in Rajasthan: agricultural biodiversity under threat. In: *Building a Multi-Stakeholder Platform for the Conservation of the Camel in Rajasthan. Proceedings of an International Conference held on 23–25 November 2004 in Sadri.* Lokhit Pashu-Palak Sansthan, Sadri, Rajasthan, India, pp. 6–17.

Little, P.D., Teka, T. and Azeze, A. (1998) Research methods on cross-border livestock trade in the Horn of Africa: further observations. *Research Report of the Broadening Access to Markets and Input Systems-Collaborative Research Support program (BASIS-CRSP) and OSSREA Project on Cross-Border Trade and Food Security in the Horn of Africa, October 1998*, Addis-Abeba, Ethiopia, p. 14.

Nin Pratt, A., Bonnet, P., Jabba, M., Ehui, S. and de Haan, C. (2005) Benefits and costs of compliance of sanitary regulations in livestock markets: the case of Rift Valley Fever in the Somali region of Ethiopia. Report from the International Livestock Research Institute, Nairobi, Kenya, and The World Bank, Washington, DC.

Saparov, G. and Annageldiyev, O. (2005) Meat productivity of the camel Arvana breed and ways to increase it. In: Faye, B. and Esenov, P. (eds) *Proceedings of the International Workshop, Desertification Combat and Food Safety: The Added Value of Camel Producers*, Ashkabad, Turkmenistan, 19–22 April 2004. In: Vol. 362 NATO Sciences Series, Life and Behavioural Sciences. IOS Press, Amsterdam, The Netherlands, pp. 224–227.

3 Camel Nutrition for Meat Production

Rafat Al Jassim[1] and Jim Hogan[2]
[1]*The University of Queensland, School of Agriculture and Food Sciences, Faculty of Science, Australia;* [2]*81 Boscombe Road, Brookfield, Australia*

3.1 Introduction

As a pseudo-ruminant, the camel has evolved as a browsing herbivore with the ability to utilize forages that are rich in fibre but contain a range of anti-nutritional factors. Its ability to survive and thrive under the harsh desert conditions of arid and semi-arid regions is believed to be due to its distinctive feeding behaviour, anatomical and physiological characteristics of its digestive system, and other unique adaptation features. Camels belong to the suborder Camelidae, which is one of three suborders that fall under the order Artiodactyla of the class Mammalia in the animal kingdom. Camelids are not ruminants; they differ from ruminants taxonomically, anatomically, physiologically and behaviourally. Camelids and ruminants also differ in their susceptibility to infectious and parasitic diseases. Some might argue that camelids fit the criteria of the suborder Ruminantia (ruminants) because they chew the cud, are cloven hoofed and have a compartmental stomach. These anatomical and physiological features are, however, not limited to ruminants. Kangaroos, for instance, are a good example of non-ruminant herbivores that benefit from microbial fermentation of feed in the stomach. Although the digestive systems of camels are anatomically different from those of true ruminants, functionally the digestive systems play the same role. Their dependence on microorganisms for the degradation of plant fibre is similar to that of true ruminants. In the foregut of the camel the ingested feed undergoes extensive fermentation processes that result in the breakdown of the structural components of the plant cell wall, synthesis of essential products such as amino acids and vitamins and detoxification of the anti-nutritional compounds in their forage plants.

In contrast with cattle, the camel's foregut contains three compartments (i.e. rumen, reticulum and abomasum) instead of four and the mucosal surface of the reticulo-rumen lacks papillation. It also contains glandular sacs in a pattern similar to that of honeycomb that form pouches that are deeper and bigger. Similar to true ruminants (i.e. cattle and sheep), the camel has developed a symbiotic relationship with a vast number and diverse microbial population that inhabit the foregut. The camel provides the physiological conditions required by these microbes to grow and in return these microbes play a major role in the digestion of feed and supply of nutrients to the camel. Overall digestion and utilization of feed by the camel and the microbial ecosystem of its foregut has not been fully investigated and therefore is not completely understood.

This chapter aims to shed some light on the different aspects of nutrition and utilization of dietary nutrients by the camel. Special attention will be paid to the nutrient requirements for maintenance and growth and the bacterial community of the camel's foregut.

3.2 Anatomy of the Camel's Digestive System

The digestive system of the camel starts with the oral cavity of the mouth followed by the oesophagus that opens into the forestomach, which is composed of the rumen-like compartment, reticulum and the abomasum. The abomasum empties into the duodenum, which is the first segment of the small intestine. The small intestine connects to the large intestine that finally opens to the outside via the rectum. In this chapter the focus will be on the different features of the forestomach in relation to digestion in the camel.

3.3 The Forestomach

The forestomach of the camel consists of two large compartments (see Fig. 3.1), referred to by Engelhardt *et al.* (1988) as compartment 1 (C1) and compartment 3 (C3), a relatively small compartment 2 (C2) and a rudimentary omasum. The large compartment (C1), which is often referred to as the rumen, is divided by a strong transversal muscular ridge into a cranial and a caudal portion. The relatively small compartment C2 is not separated from C1 and resembles the reticulum in the ruminants' compartmental stomach. The ventral parts of C1 and C2 are in the form of a glandular sac area with pouch-like sacs being separated by muscular ridges which start from the cardiac opening and connect to C3. The C3 compartment is a long tubiform and intestine-like shape with the distal part being the only hydrochloric acid secretory region (Lechner-Doll *et al.*, 1995; Engelhardt *et al.*, 2007).

In camels, only the dorsal parts of C1 and C2, the strong rib-like muscular ridge that separates the caudal and cranial compartments, and the ridges of the glandular sacs, are covered with a stratified epithelium. The remaining parts are covered with columnar surface epithelium of approximately 40 μm in height and deep tubular glands (Engelhardt *et al.*, 2007).The histological features of the mucosa of the ventral parts of C1 and C2 and of C3 show similarity with that of the abomasum in ruminant animals (Lechner-Doll *et al.*, 1995).

3.4 Motility of the Forestomach in Camels and Cattle

Motility in the camelids' forestomach compartments is different from that in cattle (Engelhardt *et al.*, 1988). The cyclical pattern of motility of C1 (rumen) and C2 (reticulum) is categorized as A- and B-contractions (Heller *et al.*, 1986a). The contraction cycle starts with a contraction of the canal connecting C2 and C3, followed by a contraction of C2 and relaxation of the canal. This is followed by a contraction of the canal, relaxation of C2 and contraction of the caudal part of C1. The B-contractions begin with the contraction of the cranial portion of C1, followed by C2, and end with the contraction of the caudal portion of C1. The numbers of A- and B-contractions vary between the camelid species. In the camel, each cycle consists of seven A- and five B-contractions. Regurgitation in camels follows the contraction of the cranial portion of C1, whereas eructation of gases occurs simultaneously following the contraction of the caudal portion of C1 and the relaxation of the cranial portion (Engelhardt *et al.*, 1988). The frequency of motility in the camel's forestomach is high during feeding and pauses are not visible. Kaske *et al.* (1989) recorded up to 130 A- and B-contractions per hour during feeding time and 80–100 contractions when feed was removed. In contrast, contractions of the reticulo-rumen in cattle begin with primary contractions, also called the A-wave of contractions, that commence in the reticulum and move caudally across the rumen. This is followed by the secondary contractions, which occur in

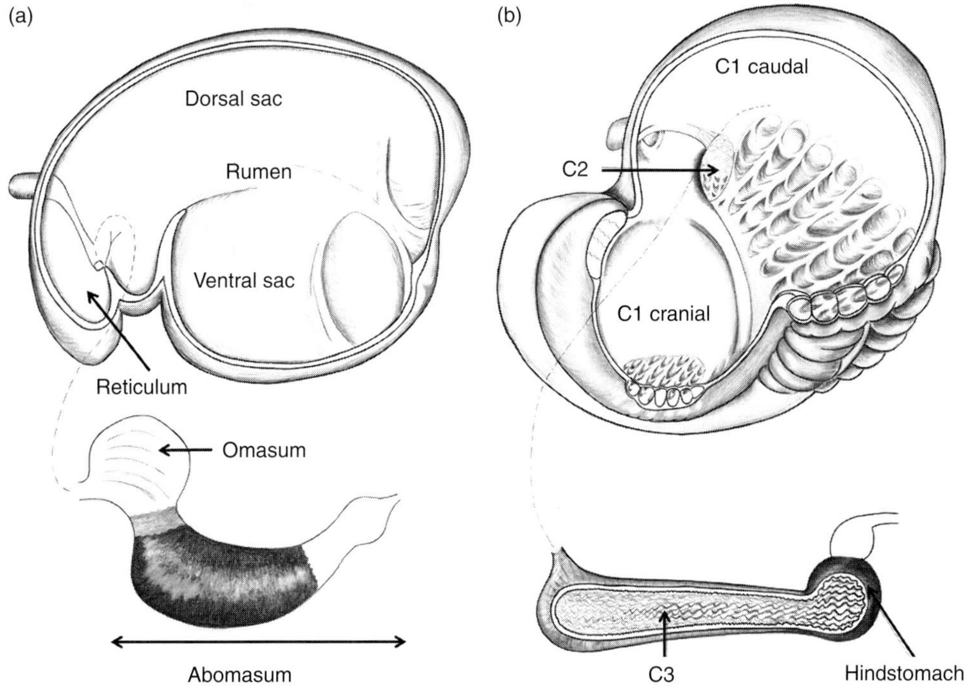

Fig 3.1. The compartmental forestomach of (a) the ruminants and (b) the camelids. Sketched by Kate Andrews, University of Queensland Gatton Campus, after Lechner-Doll *et al.* (1995).

only part of the rumen and are usually associated with eructation (Wyburn, 1980). The primary contractions start with a biphasic reticular contraction followed by a contraction of the adjacent cranial sac and the cranial pillar. These contractions push the contents of the reticulum and the cranial sac into the dorsal sac of the rumen, which contracts and pushes the digesta toward the ventral sac. Finally the contraction of the ventral sac and the relaxation of the cranial pillar allow digesta to move forward into the reticulum to complete the sequence. The second set of contractions, also called secondary contractions or the B-wave of contractions, starts the same as in the A-wave with biphasic contractions of the reticulum followed by contractions of cranial sac, cranial pillar, dorsal sac and ventral sac. Unlike in the A-wave, this time the cranial pillar contracts following the contraction of the ventral sac and pushes the digesta back into the dorsal sac. Contraction of the dorsal sac will force gases toward the cardiac sphincter for eructation. Finally the ventral sac contracts and the cranial pillar relaxes allowing digesta to move forward toward the cardia. As we see, the main function of the secondary contractions is eructation, whereas mixing and inoculation of digesta occur in both sets of contractions. Rumination is, however, facilitated by a triphasic contraction, in which an extra complete contraction follows a biphasic contraction of the reticulum forcing digesta toward the cardiac sphincter before rumination. An excellent animation of the motility of the rumen can be found on the Website of the College of Veterinary Medicine, North Carolina State University (http://www.ncsu.edu/project/cvm_gookin/rumen_motility.swf).

The reticulo-omasal orifice in the cow's compartmental stomach opens during contractions of the reticulum and closes when it dilates, allowing digesta to enter the omasum. Passage rates of both ruminal fluid and particulate matter are more closely related to duration (R^2 = 0.76) than amplitude

($R^2=0.56$) or frequency ($R^2=0.15$) of reticular contractions. The duration of reticular contractions was also found to be linearly related to the proportion of large particles ($r=0.90$) and geometric mean size of faecal particles ($r=0.61$) (Okine *et al.*, 1990). The addition of frequency and amplitude of reticular contractions adds very little to the equation ($r=0.93$), which confirms the key role of the duration of contractions in controlling passage rate from the forestomach (Okine *et al.*, 1989, 1990). There is a need for similar studies in the camel.

When camels are fed ad libitum they spend 8 h/day eating, 11 h/day ruminating and 5 h/day resting (Kaske *et al.*, 1989). When camels were fed at 08:00 h, rumination activity was recorded mainly during the night, starting after midnight until the next feeding. Little rumination activity was recorded during the day.

3.5 Retention Time and Regulation of Outflow Rate of Digesta from the Forestomach

In the ruminant animal, the outflow of solid particles from the forestomach into the intestine through the reticulo-omasal orifice is a function of flow rate and load of particles in the reticulo-rumen (Kennedy and Doyle 1993). For feed particles to pass through the reticulo-omasal orifice, their size must be reduced to a level that allows them to pass through the orifice following the contraction of the reticulum and the opening of the orifice. An increase in the frequency of contraction of the reticulum during feeding time leads to an increase in outflow rate of solid particles through the reticulo-omasal orifice, which only opens for a short time following each contraction. In sheep and cattle, sequestration of feed particles within the raft in the rumen delays the outflow of these particles from the reticulo-rumen. The outflow rate is expected to be higher when cattle consume low density diets. After the initiation of digestion, feed particles will develop voids that will be filled with gases and float on the surface, joining the raft particles and delaying their removal. There is some evidence that the smaller the particle size, the denser it is and the more likely it is to reside in the ventral sac, cranial sac and reticulum, which makes it susceptible to expulsion through the reticulo-omasal orifice following the contraction of the reticulum.

Feed particles are separated at the omasum. Some, carried in reticulo-rumen fluid, pass rapidly along the oesophageal groove to the abomasum. Others pass between the leaves of the omasum and are slowly propelled to the abomasum by mechanisms not well understood. The camel must also have a mechanism to regulate the passage of feed particles from the C1 and C2 compartments. The entrance to the C3 compartment is probably involved in this mechanism but this aspect of nutritional physiology is obscure.

3.6 Metabolites in the Forestomach of Camels

The fermentation of ingested feed in the forestomach of camelids is carried out by a large and diverse microbial population in a similar fashion to that occurring in the ruminant compartmental stomach. There, the carbohydrates are converted to short-chain fatty acids (SCFAs), mainly acetic acid, propionic acid and butyric acid, whereas proteins from the feed are broken down and rebuilt as microbial protein. That protein is later digested to release amino acids in the small intestine, and they and short- and long-chain fatty acids reach the liver for further metabolism.

Differences occur, however, between the ruminant animals and camelids in the rate and extent of fermentation, which impact on fermentation end products and the efficiency by which energy is extracted from feed. From as early as 1963, Williams' description of the forestomach contents of camels in Australia during drought conditions showed a relatively high concentration of SCFAs (133.8 mM/l) and low total nitrogen (726 mg/l).The molar proportions of SCFAs were 77.1% acetic acid, 16.4% propionic acid and 6.5% butyric acid. Williams (1963) described the rumen contents as being dry, and botanical analysis of the contents indicated a predominant presence of leaves of *Eucalyptus gamophylla* with small quantities of *Acacia aneura* and *Acacia rempeura*. A diet of this composition would be very low in crude protein content (6–9%). Measurements of SCFA concentration in the forestomach of camels grazing in the Kenyan thornbush savannah during the green and dry seasons were made using forestomach-fistulated camels (Höller *et al.*, 1989). A similar concentration profile was reported for acetate, propionate and butyrate during the wet and dry season with the mean molar percentage ratio of acetate: propionate: butyrate being 70:17:13 in the green season and 72:16:12 in the dry season. Although acetate concentration (mM/l) was similar between the wet and the dry season (122 versus 111, respectively), acetate production rate (mM/h) during the wet season was 2.7 times that during the dry season (2234 versus 816). The same applies to propionate and butyrate. It was concluded by Lechner-Doll *et al.* (1995) that concentrations of SCFAs in the fermentation chambers do not reflect production rates. Differences between the animal species used in their study were caused by a number of factors including dilution rates, rumen fluid volume and absorption rates. Absorption of SCFAs was also studied in a temporarily isolated forestomach. Engelhardt *et al.* (2007) used five fistulated camels in an experiment designed to measure the absorption of SCFAs, sodium and water from compartments 1 and 2. The two compartments (C1 and C2) were emptied through the fistula, rinsed about ten times with saline solution (0.9% NaCl) before being filled up with buffer solution containing sodium, potassium, chloride, acetate, propionate, butyrate and carbonate. The pH of the buffer was 6.6 and osmolarity was approximately 300 mosm/kg. A solute marker was also added to allow the estimation of volume and water absorption. Results of that experiment showed extensive absorption of SCFAs and that the absorption rate of the three SCFAs depended on their concentrations in the buffer solution. Absorption of SCFAs was higher when pH was lowered from 6.7 to 6.0 and, despite the difference in chain length and solubility between the three SCFAs, clearance rates across the membrane were similar. This indicates that the diffusion of SCFAs in the non-ionized, lipid-soluble form across the C1 and C2 epithelia is of minor importance (Engelhardt *et al.*, 2007). It is not known, however, whether it was the use of a different technique or the species factor that was the reason for the different trend observed in guanacos (Rübsamen and Engelhardt, 1978). By using the Pavlov pouches techniques in the C1 of llamas, Rübsamen and Engelhardt (1978) reported that clearance rates of Pr^-/HPr (propionate) were 50% higher than those of Ac^-/HAc (acetate), and clearance rates of Bu^-/HBu (butyrate) were 50% higher than those of Pr^-/HPr. Engelhardt *et al.* (2007) commented on this by raising the issue of maintaining a normal functioning epithelium of these Pavlov pouches for several weeks after surgery. It seems that the temporarily isolated forestomachs technique followed by Engelhardt *et al.* (2007) is less invasive and would therefore produce more reliable estimates.

It is important to note that, for browsing, the camel feed selectivity is higher than that of other species of herbivores because of greater height, mobility and adaptive features such as large, mobile lips and a prehensile tongue. During the wet season, camels browsed on 29 different plant species compared with 7 plant species during the dry season. Time spent by camels on browsing for specific plant species also varied during the two seasons with 77% of the total feeding time spent on browsing seven plant species in the wet season and 96% of the total feeding time browsing only three species during the dry season (Höller *et al.*, 1989). In comparison with cattle, sheep and goats, the adaptive feeding behaviour of camels gives them an advantage over the other species and makes them less constrained by the seasonal changes in the quantity and quality of plants on which they browse (Rutagwenda *et al.*, 1989). Although cattle and sheep spend 70% of their total feeding time grazing on grasses, herbs and small shrubs at ground level, camels and to a lesser extent goats show a preference for trees and shrubs that suffer little change in the quality of their leaves during the dry season. Camels can reach plants as high as 3 m, which means that during the dry season, they continue to reach the highly nutritious leaves and benefit greatly compared with cattle and sheep (Rutagwenda *et al.*, 1989). When plants of preference to camels are scarce, however, camels start to lose weight during the dry season, which indicates compromised adaptation behaviour (Abbas *et al.*, 1995). In the presence of diverse plant species, camels tend to select dicotyledons, which are highly nutritious and show less variability in their nutritive value between the wet and dry seasons (Rutagwenda *et al.*, 1989; Abbas *et al.*, 1995). Research on the Australian feral camels in arid Australia revealed a diverse selective feeding behaviour on more than 80% of the vegetation species in arid Australia (Phillips *et al.*, 2001; Dörges *et al.*, 2003). In contrast, when only limited plant species are available, selectivity is compromised and camels start to feed on plants that are less preferable to them during the wet season and that are of lower nutritive value. The digestibility of diets selected during the green, dry and extremely dry seasons in the Butana area of the Sudan were 48.6%, 34.5% and 33.2%, respectively (Abbas *et al.*, 1995). During the dry season, however, when the quality of the diet is poor, the retention time of feed particles in the forestomach of camels increases (Table 3.1), resulting in higher digestibility (Rutagwenda *et al.*, 1989, Lechner-Doll *et al.*, 1990).

It was estimated that the digestibility in the forestomach of camels of the slowly digestible, poor-quality plants increased by 5% owing to an increase of particle retention time from 25 h during the green season to 29 h in the dry season (Rutagwenda *et al.*, 1989). The increase of particle mean retention time during the dry season (expressed as a percentage of the value in the green season), was highest for sheep (46%), followed by cattle (27%), goats (22%) and camels (18%). Forestomach volumes were also greater in the dry than in the green season. The increase followed the same trend as in retention time, being highest in sheep (55%), followed by cattle (31%), goats (29%) and camels (28%) (Lechner-Doll *et al.*, 1990). The finding that the increases in mean retention time of particles and forestomach volume was smaller in camels reflected the smaller changes in the quality of their diet between the green and dry seasons. Camels are therefore able to take

Table 3.1. The mean retention time (MRT) of the solid particulates (h) and fluids or outflow rate of fluids (l/h) from the forestomach of camelids and the rumen of the ruminant animals.

Species	Diet	Solid particles	Fluids	Reference
Llamas	Hay + concentrate	27.0–32.5 h[a]	15.3 h	Heller *et al.*, 1986b
Camels	Browsing, dry season		3.7 l/h	Höller *et al.*, 1989
Camels	Browsing, wet season		6.1 l/h	Höller *et al.*, 1989
Cattle	Freshly cut ryegrass		10.2 l/h[b]	Waghorn *et al.*, 1989
Cattle	Freshly cut lucerne		12.7 l/h[b]	Waghorn *et al.*, 1989
Cattle	Freshly cut lucerne		6.0 l/h[c]	Waghorn *et al.*, 1989
Camels	Browsing, dry season	28.9 h	11.6 h	Lechner-Doll *et al.*, 1990
Camels	Browsing, wet season	24.6 h	8.6 h	Lechner-Doll *et al.*, 1990
Cattle	Grazing, dry season	35.7 h	15.1 h	Lechner-Doll *et al.*, 1990
Cattle	Grazing, wet season	28.1 h	9.7 h	Lechner-Doll *et al.*, 1990
Sheep/goats	Grazing, dry season	28.8 h	15.1 h	Lechner-Doll *et al.*, 1990
Sheep/goats	Grazing, wet season	20.1 h	9.7 h	Lechner-Doll *et al.*, 1990

[a]The range represents lower values for small particles (0.2–1.0 cm) and higher values for large particles (2.5–4.0 cm); [b] during and 2 h after feeding; [c] 2 h after feeding.

advantage of both strategies of adaptation, selective feeding when they are able to select, and utilization of slowly digestible structural cell wall constituents if no other feed of better quality is available (Lechner-Doll *et al.*, 1995). Differences in feed intake were also reported between alpacas and sheep fed two different roughage diets (Liu *et al.*, 2009). In a digestion and forestomach fermentation experiment, Liu *et al.* (2009) reported an interaction between animal species and forage source on total SCFA but not on the molar proportion of SCFA. Total SCFA concentration was higher in sheep than in alpaca and the substitution of sorghum with alfalfa decreased the concentration of SCFA in both species. The magnitude of the reduction was, however, smaller in alpacas (-17%) than in sheep (-34%). The total tract apparent digestibilities of digestible matter (DM), organic matter (OM), crude protein (CP), ether extract (EE), neutral detergent fibre (NDF) and acid detergent fibre (ADF) were similar between the two species and there was no roughage source × animal species interaction. It was suggested that discrepancy between results of this study and those reported elsewhere could be due to the differences in feed quality because alpacas are expected to perform better on a low-quality high-fibre diet (San Martin, 1987; Liu *et al.*, 2009).

3.7 Nitrogen Economy in the Camel

Camelids are known to be very efficient nitrogen utilizers. An investigation into urea excretion in urine was first made in 1925. Read (1925) reported that camel's urine contains no urea and most of the 8.7g of N excreted daily by the camel was in form of hippuric acid (35.1%), creatinine (39.4%) and purine bases (19.9%). This was later found to be incorrect and both analytical and mathematical errors were suspected to have caused such error (Petri, 1927). Camels on a low-nitrogen diet excrete less nitrogen in their urine and most of the plasma urea is recycled into the forestomach. During a normal nitrogen intake (about 33g N/day; 206g CP/day) about 40% of the urea filtered in the glomeruli was excreted in the urine, whereas during a low nitrogen intake (about 15g N/day; 94g CP/day) only 1–2% was excreted (Schmidt-Nielsen *et al.*, 1957). When camels were on a low-nitrogen diet, giving an injection of urea into the blood did not increase nitrogen excretion in the urine and all infused nitrogen was retained by the camel. Another investigation into urea degradation using the ^{14}C-urea dilution technique in camels fed diets containing graded levels of crude protein (6.1, 9.6 and 13.6%) by Emmanuel *et al.* (1976) showed that the urea excretion rate increased as dietary nitrogen intake increased. Plasma urea concentration was linearly correlated with dietary protein intake. The effect of water restriction and complete water deprivation on nitrogen balance and urea cycling in camels, desert sheep and desert goats fed a low-quality desert grass containing 3.2% crude protein have also been studied (Mousa *et al.*, 1983). All animals lost weight during both the water restriction and complete deprivation. Although camels experienced a decreased dry matter intake (g/kg$^{0.75}$), and decreased apparent dry matter digestibility, they had a smaller weight loss (% of initial weight) and increased urea recycling rate. The increased rate of urea recycling in camels under this low dietary nitrogen intake and negative nitrogen balance trial was also accompanied by a reduced urine and faecal nitrogen output (mg/kg$^{0.75}$) in comparison with sheep. Results of this experiment (Mousa *et al.*, 1983) and others (Emmanuel *et al.*, 1976; Emmanuel and Emady, 1976) all support that camels fed low nitrogen diets are more efficient than ruminant species in urea recycling and degradation of urea in their foregut. This partly explains why camels continue to thrive under the very harsh desert conditions and have an advantage over the ruminant species in survival.

3.8 Methane Emission in Camels

The 2006 Intergovernmental Panel on Climate Change (IPCC) guidelines for national greenhouse gas inventories has been the key reference for methane emission from enteric fermentation in livestock. In dealing with the various species of animals and production systems and regions, the advisory panel has chosen to follow different approaches to account for such differences. Three tiers were set with different levels of complexity to estimate methane emission. In Tier 1, basic characterizations including livestock species and categories, annual population, and, in the case of dairy cows, milk production are required and were found to be sufficient. In Tiers 2 and 3, however, more details on feed characteristics and specific information on the targeted species of animals were required. Details such as diet quality and nutritional management of the particular species of animal were added to obtain a more accurate estimate of feed intake for use in estimating methane production from enteric fermentation.

In the 2006 IPCC report, camels were considered as ruminant animals and therefore were grouped with cattle, buffalo, sheep and goats. Because of the lack of information on camels' nutrition and digestion processes, the Tier 1 method was used, however, and approximate enteric emissions were derived by extrapolation from main livestock categories that are considered to have a similar digestive system. In using Tier 1, the report stated that 'A simplified approach that relies on default emission factors either drawn from the literature or calculated using the more detailed Tier 2 methodology'. In this report the estimate of enteric methane emission was 46 kg of CH_4/head/year for a camel weighing approximately 570 kg. This weight corresponds to a metabolic body weight ($kg^{0.75}$) of 116.7 kg, making the estimated emission value of methane to be 0.3942 kg $CH_4/kg^{0.75}$/year or 1.08 mg/$kg^{0.75}$/day. This figure was derived from an earlier report by Gibbs and Johnson (1993), who extrapolated methane emission figures for camels from cattle measurements. Methane emission from cattle was found to vary between regions and production systems. It was estimated to be in the 50s for North America, Western Europe, Eastern Europe, Oceania and Latin America (ranging between 53 and 60 kg CH_4/head/year). The estimates were lower for cattle from Asia (47 kg), Africa and the Middle East (31 kg) and the Indian subcontinent (27 kg). Methane emission for lactating cows was a function of milk yield and varied between the different regions with the highest reported for the North American cows (128 kg CH_4/head/year) that produced 8400 kg of milk/head/year. The report was detailed, based on an extensive search of the literature and utilized available resources, but ignored the fundamental differences between camels and the true ruminant species of animals, which led to the use of a default value of 46 kg CH_4/head/year for camels. Although the report acknowledges indirectly the lack of information on camels, it has accepted a methane emission figure for camels that was extrapolated from cattle experiments without any adjustment to allow for differences in intake, feeding behaviour, fermentation processes and production between camels and cattle.

Calorimetric estimates of methane emission from camels fed different levels of a diet consisting of barley grain and wheat straw and during fasting were reported by Guerouali and Wardeh (1998). Methane emission was estimated to be 0.999, 0.285 and 0.642 mg/$kg^{0.75}$/day, during the periods of fasting, feeding and re-feeding, respectively. These values correspond to a total 26.3, 32.6 and 38.6 kg CH_4/year for 300, 400 and 500 kg live weight camels during the feeding period; 7.5, 9.3 and 11.0 kg CH_4/year during fasting; and 16.9, 21.0 and 27.3 kg CH_4/year during re-feeding periods. This calorimetric measurement of methane represents total methane emission from the camel. It is important to mention that concentrate supplementary feeding does not reflect the normal feeding situation for this herbivore. On the other hand, measurements of methane emission by alpaca (*Lama pacos*) and sheep by Pinares-Patino *et al.* (2003) using the sulfur hexafluoride tracer techniques showed no differences in CH_4 emission (% gross energy intake) between the two species when fed alfalfa hay indoors (5.7% versus 4.7%). Alpaca had a higher CH_4 emission, however, when fed the improved perennial ryegrass/white clover pasture (9.4% versus 7.5%) and Lotus (6.4% versus 2.7%). It was suggested that differences between alpaca and sheep in particulate fractional outflow rate might have been the underlying physiological mechanism responsible for the differences in CH_4 emission. Other differences, such as dietary selection and voluntary feed intake, might have also contributed to such differences. In general, information on methane emission by camelids is limited and more work is required, especially under normal feeding conditions.

3.9 Nutrient Requirements of the Arabian Camel

Little information is available on the appropriate nutritional requirements of camels for different purposes. Nutrient requirements for camels have not been determined and only few recommendations are available, but unfortunately these estimates were mainly derived from cattle requirements. Our knowledge of the anatomical, physiological and feeding differences between camels and cattle make the reliability of such estimates far from being realistic or accurate. There is an urgent need to start a structured programme that involves different research groups from different countries to measure the requirements for energy, protein and other nutrients for breeding, growing and racing camels. Emphasis in this chapter will be on energy requirements because of the availability of some experimental data derived from experiments carried out under defined conditions.

3.10 Energy Requirements

Little experimental data on the nutrient requirements of the camel are available and most of the available information is extrapolated from beef cattle data. In a calorimetric study an attempt was made, however, by Guerouali and Wardeh (1998) to measure maintenance energy requirements of the Arabian camel. In this study the energy requirements for maintenance was estimated at an average 0.314 MJ/kg $W^{0.75}$/day and the efficiency at which metabolizable dietary energy was utilized for maintenance was 73% during fasting and 61% at a feeding level above maintenance. The latter figure is higher than those reported for other species of animals. It might be worthwhile repeating such measurements with wholly roughage diets. It is also important to consider the energy content of weight gain. A report from Australia showed that the energy content of gain increases as the animal matures. For example, the value increases from about 9.2 MJ/kg in young lambs to 26.0 MJ/kg in the adult sheep (Weston and Hogan, 1986). Information of this kind is lacking in camels.

Guerouali and Wardeh (1998) estimated the fasting heat production for the Arabian camel during the fourth day as 0.213 MJ/kg $W^{0.75}$/day, a value closer to that reported for sheep than cattle. Exposure of camels to heat stress (40°C, for 12 h/day) resulted in only a slight change in total heat production (16.9 versus 19.6 MJ/day), whereas energy balance increased from 4.3 MJ/day to 6.8 MJ/day, indicating a well-developed adaptability to the hot climate.

3.11 Measurements of Metabolizable Energy (ME) for Growth in Camels

Meat production in animals is measured as the amounts of energy, protein, fat and minerals stored in the body of the animal. Such measurements can be made by slaughter techniques involving the analysis of the body composition of a group of animals killed at the start of an experiment and a corresponding group killed after a period of feeding. The slaughter technique also permits the establishment of a direct relationship between body water and protein and an indirect relationship between body water and body fat. In subsequent studies body composition can be predicted from the knowledge of total body water estimated from the dilution of injected tritiated water or deuterium oxide. However, most of the data on which the modelling of nutrition of sheep and cattle is based, have been derived from balance studies, i.e. estimating the amounts of nutrients stored as the difference between their intake and excretion.

The second edition of the Bulletin No. 433 (MAFF, 1987) reported techniques used for measurements of metabolizable energy of feed and the energy requirements of cattle and sheep. Common techniques used were:

1. Use of respiratory chambers or a calorimeter requires measurements of the animal's heat production as well as the energy in feed, faeces, urine and methane.
2. Use of data from a metabolism trial, if faeces and urine energy losses are known; the metabolizable energy of the feed can be calculated because the methane loses are assumed to be 0.08 of the energy value of the feed.
3. If only digestible energy is known, metabolizable energy (ME) can be calculated using the following relationship: ME = 0.81 DE; where DE is digestible energy.
4. Use of constants (factors) to convert the digestible nutrients of feed to ME. One example for the use of factors derived from digestibility experiments is the following:

$$\text{ME (MJ/kg)} = 0.0152\ \text{DCP} + 0.0342\ \text{DEE} + 0.028\,\text{DCF} + 0.0159\,\text{DNFE}$$

where DCP is digestible crude protein (g/kg); DEE is digestible ether extract (g/kg); DCF is digestible crude fibre (g/kg); DNFE is digestible nitrogen free extracts (g/kg).

It is obvious that such an approach would be more suited for concentrate feed due to the use of the Weende system, which divides carbohydrates into CF and NFE. For roughage diets, MAFF (1987) proposed the use of prediction equation that considers a factor related to the value of modified acid detergent fibre (MADF). In this approach the prediction equation is:

$$\text{ME (MJ/kg DM)} = 16.37 - 0.0205\ \text{MADF (g/kg DM)}$$

Metabolizable energy supplied to the animal is used for maintenance (ME_m), growth (ME_g) and production (ME_p). The efficiencies at which dietary ME is utilized for the different purposes are influenced by animal, dietary and environmental factors. Efficiency of utilization of ME for maintenance (K_m) is higher than that for growth (K_g) or production (K_p). The concentration of ME in the diet (M/D) is the main determinant of the efficiency of its use. For diets with an ME concentration of between 8 and 14 MJ/kg DM, an average K_m of 72% is adopted (MAFF, 1987). Total dietary ME required for maintenance (ME_m) is a function of body weight and for cattle is simply estimated from fasting metabolism (FM) as:

$$\text{FM (MJ/day)} = 5.67 + 0.061\ W$$

Where W = live weight in kg, and $ME_m = FM/0.72$. With regard to the dietary ME requirement for gain (ME_g), it is simply a function of the amount of gain (live weight gain, LWG), net energy requirement for gain (E_g), the energy value of the gain (EV_g) and the efficiency at which ME is utilized for gain (K_g). Similar to K_m, the K_g varies according to the concentration of energy in the diet and can vary from 0.30 to 0.60. It can be estimated using the equation: $K_g = 0.0435$ M/D, where M/D is the concentration of energy in the diet (MJ/kg DM). An average value of 0.435 was recommended (MAFF, 1987). If LWG is not known it can be predicted as: LWG (kg/day) = E_g/EV_g.

Accordingly, estimates for growth, energy value of gain and energy requirement for gain can be made for camels using constants reported by Gueraouali *et al.* (1989).

In order to account for different nutritional situations, three possibilities can be envisaged:

1. The animal is studied as it exists in its current environment with minimal treatment of health issues.
2. The animal is removed from its environment, treated for disease and for ecto- and endoparasites and studied in a different nutritional environment.
3. The animal is given favoured nutritional treatment before and after birth to establish its potential productivity.

Each situation provides valid research opportunities designed to answer different questions. This becomes particularly important when comparing different species such as camels and cattle. Furthermore, different species mature at different ages and valid comparisons need to be made on animals at the same physiological stages of development. In many published research accounts on camels, insufficient attention has been paid to these details and the interpretation of data can be limited. A further complication arises when animals that have lost weight are subsequently well fed; they exhibit compensatory growth, which results in the deposition of a greater proportion of protein and water and a lower proportion of fat in weight gain than would occur in a continuously well-fed animal. For these reasons it seems that claims of greater feed conversion efficiency in camels than cattle, though valid for a particular experiment, are not sufficiently well established to be accepted as general statements.

This difficulty can be illustrated by reference to papers by Gueraouali *et al.* (1989) and Mohamed (2007). The former authors reported that the efficiency of use of metabolizable energy (ME) for maintenance (K_m) was 0.73, whereas the efficiency of use of surplus ME for fattening (K_f) was 0.61. These values can be applied to the growth studies of Mohamed (2007), assuming that the mean weight of the camels in that experiment were 291.7 kg, and that digestible organic matter (DOM) contains 18 MJ/kg DE and that ME/DE was 0.82. According to these calculations, the intake of 5.9 kg DOM/day supplied 87.2 MJ ME of which 29.9 MJ was required for maintenance. Of the surplus 57.3 MJ ME, 35 MJ was stored. With a weight gain of 0.886 kg/day, the implication is that weight gain in the camel was 39.6 MJ/kg. This would further imply the deposition of appreciable amounts of fat in weight gain, which is contrary to numerous reports that camel meat is leaner than beef. In order to estimate total energy deposition, the energy value of gain including the hump fat should be accounted for. The alternative, that weight gain in these camels contained energy of about 25 MJ/kg, a typical level in beef would require a K_f value of 0.41. Studies by Kadim *et al.* (2006) confirmed that camel meat is generally leaner and that the younger the camels are, the leaner their meat is. However, the measurement carried out by Kadim *et al.* (2006) involved a specific muscle, *Longissimus thoracis*, which is the muscle between the 10 and 13 ribs and the authors did not look into fat content of the total weight gain. Abouheif *et al.* (1990a) showed that between 8 and 26 months of age, well-fed camelsgained 279.6 kg in body weight equivalent to 0.518 kg/day or 261.9 kg empty body weight equivalent to 0.485 kg/day. Of this an additional 12.3 kg (23 g/day) dissected fat was deposited in the hump, compared with 1.5 kg (3 g/day)in the kidney, pelvic and heart regions. Physical dissection of carcasses was also reported (Abouheif *et al.*, 1990b). The increase in lean weight during the same period was about 113.1 kg chilled weight (56% of the gain), which was equivalent to 0.209 kg/day. There was no difference in lean percentages between camels slaughtered at 8, 16 or 26 months of age (Abouheif *et al.*, 1990b). It seems that at this low level of live weight gain of 0.518 kg/day changes in the different proportions of the soft tissue remain unchanged. There is clearly a need for more estimates to be made of energy transactions in situation 3 mentioned in the three-point numbered list above to establish the extent of similarities and differences between camels and cattle and to assist in the interpretation of studies made in situations 1 and 2.

Although energy is regarded as the main nutrient, the proper nutrition of the animal requires adequate amounts of other nutrients such as ammonia and sulfide for the microbes in the rumen and essential amino acids, i.e. the amino acids that the animal tissues cannot synthesize in sufficient quantities. Balance studies need to be supported by quantitative studies of fermentation in the rumen and of digestion in the remainder of the tract to ensure that energy use is not limited by the supply of one or more nutrients.

The ME system can function properly only when the rumen microbes are adequately nourished to ensure the fermentation of feed and the provision of enough intestinally digested protein to meet the needs of the tissues for essential amino acids. A useful indication of microbial nutrition is provided by data on the intake of digestible OM (DOM), because, of the DOM consumed, a reasonably constant proportion is lost in the rumen. Hence DOM intake has provided a useful predictor, not only for the level of ammonia in rumen digesta, but also for the amount of microbial protein that passes to the intestines (Hogan, 1996). With temperate forages, a DOM:CP ratio of 7 (50% OM digestibility/7.1% CP) should indicate adequacy of rumen ammonia for microbes. By contrast, with tropical grasses, that is, those grasses that follow the C4 rather than the C3 pathway of photosynthesis, the ratio is probably 5:1. Protein flow (g/day) from the rumen with temperate forages follows the relationship y=0.36 CP intake + 0.16 DOM intake. The mixed CP passing from the stomach comprises about 80% amino acids, which seems to be well balanced for the nutritional needs of the tissues and of which there is a net loss of about 75% from the small intestine. Similar data seem to be lacking for the camel. The adequacy of sulfur in sheep is indicated by a ratio of N:S of 10:1 in the feed. In sheep, a portion of the daily sulfur intake is sequestered as cysteine in wool and hence is not available for recycling in the tissues. A similar situation presumably exists with camels. Of minerals in the soil, some such as Na, Cl, I, Co and Se are not readily taken up by the plant and hence these plants are potentially deficient for feeding animals. The situation is exacerbated in an animal such as the camel that seems to have higher requirements for Na and Se than sheep or cattle.

It is now clear that it is important to determine the levels of these nutritional variables in the feed and in rumen contents that permit normal feed intake and utilization, not only to remove artefacts from studies of ME use, but also to provide a sound basis for the provision of dietary supplements.

3.12 Meat Quality

Meat quality is usually judged on specific cuts of meat taken from carcasses that have been subjected to specific post-slaughter hanging and storage procedures. The meat is appraised for size of muscle, proportions of meat and fat, colour, pH and measures of tenderness such as resistance to standardized shearing forces. There do not seem to be similar studies with camels. The appraisal of meat quality by the consumer generally refers to tenderness, which is the sensation perceived by the consumer in the few seconds that a piece of meat spends between the teeth (Harper, 1999). Tenderness is equated by consumers with marbling, that is, the distribution of intra-muscular fat. The control of marbling has a genetic basis that in cattle is currently being intensively studied. Tenderness is also influenced, however, by the physiological age of the animal, by sex and plane of nutrition and by pre- and post-slaughter treatment. Post-slaughter, enzymes in the muscle anaerobically convert glycogen into lactic acid and pH falls. It is desirable for the ultimate pH to be below 5.7. Stressful conditions under which an animal is held in the period leading up to slaughter can reduce the amount of glycogen and hence the ultimate pH of meat and its tenderness. There are two different hormone systems involved (Ferguson *et al.*, 2001). Chronic stress leads to the release of glucocorticoids especially cortisol from the adrenal cortex. The reduction in glycogen in response to cortisol results in meat that is dark red in colour, dry and undesirably tough. This meat differs from the bright red, tough meat produced following adrenalin release in response to acute stress such as fighting. No studies of this nature seem to have been reported for camels but presumably the minimizing of stress to camels awaiting slaughter would benefit meat tenderness.

Meat quality also has implications for human health. Enteropathogenic bacteria are probably often ingested by ruminants, but bacteria in the normally functioning rumen seem to exert some control over the invaders. This control is weakened, however, when animals are held without food while awaiting slaughter. With dung accumulated in the holding yard providing a rich source of inoculum, the populations of bacteria such as *Salmonella* spp. and

Escherichia coli expand in the intestines (Brownlie and Grau, 1967) and increase the risk of contamination of meat following slaughter.

3.13 The Microbial Ecology of the Camel's Digestive Tract

Studies into the microbial communities of the forestomach of the Arabian camel have been carried out at the Gut Microbiology Laboratory, The University of Queensland, Gatton Campus since 2003. Our research programmes have focused mainly on the diversity of bacteria and protozoa and the function of major bacterial groups. Major lactate-producing bacteria (LAB) and lactate-utilizing bacteria (LUB) were cultured, isolated and characterized. The most efficient LUB were tested for their probiotic characteristics and for their ability to prevent acidosis in the rumen of animals. Similarly, cellulose-degrading bacteria were tested for their ability to degrade different sources of cellulose.

3.14 Protozoa

Studies into the protozoan population were carried out by Ghali (2005), who used microscopic identification protocols and the Hawksley chamber for enumeration. The protozoa count when the camels were fed a roughage diet was always higher than the roughage plus grain diet at roughage to concentrate ratio of 40:60. When the camels were fed roughage, the dominant species of protozoa were the *Entodonium* spp., representing 83–92% of the protozoa. Following the inclusion of grain in the diet the percentage of this species dramatically decreased, however (Table 3.2).

On the other hand the percentage of *Epidinium* spp. increased and became the dominant species when camels were fed roughage and a grain supplement. The increase in the percentage from only roughage to a roughage and grain diet was 18.3-fold at 0 h, 13.3-fold at 8 h and 14.7-fold at 16 h after feeding. Although the *Eudiplodinium* spp. represented a smaller percentage than *Epidinium* spp. when the camels were fed a roughage and grain diet, approximately 16%, there was also an increase in the percentage of *Eudiplodinium* spp. by at least 1.5-fold when the camels' diet changed from roughage only to a roughage and grain diet. There were also some other species of protozoa in the rumen of the camel that haven't been reported before. Those were the *Dasytricha* spp., the *Oligoisotricha* spp. and the *Buetschilia* spp., but they represented less than 2% of the total protozoa population. These species were all present when the camels were fed roughage diet but

Table 3.2. The numbers and proportions of the different types of protozoa found in the forestomach of the Arabian camel (*Camelus dromedarius*) when camels were fed roughage and roughage + grain diets (source: Ghali, 2005).

	Roughage diet[a]	Roughage + grain[a]
Total number of protozoa ($\times10^4$/ml)[b]	9.9 ± 2.75	2.5 ± 2.1
Forestomach pH[c]	6.37	5.35
Protozoa species (% population)		
Entodinium spp.	86.3	11.7
Epidinium spp.	4.6	70.4
Eudiplodinium spp.	7.27	16.67
Diplodinium spp.	0.83	1.65
Dasytricha spp.	0.40	0.60
Oligoisotricha spp.	0.75	ND
Buetschilia spp.	0.10	ND

[a]Diets were Rhodes grass (*Chloris gayana*) or Rhodes grass + steam-flaked barley; [b]the difference in protozoa numbers between diets is significant ($p<0.05$); [c]the effect of pH on protozoa number is significant ($p<0.05$). ND, not detectable.

only the *Dasytricha* spp. was found when the camels were fed roughage and grain diet at 0 h sampling.

Camels fed a roughage diet had an average forestomach pH of 6.3 to 6.5 and the dominant species of protozoa under these conditions were the *Entodinium* spp. The *Epidinium* spp. and *Eudiplodinium* spp. were only a small percentage (Table 3.2). In contrast, supplementation of the roughage diet with steam-flaked barley decreased the pH to 5.3. The *Entodinium* spp. dramatically decreased in number and the *Epidinium* spp. increased in number, becoming the dominant protozoa by percentage. The percentage of the *Eudiplodinium* also increased, but not as much as the *Epidinium*. Also the *Dasytricha*, *Oligoisotricha* and *Buetschilia* spp. disappeared completely when the pH dropped below 6.0 except for one camel where some *Dasytricha* spp. were found when the pH was 5.8.

3.15 Bacteria

To our knowledge, studies on the bacterial community diversity in the foregut of dromedary camels are not well documented. Only three reports have been published on the microbial community of the digestive tract of camels: one using culture-dependent methods, the other on the predominant lactic-acid producing and utilizing bacteria associated with diet change, and the third one on the rumen bacteria with regards to tannin toxicity (Hungate *et al.*, 1959; Tjakradidjaja *et al.*, 1999; Ghali *et al.*, 2004).

The bacterial community of the forestomach of the Arabian camel has been the focus of several investigations at The University of Queensland since 2002. Early investigation aimed at identifying the predominant lactic acid-producing (LAB) and utilizing (LUB) bacteria using culture-dependent and culture-independent techniques. Biochemical characteristics and the effect of diet on their populations have also been investigated (Ghali *et al.*, 2004; Ghali *et al.*, 2011). Recently we investigated the bacterial diversity using a clone library approach based on the 16S rDNA sequence analysis (Samsudin *et al.*, 2011). Results of our work revealed the identity of the key LAB and LUB in the forestomach of the Arabian camel. The key LAB were closely related to strains of *Streptococcus bovis* and *Selenomonas ruminantium*. Isolates closely related to *S. ruminantium* were also able to utilize lactic acid at variable rates. Other isolates of LAB were closely related to *Lactococcus garvieae*, *Butyrivibrio fibrosolvens* and *Prevotella ruminicola*. Among the LAB the most commonly (~30%) isolated were closely related to *Lachnospira pectinoschiza*. These isolates produced only small amounts of lactic acid, however (Ghali *et al.*, 2004).

The main L-lactate producers were those isolates closely related to *S. bovis*, *S. ruminantium* and *L. garvieae*, whereas the efficient lactate utilizers were *S. ruminantium* related isolates. D-Lactate was produced by isolates closely related to either *Lachnospira pectinoschiza* or *S. ruminantium*. In addition to *S. bovis* being the key LAB in the foregut of the Arabian camel, we have established the similarity of camel *S. bovis* isolates to those from cattle, sheep and deer (Ghali *et al.*, 2004).The predominant bacteria isolated and characterized in these studies were identical and/or closely related to those typically found in true ruminants (e.g. *S. ruminantium*, *B. fibrisolvens* and *S. bovis*). In addition, some of the bacteria isolated represent novel species of *Lachnospira* and *Clostridium* in the context of lactic acid bacteria from a large herbivorous host. The report has also identified the bacterium *S. ruminantium* as being predominant in the camel foregut in response to grain feeding. Findings from this study have contributed to our understanding of microbial population and would provide opportunities to reduce foregut acidosis in the camel.

Recently the molecular diversity of the foregut bacterial community in the Arabian feral camel (*Camelus dromedarius*) in central Australia was investigated through comparative analyses of 16S rRNA gene sequences prepared from the foregut contents of 12 adult feral camels fed on native vegetation (Samsudin *et al.*, 2011). The majority of cloned bacteria were affiliated with the bacterial phylum Firmicutes (67% of total clones) and were related to the classes Clostridia, Bacilli and Mollicutes, followed by the Bacteroidetes (25%) that were mostly represented by the family Prevotellaceae. The remaining phyla were represented by Actinobacteria, Chloroflexi, Cynophyta, Lentisphaerae, Planctomycetes, Proteobacteria and Sphirochaetes. Moreover, a small number of clones of cultivated bacteria were identified as *Brevundimonas* sp., *Butyrivibrio fibrisolvens*, *Prevotella* sp. and *Ruminococcus flavefaciens*. Sequence data from Samsudin *et al.* (2011) represent novel bacterial sequences representing new species, several new genera and probably a new family. The novelty in this foregut environment is remarkable, where 97% of the operational taxonomic units (OTUs) were distantly related to any known sequence in the public database. Furthermore, this research should not only contribute to our knowledge about the poorly understood microbial ecosystem of the camel gastrointestinal tract, but also should enable an understanding of the inter-relationships of these microorganisms to the animal's productivity and performance. To achieve these objectives, Samsudin *et al.* (2011) suggested that future studies should be carried out focusing on identifying a variety of active members and understanding their functional role within the gut environment. Further studies using a metagenomic approach will help in achieving these objectives.

3.16 Archaeal Microorganisms

A preliminary study into the population structure of faecal methanogenes in the Bactrian camel (*Camelus bactrianus*) from two zoos in the USA was carried out using separate 16S rRNA gene libraries for each zoo (Turnbull *et al.*, 2012). The two zoos, the Southwick Zoo (Mendon, MA) and the Potter Park Zoo (Lansing, MI) are 1220 km apart.

Although methanogen sequences belonging to the genus *Methanobrevibacter* were dominant in both libraries, they showed significant differences in diversity and structure. Population structure analysis revealed that only two OTUs were shared between libraries, whereas two OTUs were unique to the Southwick Zoo library and seven OTUs were unique to the Potter Park Zoo library. It was concluded that these preliminary results highlight how methanogen population structures can vary greatly between animals of the same species maintained in captivity at different locations. Authors also suggested the need to carry out additional studies using alternative techniques such as next generation sequencing to analyse a larger group of animals under controlled diets to gain further insights into the diversity of gastrointestinal methanogens in captive and wild Bactrian and dromedary camels (Turnbull *et al.*, 2012).

3.17 Conclusion

Camelids evolved as herbivores with different anatomical, physiological and behavioural adaptations that are most suited for the arid and semi-arid environments. Their digestive systems developed through a long history of feeding on native range vegetations that are low in digestibility and contain anti-nutritional factors that might act as inhibitors for some of the forestomach microbes. Although the digestive systems of the camelids are anatomically different from that of the true ruminant animals (i.e. cattle, sheep and goats), physiologically they act the same and camelids, like cattle, sheep and goats, rely on vast and diverse numbers of microorganisms to digest food and extract energy. Fermentation processes in camelids share similarities and differences to those in ruminant animals and require further investigation to improve our understanding of the role and function of the forestomach and the whole digestive tract.

References

Abbas, A.M., Mousa, H.M., Lechner-Doll, M. and von Engelhardt, W. (1995) Nutritional value of plants selected by camels (*Camelus dromedarius*) in the Butana area of the Sudan. *Journal of Animal Physiology and Animal Nutrition* 74, 1–8.

Abouheif, M.A., Basmaeil, S.M. and Bakkar, M.N. (1990a) A standard method for jointing camel carcasses with reference to the effect of slaughter age on carcass characteristics in Najdi camels. I. Wholesale cut weight. *Asian-Australasian Journal of Animal Science* 3, 97–102.

Abouheif, M.A., Basmaeil, S.M. and Bakkar, M.N. (1990b) A standard method for jointing camel carcasses with reference to the effect of slaughter age on carcass characteristics in Najdi camels. II. Variation in lean growth and distribution. *Asian-Australasian Journal of Animal Science* 3, 155–159.

Brownlie, L.E. and Grau, H. (1967) Effect of food intake on growth and survival of Salmonellas and *Escherichia coli* in the bovine rumen. *Journal of General Microbiology* 46, 125–134.

Dörges, B., Heucke, J. and Dance, R. (2003) Observations on the effect of camels grazing the vegetation of central Australia. Technote No. 118, 1–5. (http://www.primaryindustry.nt.gov.au)

Emmanuel, B. and Emady, M. (1976) A comparative study of endogenous urea degradation in ruminants. *Comparative Biochemistry and Physiology Part A: Physiology* 59A, 215–217.

Emmanuel, B. Howard, B.R. and Emady, M. (1976) Urea degradation in the camel. *Canadian Journal of Animal Science* 56, 595–601.

Engelhardt, W.V., Lechner-Doll, M., Heller, R., Schwartz, H.J., Rutagwenda, T. and Schultka, W. (1988) Physiology of the forestomach in camelids with particular reference to adaptation to extreme dietary conditions. A comparative approach. *Animal Research and Development* 28, 56–70.

Engelhardt, W.V., Dycker, C. and Lechner-Doll, M. (2007) Absorption of short-chain fatty acids, sodium and water from the forestomach of camels. *Journal of Comparative Physiology B* 177, 631–640.

Ferguson, D.M., Bruce, H.L., Thompson, I.M., Egan, A.F., Perry D. and Shorthose, W.R. (2001) Factors affecting beef quality-farm gate to chilled carcass. *Australian Journal of Experimental Agriculture* 41, 879–891.

Ghali, M.B. (2005) The bacterial and protozoal diversity of the gastro-intestinal tract of the dromedary camel. PhD thesis, The University of Queensland, Gatton, Australia.

Ghali, M.B., Scott, P.T. and Al Jassim, R.A.M. (2004) Characterization of *Streptococcus bovis* from the rumen of the dromedary camel and rusa deer. *Letters in Applied Microbiology* 39, 341–346.

Ghali, M.B., Scott, P.T., Alhadrami, G.A. and Al Jassim, R.A.M. (2011) Identification and characterisation of the predominant lactic acid producing and utilising bacteria in the foregut of the feral camel (*Camelus dromedarius*) in Australia. *Animal Production Science* 51, 597–604.

Gibbs, M.J. and Johnson, D.E. (1993) Livestock Emissions. In: *International Methane Emissions.* US Environmental Protection Agency, Climate Change Division, Washington, DC.

Guerouali, A. and Wardeh, M.F. (1998) Assessing nutrient requirements and limits to production of the camel under its simulated natural environment. In: *Proceedings of the Third Annual Meeting for Animal Production under Arid Conditions.* United Arab Emirates University Publishing Unit, 36, pp. 36–51.

Guerouali, A., Zine Filali, R., Vermorel, M. and Wardeh, M.F. (1989) Maintenance energy requirements and energy utilisation by the dromedary at rest. *CIHEAM-Options Mediterraneennes* 59–69.

Harper, G.S. (1999) Trends in skeletal muscle biology and the understanding of toughness in beef. *Australian Journal of Agricultural Research* 50, 1105–1129.

Heller, R., Lechner, M. and Engelhardt, W. (1986a) Forestomach motility in the camel (*Camelus dromedarius*). *Comparative Biochemistry and Physiology Part A: Physiology* 84A, 285–288.

Heller, R., Cercasov, V. and Engelhardt, W.V., (1986b) Retention of fluid and particles in the digestive tract of the llama (*Lama guanacoef. glama*). *Comparative Biochemistry and Physiology Part A: Physiology* 83A, 687–691.

Hogan, J.P. (1996) Options for manipulating nutrition if feed supply is immutable. *Australian Journal of Agricultural Research* 47, 289–305.

Höller, H., Breves, G., Lechner-Doll, M. and Schulze, E. (1989) Concentrations of volatile fatty acids and acetate production rates in the forestomachs of grazing camels. *Comparative Biochemistry and Physiology* Part B: *Comparative Biochemistry* 93B, 2, 413–416.

Hungate, R.E., Philips, G.D., McGregor, A., Hungate, D.P. and Buechner, H.K. (1959) Microbial fermentation in certain mammals. *Science* 130, 1192–1194.

Kadim, I.T., Mahgoub, O., Al-Marzooqi, W., Zadjali, S., Annamalai, K. and Mansour, M.A. (2006) Effects of age on composition and quality of muscle *Longissimus thoracis* of the Omani

Arabian Camel (*Camelus dromedarius*). *Meat Science* 73, 619–625.

Kaske, M., Osman, T., Lechner-Doll, M., Larsson, M. and Engelhardt, W.V. (1989) Circadian changes of forestomach motility and of rumination in camels. *Asian-Australian Journal of Animal Science* 2, 301–302.

Kennedy, P.M. and Doyle, P.T. (1993) Particle-size reduction by ruminants - effects of cell wall composition and structure. In: Jung, H.G., Buxton, D.R., Hatfield, R.D. and Ralph, J. (eds) *Proceedings of the International Symposium on Forage Cell wall Structure and Digestibility.* USDA-Agricultural Research Service and the U.S. Dairy Forage Research Center, Madison, Wisconsin, 7–10 October 1991, pp. 499–534.

Lechner-Doll, M., Rutagwenda, T., Schwartz, H.J., Schultka, W. and Englehardt, W.V. (1990) Seasonal changes in ingesta mean retention time and forestomach fluid volume in indigenous camels, cattle, sheep and goats grazing a thornbush savannah pasture in Kenya. *Journal of Agricultural Science, Cambridge* 115, 409–420.

Lechner-Doll, M., von Engelhardt, W., Abbas, A.M., Mousa, H.M., Luciano, L. and Reale, E. (1995) Particularities in forestomach anatomy, physiology and biochemistry of camelids compared to ruminants. In: Tisserand, J.L. (ed.) Elevage imentation du dromadaire – Camel production and nutrition. *Options Méditerranéennes, Serie B: Etudes et Recherches.* CIHEAM, Paris 13, 19–32.

Liu, Q., Dong, C.S., Li, H.Q., Yang, W.Z., Jiang, J.B., Gao, W.J., Pei, C.X. and Liang, Z.Q. (2009) Forestomach fermentation characteristics and diet digestibility in alpacas (*Lama pacos*) and sheep (*Ovis aries*) fed two roughage diets. *Animal Feed Science and Technology* 154, 151–159.

MAFF (1987) Energy Allowances and Feeding Systems for Ruminants. Tech. Bull. No. 433. Her Majesty's Stationary Office, London, pp. 11–17.

Mohamed, I.M. (2007) Evaluation of growth performance for growing Maghraby camel fed on un-conventional feed. *International Journal of Agriculture & Biology* 9, 18–21.

Mousa, H.M., Ali, K.E. and Hume, I.D. (1983) Effect of water deprivation and urea metabolism in camels, desert sheep and desert goats fed dry desert grass. *Comparative Biochemistry and Physiology Part A: Physiology* 74A, 3, 715–720.

Okine, E.K., Mathison, G.W. and Hardin, R.T. (1989) Effect of changes in frequency of reticular contractions on fluid and particulate passage rate in cattle. *Journal of Animal Science* 67, 3388–3396.

Okine, E.K., Mathison, G.W. and Hardin, R.T. (1990) Effect of changes in attributes of reticular contraction on fecal particle sizes in cattle. *Canadian Journal of Animal Science* 70, 159–166.

Petri, J. (1927) Notiz iiber den Harnstoffqehalt des Kamelharns. *Zeitschrift für Physikalische Chemie* 166, 125–127.

Phillips, A., Heucke, J., Dörges, B. and O'Reilly, G. (2001) Co-grazing cattle and camels. A report for rural industries research and development corporation, RIRDC Publication No 01/092 Barton, ACT, Australia.

Pinares-Patino, C.S., Ulyatt, M.J., Waghorn, G.C., Lassey, K.R., Barry, T.N., Holmes, C.W. and Johnson, D.E. (2003) Methane emission by alpaca and sheep fed on Lucerne hay or grazed on pastures of perennial ryegrass/white clover or birds foot trefoil. *Journal of Agricultural Science, Cambridge* 140, 215–226.

Read, B.E. (1925) Chemical constituents of camel's urine. *Journal of Biological Chemistry* 64, 615–617.

Rübsamen, K. and Engelhardt, W.V. (1978) Bicarbonate secretion and solute absorption in the glandular pouch in the forestomach of the llama. *American Journal of Physiology Endocrinology and Metabolism* 235, E1–E6.

Rutagwenda, T., Lechner-Doll, M., Kaske, M., Engelhardt, W.V., Schultka, W. and Schwartz, H.J. (1989) Adaptation strategies of camels on a thornbush pasture: comparison with other domestic animals. *Options Méditerranéennes, Serie Seminaires*, CIHEAM, Paris 2, 69–73.

Samsudin, A.A., Evans, P.N., Wright, A-D.G. and Al Jassim, R. (2011) Molecular diversity of the foregut bacteria community in the dromedary camel (*Camelus dromedarius*). *Environmental Microbiology* 13, 3024–3035.

San Martin, F. (1987) Comparative forage selectivity and nutrition of South American camelids and sheep. PhD thesis. Texas Tech University, Lubbock, TX.

Schmidt-Nielsen K., Schmidt-Nielsen B., Jarnum, S.A. and Houpt, T.R. (1957) Body temperature of the camel and its relation to water economy. *American Journal of Physiology* 188, 103–112.

Tjakradidjaja, A.S., Brooker, J.D. and Bottema, C.D.K. (1999) Characterisation of tannin resistant bacteria from the rumen fluid of feral goats and camels with restriction analysis of amplified 16S rDNA. In: Brooker, J.D. (ed.) *Tannins in Livestock and Human Nutrition.* Australian Centre for International Agricultural Research, Adelaide, pp. 161–165.

Turnbull, K.L., Smith, R.P., Benoit St-Pierre, B. and Wright, A-D.G. (2012) Molecular diversity of methanogens in fecal samples from Bactrian camels (*Camelus bactrianus*) at two zoos. *Research in Veterinary Science* 93, 246–249.

Waghorn, G.C., Shelton, I.D. and Thomas, V.J. (1989) Particle breakdown and rumen digestion of fresh ryegrass (*Lolium perenne* L.) and lucerne (*Medicago sativa* L.) fed to cows during a restricted feeding period. *British Journal of Nutrition* 61, 409–423.

Weston, R.H. and Hogan, J.P. (1986) Nutrition of Herbage-fed Ruminants. In: Alexander, G. and Williams, O.B. (eds) *The Pastoral Industries of Australia, Practice and Technology of Sheep, Cattle, Goat and Deer Production.* Sydney University Press, Sydney, pp.183–210.

Williams, V.J. (1963) Rumen function in the camel. *Nature* 197, 1221.

Wyburn, R.S. (1980) The mixing and propulsion of the stomach contents of ruminants. In: Ruckebusch, Y. and Thivend, P. (eds) *Digestive Physiology and Metabolism in Ruminants.* AVI Publishing Company, Inc., Westport, Connecticut, pp. 35–53.

4 Camel Body Growth

Isam T. Kadim and Osman Mahgoub
Department of Animal and Veterinary Sciences, College of Agricultural and Marine Sciences, Sultan Qaboos University, Muscat, Sultanate of Oman

4.1 Introduction

The birth weight and growth physiology of camels are important parameters to meat producers and animal scientists. Camel body growth and development, which are linked to increases in the size and weight of the body tissues to increase body tissue cell mass, are the governing factors of camel meat production. The net result of this growth and development is an increase in the size and weight of muscle, bone, fat and other body tissues. Changes in the shape and composition of the animal body are due to cellular differentiation, which is of significance in meat production and quality. Generally, some parts of the animal body are more preferable to producers and consumers than others.

Birth weight and growth rate of camels are affected by sex, genetics, nutrition and the health status of the animal. The growth rate of camels varies according to the availability of feed and may be altered seasonally, especially in the outdoor feeding systems. On the other hand, camel growth can be partially attributed to the availability of browse throughout the year, regardless of good or bad years for grazing animals. Animal body development tends to be affected to a lesser extent by the above factors. The camel growth curve relates live weight to age and it has an S-shape, which it is similar to other livestock. The growth and development of camel has three phases like other livestock: (i) increasing live body weight with increasing age; (ii) explosive body growth and development; and (iii) low body growth. Hammond (1940) stated that the first wave of animal growth begins at the head and moves towards the chest, and the second wave starts at the limbs and moves upwards. Although, the rib and loin regions are the most preferred parts of the carcass for meat consumption, they are the last regions to develop. The growth and development of muscles in different locations reflects their functions and the animal's needs. For example, the early development of the muscles of the distal limbs reflects the need for mobility required to forage for food, whereas the development of the jaw muscles promotes effective mastication of the food (Berg and Butterfield, 1976).

Although the camel female matures earlier, the male is larger and heavier than the female at a later age. Different parts of the body tissues grow at different rates, and this causes differences in size between males and females. This chapter reviews information on birth weight, growth rate, mature body weight and the estimation of live weight using various regression equations.

4.2 Birth Weight

Heritability of birth weight is numerically higher in camels than in other meat species, which could lead to rapid improvement, largely through the adoption of a good genetic selection programme. Improvement of camel nutrition and applying efficient management intensive rearing systems are appropriate tools to increase calf birth weight. Heavier camel birth weights should result in better calf survival and improve meat production. The birth weight is influenced by the sum of factors contributing to the nutrition of the fetus in the uterus. Hansard and Berry (1969) summarized the factors influencing the birth weight of the animals and estimated that the largest component of variation (36%) is attributed to the combined genotypes of the dam (20%), fetus (17%), parity (7%), nutrition (6%), sex (2%) and the maternal age (1%). The exact role of these factors in the camel has not been investigated. Progeny records for 383 Saudi camels were genetically evaluated for growth performance of body weights at birth and bimonthly up to 12 months of age (Al-Sobayil *et al.*, 2007). The authors found that the phenotypic variations and heritabilities for most growth traits were moderate or slightly high, ranging from 7.0% to 35.2% and from 0.24 to 0.40, respectively.

A female camel usually bears a single calf and more rarely twins after a gestation period of 13 months (385–400 days). The lifetime output of the female is limited because of low calving rates, long milk-feeding periods and long gestation periods. The birth weight of the calf is the basis of meat production. The higher birth weight gives advantage for calves to grow and to have a higher body weight at weaning and maturity. The newborn camel walks within a few hours of birth (Fig. 4.1), but remains close to its mother until 5 years of age (Bhargava *et al.*, 1965). It has been reported for dromedary camels, however (Hammadi *et al.*, 1998), that the duration of the post-partum interval in lactating females is positively correlated with daily milk production.

Shalash (1983) stated that heredity is one of the main factors influencing camel fetus growth, either directly via the genotype of the fetus or indirectly through the genotype of the dam. There are also other factors that influence camel birth weight, including nutritional status of the dam (Hammadi *et al.*, 2001), sex (Burgemeister, 1975; Zhao *et al.*, 1999), health status of the dam, season of year and parity (Mutairi,

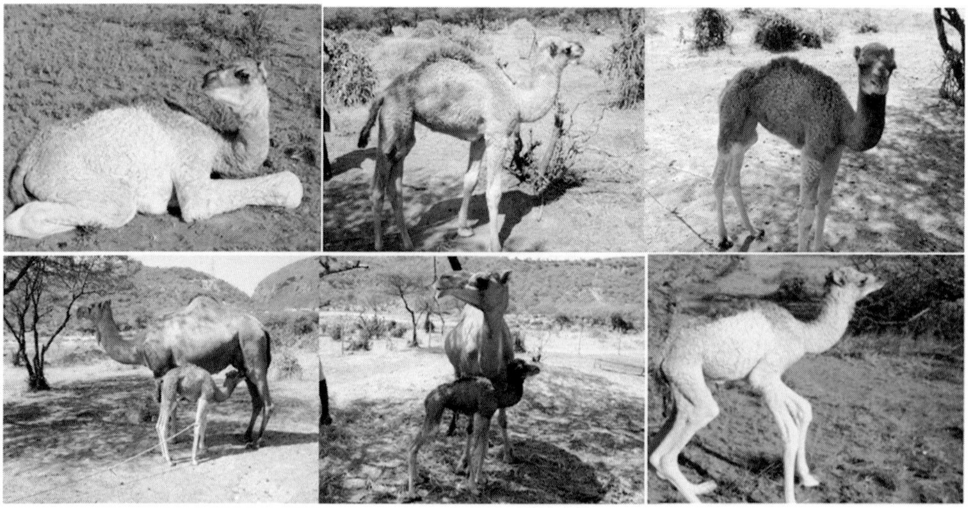

Fig. 4.1. Camel calves at birth.

1999; Khan *et al.*, 2003). Musa (1984) and Russel (1975) found that the development of the camel fetus and its associated growth curve are strikingly similar to that of cattle. The nutritional status of the dam had a direct effect on fetal growth and development, a factor that would be important in camel meat production. Poor nutritional levels during gestation could lead to increased prenatal mortality. Undernutrition during the last months of gestation could result in low rates of body weight gain, loss of body weight in dams and abortion in primiparous females (Hammadi *et al.*, 2001). The season has a significant effect on the birth weight of camel calves and under traditional management the season exerts its effect through the availability of feed and management conditions of the camels. In Kenya, young dams that received basic veterinary care gave birth to calves weighing 29.0 kg at birth, whereas the non-treated ones gave birth to calves weighing 25.8 kg (Wilson, 1998). Malnutrition of the dam during the last phase of gestation and the beginning of lactation reduced the birth weight of camel calves and resulted in a low growth rate of both young and adult camels and probably abortions in females (Zeleke and Bekele, 1999). In general, during the first month postpartum, nutritional requirements for newborn calves are limited and the milk quantity of dams is insufficient for a moderate gain.

The birth weight of the calf was significantly affected by the prepartum nutritional level of the dam (Hammadi *et al.*, 2001). Pregnant dromedary females fed either supplemented or unsupplemented diet had calves weighing 30.3 and 23.4 kg at birth, respectively. Early postnatal growth is also influenced by the accessibility to the dam's milk because a negative weight gain occurred when milk production decreased (Russel, 1975; Musa, 1984). Among 242 calves, 38 (13.7%) died between birth and 6 months of age (Mutairi, 1999). A higher mortality rate (14.9–20.3%) was reported by Megersa *et al.* (2008) among Ethiopian camels. This level of mortality rate was attributed to poor management systems including nutrition. However, Bakheit *et al.* (2009) reported no significant difference in birth weight of calves raised under semi-intensive and traditional systems in Sudanese camels. Figure 4.2 shows camel calf birth weights for 239 calves during 14 seasons as measured by Mutairi (1999). It was found that the season had a significant effect on the birth weight, which reflected feed availability.

In contrast with many other species, female camel calves are usually as heavy as males at birth (Yagil, 1985; Ouda, 1995; Wilson, 1998; Mutairi, 1999; Bakheit *et al.*, 2009). Female camel calves (37.2 kg) were slightly, but not significantly, lighter than males (38.2 kg) as reported by Yagil (1985). The birth weight of male Sudanese calves (39 ± 0.31 kg) was significantly heavier than the females (36 ± 0.34 kg) (Bakheit *et al.*, 2009). Similarly, Harmas *et al.* (1990) reported average camel birth weights of 35 ± 95 kg and 34.05 ± 0.46 kg for male and females, respectively, with significant sex effects. Mutairi (1999) found that the average birth weight of male and female calves were 37.45 ± 0.55 kg and 37.27 ± 0.41 kg, respectively. Among 136 deliveries of male calves, 70 (51.5%) were in the range 36–40 kg (Mutairi, 1999). The number of female calves in the same range was 57 (55.3%). More than 90% of the birth weights of male and female calves ranged from 31 kg to 45 kg. Gitao (2006) stated, however, that male calves usually weighed more than females but the difference is not significant. This discrepancy in birth weights between the sexes is retained during the first year of age.

The season of birth and age of dam had a significant effect on the birth weight of the camel. Harmas *et al.* (1990) reported mean birth weights of 31 kg for dam camels at the age of 5–6 years and 36 kg for dam camels at the age of 7–10, 11–15 and 15 years or more.

The birth weight of the camel calf is significantly affected by the age of the dam. The means of calves' birth weights were 30.83 ± 0.76 kg for camels at the age of 5–6 years, 35.82 ± 0.56 kg for camels at the age of 7–10 years, 36.26 ± 0.68 kg for camels at the age of 11–15 years, and 35.46 ± 0.72 kg for camels at the age of 15 years or more (Harmas *et al.*, 1990). Mutairi (1999) found that the

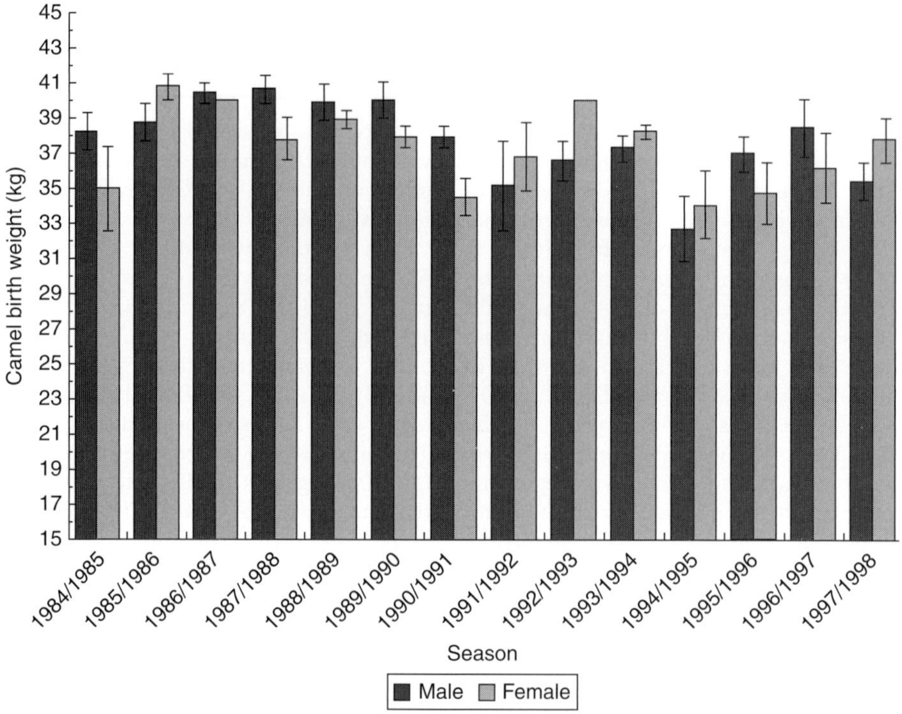

Fig. 4.2. Average birth weights of male and female camels in different seasons (Mutairi, 1999).

average male and female weights of camel calves in the first delivery age of dam (4–5 years) were 34.48 kg and 34.66 kg, respectively (Table 4.1). Low calves' weight at the first delivery might be attributed to the small size of the dam. The correlation between the mother's age and birth weight of camel calves was significantly high (0.87; Mutairi, 1999).

The heredity and geographical location are other factors affecting prenatal growth of the camel, directly via the genotype of the fetus and indirectly through the genotype of the dam or for nutritional reasons related to the availability of natural grazing (Zhao *et al.*, 1999). A positive correlation exists between the maternal body size and the prenatal growth of the fetus. The birth weights of the dromedary from different locations ranged from 26 kg to 45 kg (Table 4.2). The birth weight of the camel calves in the Indian Bikaneri breed ranged from 26.3 kg to 51.2 kg, and the average birth weight of males was 38.19 kg and females 37.19 kg, with a pooled average of 37.3 kg (Bhargava *et al.*, 1965). In Tunisia and Kenya calves are smaller (Hertrampf, 2004), weighing an average of 25.8 kg and 30.9 kg, respectively (Burgemeister, 1975), whereas the range in birth weight of Sudanese camel calves was between 30 kg and 40 kg (El-Amin, 1979). The weight of the newborn camel in Australia was between 30 kg and 40 kg. In a study of 239 Saudi camel calves over 14 years (Mutairi, 1999), birth weights of male and female camel calves were 37.45 ± 0.55 kg (19–50 kg) and 37.37 ± 0.41 kg, (23–49 kg), respectively. Birth weight varied between different breeds of Indian camels in the same region, averaging 38.8 kg, 31.8 kg and 31.0 kg for Bikaneri, Kachhi and Jaisalmaeri types, respectively (NRCC, 1990). Similarly in Pakistan, Khan *et al.* (2003) reported averages of 44.0 kg, 41.0 kg, 44.0 kg and 40.0 kg for Kachhi, Gaddi, Bagri and Dhatti types, respectively. In Tunisia and Kenya, camel

Table 4.1. The effect of dam age on the birth weight of male and female camel calves (Mutairi, 1999).

	Mean birth weight (kg ± SE)				
Dam age	*N*	Male	*N*	Female	Overall average
1st delivery (4–5 years)	37	34.48 ± 0.73	22	34.66 ± 0.69	34.57 ± 0.7
2nd delivery (6–7 years)	22	37.68 ± 0.73	15	37.44 ± 1.03	37.44 ± 0.93
3rd delivery (8–9 years)	15	36.58 ± 1.39	16	38.06 ± 0.92	37.32 ± 1.10
4th delivery (10–11 years)	7	38.13 ± 1.75	8	40.25 ± 1.48	39.19 ± 1.52
5th delivery (12–13 years)	5	40.60 ± 1.22	2	36.33 ± 0.27	38.47 ± 0.90

Table 4.2. Birth weight of dromedary camels from various locations.

Country	Number	Birth weight (kg)	References
India	525	37.3–41.0	Bhargava *et al.* (1965); NRCC (1990)
Pakistan		42.3	Khan *et al.* (2003)
Tunisia	29	35 ± 6 (male) 32 ± 5 (female)	Kamoun (1995)
Tunisia	22	27.0–28.4	Hammadi *et al.* (2001)
Libya	158	35.95 ± 0.46 (male) 34.05 ± 0.46 (female)	Harmas *et al.* (1990)
Saudi Arabia	32 49	39.94 ± 0.61 (male) 37.77 ± 0.53 (female)	Mutairi (1999)
Sudan	20	39.0 ± 0.31 (male) 36.0 ± 0.34 (female)	Bakheit *et al.* (2009)
Sudan	22	32.5–35	El-Amin (1979)
Kenya	61	27.8	Wilson (1998)
Australia		30–40	Camel Newsletter (1997)

calves were smaller (Hertrampf, 2004), weighing an average of 25.8 kg and 30.9 kg, respectively (Burgemeister, 1975), whereas Sudanese camels had higher birth weights of between 30 kg and 40 kg (El-Amin, 1979). Variations in average camel birth weights were also reported in other countries including: 27 kg for Somali camels (Field, 1979; Simpkin, 1983; Ouda, 1995), 27 kg for Tunisian camels (Hammadi *et al.*, 2001) and 39 kg for Indian camels (Bissa, 1996).

4.3 Body Weight Gain

Live body weight at a particular age is a reflection of body weight gain. Camel daily body gains widely vary between regions, breeds and within the same breed and are affected by sex, nutritional status, management system and locations. During the first month of age, nutritional requirements for the young camel remain limited and a small quantity of milk is sufficient for a moderate gain (Hammadi *et al.*, 2001). In the two subsequent months, differences in calf weight gain reflect differences in milk production between supplemented and non-supplemented females (Hammadi *et al.*, 1999). The average daily weight gain of the camel can reach 1000 g/day under the most favourable fattening conditions (Kamoun, 2004). In general, the daily growth rate of young dromedary camels ranges from 300 to 1000 g/day from birth to 1 year of age. The number of calves already borne by the dam affects birth weight and weight gain at a younger age. Compensatory growth in the camel occurs from 24–30 months of age and the effects disappear after this age. Weight gain in the young camel is governed by the individual's genetic makeup, but needs to be developed by adequate feeding and proper management. If the camel calf is allowed to suckle all of its mother's milk, its weight gain can reach as much as 650 g/day (Gitao, 2006). Several studies under various conditions have indicated the potential of the rapid growth rate during the early months of camel life (Dong, 1979; Degen *et al.*, 1987; Ismail, 1996; Iqbal *et al.*, 1999; Bakheit *et al.*, 2009). Dromedary camel calves double their birth weight within 1 month (Ismail, 1996; Iqbal *et al.*, 1999), whereas Bactrian calves have doubled at the age of 2.5 months (Chapman, 1985). Gitao (2006) stated that the average daily gain is 250 g/day up to 6 months, reaching 180–200 kg live body weight at 1 year old. A 6-month study by Iqbal *et al.* (1999) compared the efficiency of body weight gains in camel calves raised by farmers with those raised under the management systems in Pakistan (Table 4.3). The daily growth rate of camel calves from 7 days to 6 months of age under the management system and on farmer's premises were 0.75 kg and 0.82 kg, respectively, with 0.79 kg average growth rate. They attributed the difference to the better level of attention given to calves by the farmers. Similary, Bakheit *et al.* (2009) found that the average daily gains of camel calves under intensive versus traditional systems were 477 ± 10.94 g versus 352 ± 10.55 g (birth to 6 months), 542 ± 8.25 g versus 272 ± 15.98 g (6 to 12 months) and 585 ± 8.37 g versus 316.71 ± 5.46 g (12 to 18 months), with the average growth rate under intensive and traditional systems of 535 ± 9.83g and 317 ± 5.46 g, respectively. Kenyan camel calves from supplemented dams grew significantly faster than those from non-supplemented ones, gaining 441.3 g/day and 424.8 g/day versus 275.7 g/day and 307.7 g/day in Kargi and Ngurunit types, respectively (Kuria *et al.*, 2004). The growth performance of 14 camel calves weaned at 3 months of age was compared with those weaned using the traditional system of calf rearing (Saini and Singh, 2006). The first group (126.70 kg body weight) was weaned and maintained on concentrate mixture and available fodder, whereas the second group (139.40 kg body weight) was kept with the dams to suckle and graze. The average daily weight gain over 137 days was significantly higher in supplemented calves (535.03 g/day) than in non-supplemented ones (491.24 g/day). Dry matter intake in weaned calves increased with increasing body weight varying from 1.66 to 2.25 kg/100 kg body weight with an average of 1.92 kg/100 kg live body weight. The study of Iqbal *et al.* (1999) indicated that the average monthly gain weight of camel calves was 23.62 kg with a range of 21.60–25.90 kg. Hammadi *et al.* (2001) reported camel body weights of 27 kg, 48 kg, 65 kg and 79 kg at birth, 30, 60 and 90 days of age, respectively, which implies a 580 g/day average daily gain from birth to 90 days of age. Bissa (1996) reported an average body weight of 39, 119 and 171 kg at birth, 90 and 180 days, respectively, for Indian camels indicating an average growth rate of 733 g/day between birth and 180 days. These values are lower than those commonly reported for beef cattle, but it should be noted that camels are normally raised under an extensive system depending mainly on rangeland grazing rather than on feedlots. The effect of feeding camels under stall-feeding with three levels of nutrition on the Omani camel growth rate was studied by Mahgoub *et al.* (2012). They found that the average daily gain was 71 g/day, 347 g/day and 400 g/day for the camels given 1.5%, 2.0% and 2.5% body weight feed intake, respectively.

Generally, the growth curve for camel calves follows a pattern more or less similar to that of other animal species. The growth curves for males and females are given in Fig. 4.3. There was no obvious rapid phase of early growth and weight gain was maintained steadily well past sexual maturity. This more or less continuing growth is probably partially attributed to the availability of browse. Table 4.4 shows the average daily weight gain from birth to 3 years of 32 male and 49 female calves were 420 g/day and 393 g/day, respectively (Mutairi, 1999).

Table 4.3. Comparative growth rate of camel calves raised under an intensive management system or on farmer's premises (Iqbal *et al.*, 1999).

Month	Growth rate under management system (kg/month)	Growth rate with farmer (kg/month)	Gain (kg/day)	Overall (kg/month)
1	20.25 ± 1.91 (20.04 ± 1.75)	28.18 ± 1.546	0.80	24.21
2	18.85 ± 2.60 (21.9 ± 0.73)	24.29 ± 0.94	0.72	21.60
3	21.9 ± 0.73 (23.78 ± 1.16)	23.09 ± 1.17	0.75	22.50
4	26.04 ± 1.15 (22.23 ± 1.63)	25.77 ± 0.55	0.86	25.90
5	24.7 ± 2.23 (25.05 ± 1.89)	24.76 ± 1.11	0.82	24.73
6	22.45 ± 0.78 (20.82 ± 1.89)	23.11 ± 0.99	0.76	22.78
Overall	22.37 (22.30)	24.87 ± 0.49	0.79	23.62

Figures in parentheses are the actual weights of calves.

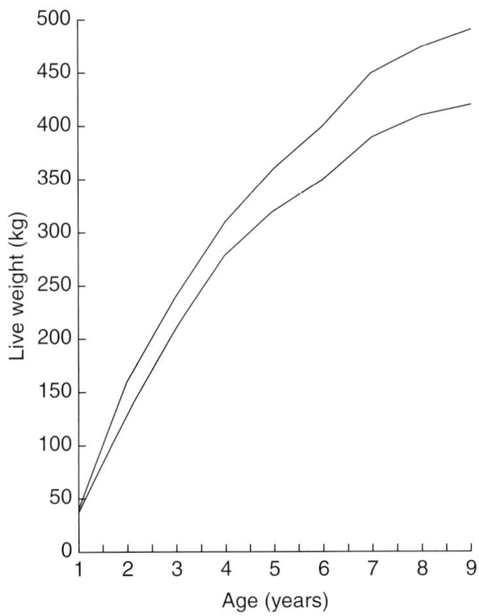

Fig. 4.3. Growth curves of Sudanese camels (Wilson, 1978).

The average daily body weight gain of Bikaneri camels in different age groups under improved management at the National Research Centre on Camels is presented in Fig. 4.4 (Tandon *et al.*, 1988). The average daily gain gradually increased from 400 g/day in the 0–1 year group to a maximum of 720 g/day in the 7–8 years group then declined to 300 g/day by 10–11 years of age. The growth rates given should, however, be considered as maximum values because a growth of 300 g/day will result in a weight gain of over 100 kg/year, which does not match the change in body weights shown in Fig. 4.4 at the ages of 6 years or more. The graph resembles the specific growth pattern in other farm animals, with an inflection point where growth rate is at a maximum at about 1 year old. This pattern is affected by many factors such as weaning age, season and nutrition.

Burgemeister (1975) and Zhao *et al.* (1999) also studied the weekly postnatal growth performance of camel calves and found that male calves tend to grow faster than females. No differences in live weight between the sexes were observed up to 2 years old (Ouda *et al.*, 1992), up to 4 years of age (Simpkin, 1983) or up to 14 years old (Mutairi,1999).

Mortality was recorded in 38 (13.7%) camel calves between birth and 3 years old (Mutairi, 1999). About 66% of mortality cases occurred in the winter season. The dam near delivery should therefore be transferred to a warm place where the calf is kept under observation for a while. The causes of mortality in dromedary camel calves aged less than a year in India were studied by Sena *et al.* (2006) who found that the highest incidence of mortality occurred at 0–3 months, followed by 6 months – 1 year and 3–6 months of age. Causes of death included heat stroke (18.367%), impaction (4.081%), encephalitis (6.122%), enteritis (26.530%), pneumonia (40.816%) and respiratory distress (2.040%). Calf mortality from birth to 3 months, 3–12 months, 12–24 months, 24–36 months and more than 36 months was 8.6, 3.3, 5.3, 4.7 and 5%, respectively

Table 4.4. Growth rate of camel calves from birth to 3 years of age (Mutairi, 1999).

Sex	*N*	Average birth weight (kg)	Average weight (3 years)	Daily gain (g/day)
Males	32	39.94 ± 0.61	489.31 ± 11.29	410
Females	49	37.77 ± 0.53	468.60 ± 9.24	393

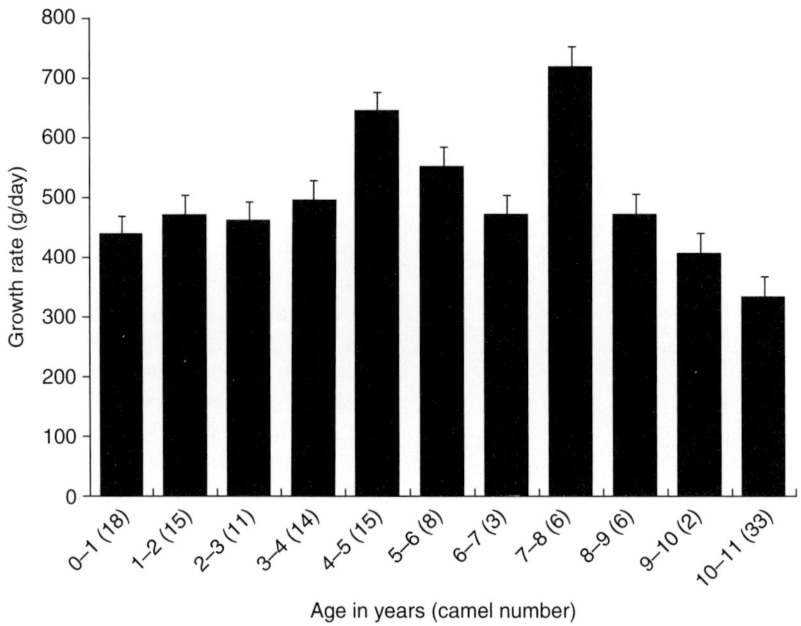

Fig. 4.4. Average body weights of camels of different ages (Tandon *et al.*, 1988).

(Bissa *et al.*, 2004). These authors concluded that the total mortality of female calves was 21.89% before they reached the age of first calving. Special care should be provided to camels during their first delivery season and neonatal care of dromedary calves at 0–3 months of age is of utmost importance in order to reduce mortality.

Daily growth rate is not consistent throughout life. The average daily weight gain of camels in different age groups is presented in Fig. 4.5. Average daily growth rate gradually increases from 400 g/day in the 0–1 year group to a maximum of 720 g/day in the 7–8 year group then declines to 300 g/day by 10–11 years of age. This pattern is obviously affected by many factors such as weaning age, season, nutrition, etc. The pre- and post-weaning growth rate has a significant effect on the final weight of camels. The pre-weaning growth rate of the camel calf is affected by milk quantity and the system of management (Babiker and Tibin, 1989). Postnatal growth in animals is characterized by continued growth of the skeleton, musculature and organs until the animal reaches approximately 50–60% of its mature weight and then skeletal and organ weight gains have slowed down and fat deposition has increased to a modest rate (Trenkle and Marple, 1983). In camels, however, this point is not reached until after 2 years of age, which is rather long compared with other animals (Zhao, 1995). The limited research work carried out on improving camel nutrition showed significant relationships between average body gain and daily feed intake of concentrates in the

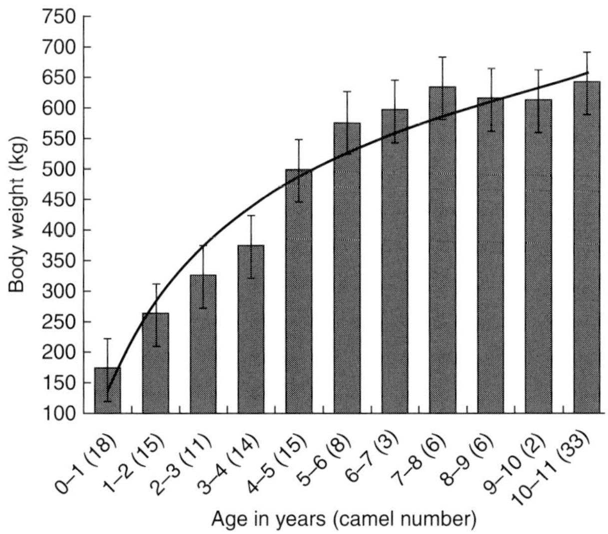

Fig. 4.5. Average daily weight gain of camels in different age groups (Tandon *et al.*, 1988).

dromedary camels. Camels fed a diet of high protein and energy gained more weight (550 g/day) than non-supplemented camels fed only mangroves (260 g/day) (Kamoun, 1995). Tribal camel calves in Kenya grew at a rate of 222 g/day up to 6 months of age in dry years and at a rate of 655 g/day in wet years (Field, 1979). Post-weaning growth rate depends mainly on husbandry practices and conditions of the vegetation (Babiker and Tibin, 1989). It is partially dependent on the availability of browse throughout the year according to Wilson (1998). The average daily gain of the Saudi male camel was between 567 g and 790 g in a fattening period over 90 days (Mutairi, 2009). Field (1984) studied the growth patterns of two groups of dromedary calves, one under pastoral conditions and another under control conditions in which the young received at least 75% of their dam's milk. The former group showed an average daily gain of 222 g and 255 g during the dry and the wet seasons, respectively, whereas gains ranged from 378 g to 655 g for the latter group.

The figures of Field's study revealed the important influence of dam milk on the growth and development of camels. This reflects the negative effect of competition for milk between calves and owners under the pastoral management system. The postnatal calf growth curves given by Field (1984) also showed a better performance by calves born during the wet season, irrespective of the breed of the camel. However, this advantage is not permanent, because calves born in the dry season seem to catch up after 9–12 months by means of compensatory growth. Eighteen young camels (6–14 months of age) were randomly assigned to three feeding groups: concentrate pellets of 18% crude protein at the rate of 1.5% body weight plus either lucerne hay, Rhodes grass hay or wheat straw treated with ammonia gas (Bakkar *et al.*, 1999). They found average body weights were 315 kg, 298 kg and 291.4 kg and average daily gains were 932 g/day, 803 g/day and 767 g/day for the three groups, respectively. Camel calves born during the rainy season had gains of 318 g/day and 289 g/day in the favourable and unfavourable seasons, respectively, typical of extensive farming systems (Pacholek *et al.*, 1999).

On average, research station calves attained a net weight gain of 135.45 ± 6.35 kg, whereas farmers' calves gained 149.20 ± 3.06 kg during the 6-month study period (Iqbal *et al.*, 1999). Degen *et al.* (1987) reported that camel calves averaged 155 kg at 180 days and the average daily gain to that age was 0.68 kg. El-Badawi (1996) also reported similar results of 150–175 kg live weight at 6 months old. A decline in the growth rate during the second month of age could be attributed to an increase in feed requirements coupled with restricted milk feeding. The growth rate of camel calves depends on the availability of feed; the enhanced growth rate might be due to the abundant supply of lush vegetation.

Although the daily weight gain is low in Ethiopian camels, Zeleke and Bekele (1999) found that female immature camels (1–4 years old) had a significantly higher daily weight gain (59.4 g) than males of the same age (33.2 g), which is in agreement with the finding of Simpkin (1985) for Kenyan camels. This could be because females mature earlier than males and reach puberty at an earlier age. Zeleke and Bekele (1999) reported that a significantly higher daily weight gain (63.1 g) was recorded in camels 1–2 years old compared with camels 3–4 years old (29.5 g). Similar results were reported by Simpkin (1985). It is a natural phenomenon that young animals of 1–2 years old grow faster than those 3–4 years old under the same management conditions.

This pattern of growth is affected by many factors such as breed, geographical location, weaning age, season and nutritional status of the animals. Weaning of the camels could contribute to a slow daily body gain which is noticeable during the second year of life. In Saudi Arabia, male and female camels had similar growth rates with a rise in daily body gain from 780 g/day in the first month after birth to 1040 g/day in the fifth month and then a fall to 400 g/day in month 12 (Wilson, 1998). For Tunisian camels, daily body gain from birth to 90 days of age was 806 g/day (Hammadi *et al.*, 2001). Camel weight at birth, 30, 60 and 90 days of age was 27 kg, 48 kg, 65 kg and 79 kg, respectively, which indicated that body weight gain was 580g/day between birth and 90 days of age (Hammadi *et al.*, 2001). For Indian camels, Bissa (1996) reported average live body weights of 39 kg, 119 kg and 171 kg at birth, 90 and 180 days of age, respectively, with an average daily body gain of 733 g/day between birth and 180 days of age. In Morocco, 10-month-old camels of the Guerzni type were compared with the Marmouri type receiving the same amount of feed. The average growth rate of the Guerzni type was significantly higher (410 g/day) than the Marmouri type (320 g/day) (Guerouali and Acharbance, 2004). These values for camel live body gains are lower than those reported for cattle. However, these differences might be because camels are normally raised under extensive systems depending mainly on rangeland grazing rather than on feedlots. The limited research carried out on improving camel productivity demonstrated significant relationships between average daily gain and average daily intake of concentrates for dromedary camels (Kamoun, 1995).

Pre- and post-weaning body condition, health and growth rates have significant effects on final weights of camels. The pre-weaning growth rate of the camel calf is affected by the milk availability of the dams, the system of management and management interventions, and the availability of browse throughout the year (Babiker and Tibin, 1989). The season in which the camel is born is an important factor affecting body weight gain because it effectively controls the amount of feed available to animals. Camel calves in Kenya grew at a rate of 222 g/day up to 6 months of age in dry years and at a rate of 655 g/day in wet years (Field, 1979).

Post-weaning growth rate depends mainly on husbandry practices and conditions of the vegetation (Babiker and Tibin, 1989). The postnatal growth performance of the camel was studied by Kamoun (1995) and the results showed that male calves grow faster than females; the average daily weight gain from birth to weaning was 760 g and 620 g for male and female camels, respectively.

Reports of weight gains in camels vary greatly. Some examples are quoted here. The camel daily weight gain can be increased by good-quality feed rather than an increase in quantity. The availability of feed and the use of an efficient management system are more effective in increasing weight when used at an early stage of life. Under open-range conditions, a live weight increase of 1 kg/day has been reported. Rates of feed intake in relation to live weight gain are generally in the range of 4–8 kg intake for each kilogram of weight gain (Wilson, 1998). On an energy basis, 22–29 MJ metabolizable energy are needed for each kilogram of weight increase for a young camel. According to Wilson (1998), camel calves gained 870 g/day in very early life at a metabolizable energy intake of 19.5 MJ/day with an average daily gain to 180 days of 680 g. It has been reported that camels fatten rapidly when fed 15–20 kg of a mixture of straw, beet pulp silage, molasses and 10–15% barley grains, whereas camels fed sugar beet tops gain as much as 1.5 kg/day and can be made ready for slaughter in 60 days (Wilson, 1998). Kamoun (1995) reported a growth rate of 280 g/day in camels on medium-quality forage (wheat straw or oat hay and concentrates). In Tunisia, 1-year-old camels fed for 175 days on oat hay *ad libitum* and a concentrate of wheat bran and olive pulp gained 326–565 g/day, eating 1.6 kg DM/100 kg live weight or 61 g DM/kg$^{0.75}$/day at a conversion ratio of 7.4:1.0 (Wilson, 1978). In Ethiopia, camels at 265 kg live weight gained 100 g/day over 90 days at lower intake levels of about 1.25 kg DM/100 kg or 50 g DM/kg$^{0.75}$. The pre-weaning growth rate of the camel calf is affected by the competition for milk, milk quantity and management (Babiker and Tibin, 1989). Post-weaning growth rate depends mainly on husbandry practices and the condition of the vegetation (Babiker and Tibin, 1989). These studies indicated that the camel is a slow grower. More studies are required, however, to assess camel growth under optimal management conditions and determine the optimum slaughter weight. Water is another factor that significantly influenced camel daily growth rate. A 1-year-old camel (200 kg) gained 430 g/day over 6 months on a daily watering system and gained 380 g/day when on a weekly watering system (Wilson, 1978). Breeds and types with lighter birth weights might gain weight more rapidly than breeds of heavier weights and thus might become physically mature at an earlier age (Wilson, 1998). In Egypt, animals fed on a high-energy diet comprising cottonseed, rice, molasses and mineral mix gained 150 kg in body weight in 6 months (almost 0.82 kg/day). Well-fed young camels under intensive conditions have gained 0.58 kg/day. In tribal situations, 222 g/day have been recorded in poor years to the age of 6 months, and 655 g/day in wet years when the calves were allowed to take all the mother's milk. Camels can utilize urea, molasses, dried sugar tar and poor-quality feedstuffs treated with ammonia. These reports suggest that feedlotting might be cost effective. It could be possible to combine a system of low-cost open-range breeding with an intensive finishing period. Reported live weight variations in camels suggest there is ample scope for genetic manipulation and development of meat type (Manefield and Tinson, 1997).

Management systems play a significant role in camel growth and production. Such systems include environmental conditions, composition and size of the herd, and the way camels are raised alone or mixed with sheep, goats and cattle (Bakheit, 1999). Camel feed management should consider production patterns of feed availability and production target, such as increased milk production, prolonged lactation, herd growth, reproduction and meat production (Hashi *et al.*, 1995).

There are many factors that influence growth rate, including heredity, nutrition, sex and health. Heredity mainly affects prenatal growth of the camel, directly via the genotype of the fetus and indirectly through the genotype of the dam (Shalash, 1983). The nutritional status of the dam might also have a direct effect on fetal growth. Poor nutritional levels during gestation could lead to increased prenatal mortality. After a gestation period of 13 months, a camel female usually bears a single calf, and rarely twins. The newborn camel walks within hours of birth but remains close to its mother until maturity.

4.4 Mature Body Weights

A camel's genetic make-up is usually the most important factor in its final mature weight. The effects of management and veterinary care can influence the time it takes to reach the final weight. Deworming, the control of ticks and supplementary feeding can improve camel mature weight. These dividends can be used to reduce the age at which camels can be sold for meat production.

There are varying estimates of camel mature live body weight in the literature. Mature body weights are mainly related to breed or type, age, sex, nutrition and general health of the camels (El-Amin, 1979). Camels attain maturity comparatively slowly, as indicated by the average body weights of camels in different age groups (Fig. 4.4), which show that camels reach a maximum live weight of about 650 kg at 7–8 years of age. The graph resembles the sigmoid-shaped growth curve of other farm animals and matches the pattern in Fig. 4.3 with an inflection point at the 7–8 year group. In general, mature weights of camels range from 400 kg to 800 kg.

Breed and type affect camel live weight. Most breeds at maturity weigh 450–550 kg with the heavy camel breeds weighing up to 660 kg when mature and in good condition (Wilson, 1984; Hertrampf, 2004). Wilson (1984) provided estimates of live weights of camels in different countries with the lightest live body weights in Somalia desert camels (350–400 kg) and the heaviest live weight (660 kg) in Indian camels. The Sudanese camel is generally larger than other north African and Somali types (Wilson *et al.*, 1978). The mean mature body weight of 14 males was 476 ± 75 kg with a range of 340–581 kg, whereas the corresponding average weight of 35 mature females was 419 ± 47 kg with a range of 307–522 kg (Wilson, 1978). The weights of Australian mature camels ranged from 514 kg to 635 kg for males and 470 kg to 510 kg for females (Camel Newsletter, 1997). Iranian camels at 5 years old ranged in weight from 340 to 430 kg (Khatami, 1970).

Nutritional history and body condition have significant effects on live weight. The ability of camels to cope with food shortage is the result of a long evolutionary process in natural conditions where food ability seasonally fluctuates. In arid conditions, all the adaptive mechanisms, and especially body fat mobilization strategies, are of considerable importance in determining camel mature body weight. Live weights of mature well-finished male desert Saudi camels were between 359 kg and 512 kg with an average of 475 kg (Babiker and Yousif, 1987). There are, however, reports of extremely high body weights in camels. For instance Herrmann and Fischer (2004) reported a range of live weights of between 350 kg and 800 kg for eight Somali × Turkana castrated male camels. They attributed the high live weight to the body condition of the camels, which was ranked as very good without any external injuries.

In Kenya, camels raised under a traditional pastoral system can reach 400–500 kg mature body weight at 9 years of age, whereas this range can be achieved at 4 years of age under ranch conditions. Benadir type camels under ranch conditions reach 600–700 kg by 5 years of age, whereas traditionally managed camels reach a similar weight at older ages.

Although there are no marked differences between the sexes in live weight earlier in life, males become heavier than females at older ages. Mature male camels were heavier than females by 38% in the study by Kurtu (2004). Wilson (1998) reported that male camels in Ethiopia weigh about 685 kg at maturity, whereas females weigh about 525 kg, supported by another report by the same author that showed higher body weights for mature males (448 kg) than females (414 kg). Among Indian camel breeds (Bikaneri, Jaisalmaeri, Kachhi and Arabi × Bikaneri), the Bikaneri breed during 3 years shows the highest average adult body weight of 617 kg for males and 578 kg for females (Khanna, 2004). In Tunisia, the mature weight of the male camel was 450 kg, whereas for the female it was 400 kg. The average live weight of camels in Algeria was 600–700 kg (Djemali, 2004).

4.5 Estimating Live Weight from Body Measurements

Live body weight measurements are an important management tool to assess the growth and development of camels for meat production, but are often unavailable in pastoral communities. Although the daily recorded of body weight gain of a camel varies widely owing to many factors, it can still be used to assess economic benefit for improvement animal productivity (Kamoun, 2004). According to Al-Hazni *et al.* (1994) live weight can be used in breed identification, classification and as a prerequisite for the management and conservation of camel genetic resources. As an alternative to actual weight, body measurements (Fig. 4.6) can be used as a useful tool in estimating the live weight of animals in a less complicated and inexpensive method (Goe *et al.*, 2001; Keith *et al.*, 2009). Equations can be derived from live body measurements to estimate the live weight of camel at different ages where a weighing machine is not available (Nasser, 1999). Actual or estimated live body weight therefore can be used to evaluate growth rate, stocking rates, feeding programmes, nutritional status, drug administration (dosage) and management practices such as selection of replacements. It is an important tool in marketing animals because farmers can get good value for their animals when prices are decided upon body weight. Body weight information can also be used in determining the value of animals and the efficiency of rearing. Weighing camels under field conditions is difficult; several equations have therefore been developed by researchers to estimate the weight of camels from body measurements.

The scaled variables live weight and time after birth were subjected to a number of mathematical transformations in order to give various sets of transformed standardized growth curves with different properties. They could be used to understand the set of curves and also could be better used for various purposes. For any transformations of the variable live weight, the growth of animals and also the mean growth curves

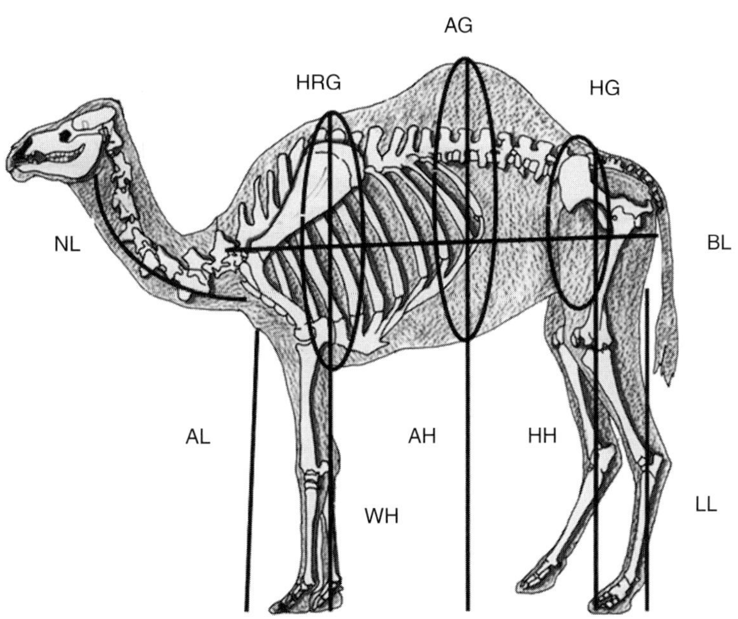

Fig. 4.6. Diagram indicating where measurements are taken on the live camel. Abdominal height (AH), wither height (WH), heart girth (HRG), hip girth (HG), hip height (HH), abdomen girth (AG), leg length (LL), arm length (AL), neck length (NL) and body length (BL).

can be represented in principle by some form of growth equation. Mathematical models to describe the growth of animals and tissues have been developed by many scientists. The analyses of components of growth curves, such as functional growth rates and estimates of live body weight, are being examined by animal geneticists as potential tools for the selection of animals with specific growth-related traits. The estimated growth curves can be used to compare animals across species, breeds or sexes, termed genetic size-scaling (Taylor and Fitzhugh, 1971; Taylor, 1980a). The growth curves can be generated for different species for various and mature body weights and metabolic age. Comparison regression lines across species showed sigmoid growth curves of a similar pattern, but significant deviations from the mean curves were found, which indicated that all species do not reach the same fraction of their mature size at a given metabolic age (Taylor, 1980b).

Some equations to predict live body weight use linear body measurements such as shoulder height (SH), heart girth (HG) and abdominal girth (AG), and others used a single measurement (HG only). With any type of equation, the mature weight of lean and tall camels will be overestimated, whereas the weight of short and stocky camels will be underestimated. The derived equations might only predict 10–15% of the actual mature weight. Camels attain maturity comparatively slowly and the benefits of meat production in camels are unfortunately not matched by high growth rates. An attempt was made to develop a regression equation of weight on girth for larger camels. Girth measurements can be taken on both standing and crouched camels round the chest with the tape behind the front legs.

Attempts have been made by researchers to determine the value of individually recorded body measurements for predicting the weight of a camel. The acquisition of such information might be of considerable practical importance, particularly in a breeding programme where improving body weight and carcass quality are major objectives. A close relationship between body weight and the sum of heart girth, abdominal girth and shoulder height in male Kenyan camels was reported by Field (1979). Also, Abouheif *et al.* (1985) recorded positive and significant correlations between carcass weight and hip girth, heart girth and abdominal girth in male Abhawi camels of Saudi Arabia. They also found a number of reliable prediction equations to determine carcass weight using these body measurements. On the other hand, Wilson (1978) found that predicting live weight from chest girth measurements in male Sudanese camels was less reliable than in the case of cattle.

Heart girth (HG), abdominal girth (AG) and wither height (WH) measurements can be taken using a tape measure. Ihuthia *et al.* (2010) measured 59 camel calves and found that AG and HG had highly significant correlation coefficients of $r = 0.957$ and $r = 0.934$, respectively, to the live weight of camel calves than abdominal height (AH) ($r = 0.432$). They suggested that AG had the greatest influence on the live weight of camel calves followed by HG. A multiple regression equation including HG, AG and AH for the live weight estimates had a coefficient of determination R^2 accounting for 92.3% of the variation, which was higher than for individual or any two combined variables. The AG coefficient of determination R^2 accounts for 92.4%, HG 87% and AH 17.2% of the body weight variation (Ihuthia *et al.*, 2010). The correlation of predicted weights and the actual live weights was high (r = 0.963) for the multiple regression equation: Live body weight (kg) = −100.6 + 101.2 AG (m) + 58.2 HG (m) + 9.91 AH (m) derived from the three linear body measurements. Heart girth, abdominal girth and shoulder height (SH) measurements were taken by Kuria *et al.* (2007) using an ordinary tape measure on 64 suckling calves aged 3 weeks to 7 months. They suggested that HG had the greatest influence on live weight ($r = 0.96$). On the contrary, Ihuthia *et al.* (2010) suggested that AG is the best single weight estimator. The differences between the two studies could be due to the time of measurement. Time of measurement is crucial for accuracy and it should be early in the morning when the animals have not yet fed to reduce the measurement variability. Among body length, shoulder height and heart girth, the shoulder height in adult camels was found by Patel *et al.* (2007) to be a reliable measure for growth from its association with important body measurements.

Schwartz *et al.* (1983) developed a linear body measurement equation from mature camels, whereas Simpkin (1998) measured young camel calves and calves of more than 1 year. Kuria *et al.* (2007) and Ihuthia *et al.* (2010) developed an equation for camel calves up to 7 and 12 months, respectively. The prediction of body weight for camels less than 1 year old is important to assess their performance during this critical period. For comparison, different models estimating linear body data were fitted into Schwartz *et al.* (1983), Simpkin (1998), Kuria *et al.* (2007) and two equations of Ihuthia *et al.* (2010) to project growth curves (Fig. 4.7). Schwartz *et al.* (1983) used AG × HG × WH × 50 to estimate the weight of camels, whereas Nasser (1999) used HG × AG × SH × 50 from 56 camels to estimate their weights at 6 months and 6 years. On the other hand multi-factors were modified by Simpkin (1998) for camel calves. The growth curves generated by the regressions of Ihuthia *et al.* (2010) are linear and gave higher estimated calf weight for the first few months with low body weight thereafter. The regressions generated by Schwartz *et al.* (1983) and Simpkin (1998) are exponential though the former depicts a faster growth. The regression model of Kuria *et al.* (2007) gave a linear growth curve, which is similar to the first model of Ihuthia *et al.* (2010) with a faster growth rate. The differences among the estimated sigmoid growth curve for live body weight of camels could be due to the small number of animals used and the number of measurements taken.

Repeated measurements are difficult in a pastoral system because of the camels' high mobility and diverse locations, and inaccuracies arise because of human error in recalling information. The camel herds are usually under different pastoral systems and migration is the better time for data collection. Bissa *et al.* (1998) showed the growth of camel calves for the first year is linear and it can be used to predict live weight using two measurements: HG and SH (weight kg = 52.17 $SH^{1.64}$ $HG^{1.71}$ + 1.35). Kamoun (2004) modelled the growth of

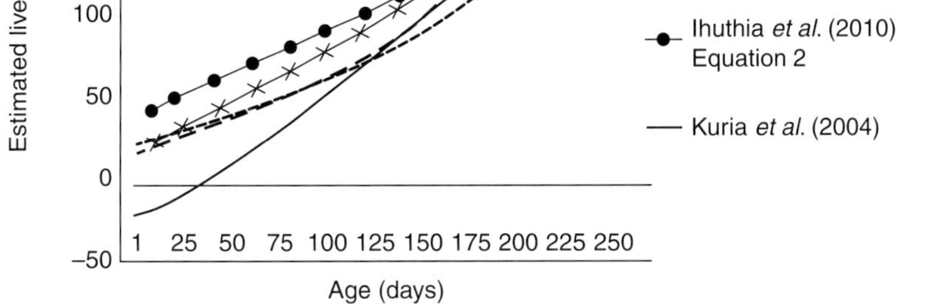

Fig. 4.7. A comparison of live weight estimates using prediction equations from four research groups (Ihuthia *et al.*, 2010).

young dromedaries using the following equation: live weight (kg) = Wo $e^{2.15\{1-e(-0.0059\,t)\}}$ (Kamoun, 2004).

Data on 227 Najdi male camels were used by Abouheif *et al.* (1986) to determine live weight using seven body measurements: neck length (NL), arm length (AL), leg length (LL), body length (BL), heart girth (HG), abdomen girth (AG) and hip girth (HG) (Fig. 4.6). They found that correlations of live body weight with HG, SG and AG were the highest among all the studied body measurements.

The monthly body weight of calves was determined by actual weighing or predicted using shoulder height × girth of shoulder × girth around the hump (Iqbal *et al.*, 1999). There was a close correlation between the two measurements with the range of body weights between 135.45 ± 6.35 kg and 149.20 ± 3.06 kg. Camel calves aged 5–180 days from 20 herds in four consecutive years were used to formulate a simple equation to estimate the body weight using height at withers and chest girth (Pacholek *et al.*, 1999). They established a barymetric formula from the measurement of heart girth (HG) of between 0.7 m and 1.5 m and found that sex had a significant effect on variation; with the same HG, females were heavier than males at birth. Predicted live weights varied between 30.56 ± 3.71 kg and 174.02 ± 3.01 kg in males and between 32.37 ± 3.67 kg and 168.80 ± 2.90 kg in females (Pacholek *et al.*, 1999).

4.6 Conclusion

The birth weight and growth physiology of camels are the principle parameters for meat producers and are affected by sex, genetics, nutrition and health status of the camel. Camel growth is linked to an increase in size and weight of the muscle, bone, fat and other body tissues to increase body tissue cell mass, which reflects their functions and the animal's needs. Body tissue cellular differentiation causes shape and composition changes of the camel body, which has a significant effect in meat production. The camel growth curve relates live weight and age and it can take an S-shape that includes three phases similar to other livestock. Weighing camels under field conditions is difficult; several equations have therefore been developed by researchers to estimate the weight of camels at different ages from linear body measurements.

References

Abouheif, M.A., Bakkar, M.N. and Basmaeil, S.M. (1985) Relationship of carcass weight and carcass measurements in Abhawi camels of Saudi Arabia. *Journal of College of Agriculture, King Saud University* 7, 51–59.

Abouheif, M.A., Basmaeil, S.M. and Bakkar, M.N. (1986) Estimation of body and carcass weights in Saudi Arabian Najdi male camels. *Arab Gulf Journal of Science Research* 4, 733–743.

Al-Hazni, M.A., Ghandour, A.M., and El-Gohar, M. (1994) A study of the biometry of some breeds of Arabian camel (*Camelus dromedaries*) in Saudi Arab. *Journal of King Abdual Aziz University Science* 6, 87–99.

Al-Sobayil, K.A., Khalil, M.H., Al-Saef, A.M., Mohamed, K.M. and Salal, S.A. (2007) Genetic evaluation of growth of calves at early stage in Saudi camels. *Journal of Camel Practice and Research* 14, 175–180.

Babiker, S.A. and Tibin, I.M. (1989) A note on desert camel meat production and characteristics. *The International Symposium on the Development of Animal Resources in the Sudan*, pp. 116–120.

Babiker, S.A. and Yousif, K.H. (1987) Carcass yield and characteristics of mature male camels of the Sudan. *Annual Report, Camel Research Unit, University of Khartoum, Sudan*, 120–124.

Bakheit, S.A.F. (1999) Studies on milk production and composition of camels (Camelus dromedarius) under nomadic system. MSc Thesis, University of Sudan, Khartoum.

Bakheit, S.A., Faye, B., Nikheila, A.M. and Majid, A.M. (2009) The impact of farming system on Sudanese camel calves growth rate. *The 2nd Conference of the International Society of Camelid Research and Development.* Djerba, Tunisia, 12–14 March, 2009, p. 81.

Bakkar, M.N., Basmaeil, S.M. and Hamam, A.A. (1999) Meat production and quality: two unique properties of young camels. *Proceedings of the International Workshop on the Camel Calf.* Ouarzazate, Morocco, 24–26 October, 1999, p. 218.

Berg, R.T. and Butterfield, R.M. (1976) *New Concepts of Cattle Growth*. John Wiley, New York.

Bhargava, K.K., Sharma, V.D. and Singh, M.A. (1965) A study of the birth and body measurement of camels (*Camelus dromedarius*). *Indian Journal of Veterinary Science and Animal Husbandry* 35, 358–362.

Bissa, U.J. (1996) Early growth and its association with certain blood parameters in different breeds of camel. MSc Thesis, Department of Animal Breeding and Genetics. College of Veterinary and Animal Science, Rajasthan Agricultural University, Bikaner. Camel Newsletter, No. 12. 1996, pp. 21–23.

Bissa, U.K., Yadav, B.S., Khanna, N.D. and Pant, K.P. (1998) Growth curve of body weight from birth to four years in Bikaneri breed of Indian camels (*Camelus dromedarius*). *The 3rd Meeting for Animal Production under Arid Conditions*. United Arab Emirates University, Vol. 2, 15–24.

Bissa, U.K., Yadav, S.B.S., Beniwal, B.K. and Sahani, M.S. (2004) Losses of female camel calves at different ages from birth to age at first calving. *Indian Journal of Animal Sciences* 74, 965–968.

Burgemeister, R. (1975) Elevage de chameaux en Afrique du Nord, 88. Eschborn, Germany, GTZ, pp. 40–48.

Camel Newsletter (1970) The Central Australian Camel Industry. *Camel Newsletter*, No 13, pp. 52–58.

Camel Newsletter (1997) The Central Australian Camel Industry. Camel Newsletter, No. 13, pp. 52–58.

Chapman, M.J. (1985) Bacterian camels. *World Animal Review* 55, 14–19.

Degen, A.A., Elias, E. and Kam, M. (1987) A preliminary report on the energy intake and growth rate of early weaned camel (*Camelus dromedarius*) calves. *Animal Production* 45, 301–306.

Djemali, M. (2004) Camel genetic resource in North Africa: a report on three Arabian countries. ICAR Technical Series, No. 11.

Dong, W.E.I. (1979) Chinese camels and their productivities. In: *Proceedings of Workshop on Camels, Khartoum*, Sudan, 18–20 December 1978, Vol. 55.

El-Amin, F.M. (1979) The dromedary camel of the Sudan. In: IFS Symposium Camels, Sudan, pp. 35–54.

El-Badawi, A.Y. (1996) The possibilities of using camels in new reclaimed lands. *In: Proceedings First Meeting of the National Committee for Camel Research in the Arab Republic of Egypt*. 16 November 1995, Cairo, Egypt.

Field, C.R. (1979) Camel growth and milk production in Marasabit district, Northern Kenya. IFS Prov. Rep. No. 6. *Camels International Science*, Stockholm, Sweden pp. 215–240.

Field, C.R. (1984) Camel growth and milk production in Marsabit District, North Kenya. In: Cockrill, W.R (ed.) *The Camelid. An All-purpose Animal*. Scandinavian Institute of Africa Studies, Uppsala, Sweden, pp. 209–230.

Gitao, G.C. (2006) *Camel Husbandry: A Practical Guide to Camel Husbandry*. Downtown Printing Works Ltd, Nairobi, Kenya.

Goe, M.R., Alldredge, J.R. and Light, D. (2001) Use of heart girth to predict body weight for working oxen in Ethiopia highlands. *Journal of Livestock Production Science* 69, 187–195.

Guerouali, A. and Acharbane, R. (2004) Camel genetic resources in Morocco: a report on three Arabian countries. ICAR Technical Series, No. 11.

Hammadi, M., Khorchani T., Khaldi, G., Abdouli, H., Slimane, N. and Renaville, R. (1998) Effect of dietary supplementation on productive and reproductive performance of dromedary camels under arid range conditions. In: *Third Annual Meeting for Animal Production and Future Perspectives*. May 1998, Abou-Dhabi.

Hammadi, M., Khorchani, T., Khaldi, G., Majdoub, A., Abdouli H., Slimane, N., Portetelle, D. and Renaville, R. (2001) Effect of diet supplementation on growth and reproduction in camels under arid range conditions. *Biotechnology Agronomy Society and Environment* 5, 69–72.

Hammond, J. (1940) *Farm Animals: Their Breeding, Growth and Inheritance*, 1st edn, Edward Arnold, London.

Hansard, S.L. and Berry, R.K. (1969) Fetal nutrition. In: Hafez, E.S.E. and Dyer, I.A. (eds) *Animal Growth and Nutrition*. Lea and Febiger, Philadelphia, Pennsylvania, pp. 238–247.

Harmas, S., Shareha, A., Biala, A. and Abu-Shawachi, H. (1990) Investigation on growth measures of Magrib camel (*Camelus dromedrius*), *Camel Newsletter* 7, 82 (abstract).

Hashi, A.M., Kamoun, M. and Cianci, D. (1995) Feed requirements of the camel. *Options Mediterraneennes (CIHEAM)*, pp. 71–80.

Herrmann, K. and Fischer, A. (2004) Method for hygienic slaughter of camel. In: Farah, Z. and Fischer, A. (eds) *Milk and Meat from the Camel; Handbook on Products and Processing*. vdt Hochschulverlag AG an ETH Zurich, Zurich/Singen.

Hertrampf, J.W. (2004) The "ship of the desert" as a meat supplier. *Fleischwirtschaft* 84, 111–114.

Ihuthia, P.M., Githaiga, W.R. and Margaret, W. (2010) Correlation of actual live weight and estimates of live weights of camel calves (Camelus dromedarius) in Samburu district of northern Kenya. *Journal of Camelid Science* 3, 26–32.

Iqbal, A., Gill, R.A., Khan, B.B., Younas, M. and Jasra, A.W. (1999) Comparative growth performance of camel calves kept under station and farmers' conditions. *Proceedings of the International Workshop on the Camel Calf.* 24–26 October 1999, Quarzate, Morocco, pp. 198–200.

Ismail, M.D. (1996) The dromedary camel as multipurpose animal. In: *International Workshop on Sustainable Use of Rangelands and Desertification Control*, 3-6 November, 1995, Jeddah, Saudi Arabia, pp. 1–14.

Kamoun, M. (1995) Dromedary meat: production, qualitative aspects and capability for transformation. *Option Mediterraneennes. Serie B, Etudes et Recherches* 13, 105–130.

Kamoun, M. (2004) Meat recording systems in camelids. In: Cardellino, R., Rosati, A. and Mosconi, C. (eds) *Current Status of Genetic Resources, Recording and Production Systems in African, Asian and American Camelids*, 30 May 2004, Sousse, Tunisia. *ICAR Technical Series* 11, 105–130.

Keith, L., Okere, C., Solaiman S. and Tiller O. (2009) Accuracy of predicting weights from body conformation and testicular morphology in pubertal Boer goats. *Research Journal of Animal Science* 3, 26–30.

Khan, B.B., Iqbal, A. and Riaz, M. (2003) *Production and Management of Camels.* Pak TM Printers, Pakistan.

Khanna, N.D. (2004) Camel breeds of India. *Journal of Camel Science* 1, 8–15.

Khatami, K. (1970) *Camel Meat: A New Promising Approach to the Solution of Meat and Protein in the Arid and Semi-arid Countries of the World.* Ministry of Agriculture, Tehran.

Kuria, S.G., Wahome, R.G., Gachuiri, C.K., Wanyoike, M.M. and Mwangi, J.N. (2007) Use of linear body measurements in estimating live weight of camel (*Camelus dromedarius*) calves in Kenya. *Journal of Camel Practice and Research* 14, 21–25.

Kuria, S.G., Gachuiri, C.K., Wanyoike, M.M. and Wahome, R.G. (2004) Effect of mineral supplementation on milk yield and calf growth of camels in Marsabit District of Kenya. *Journal of Camel Practice and Research* 11, 87–96.

Kurtu, M.Y. (2004) An assessment of the productivity for meat and carcass yield of camel (Camelus dromedarius) and the consumption of camel meat in the Eastern region of Ethiopia. *Tropical Animal Health and Production* 36, 65–76.

Mahgoub, O., Kadim, I.T., Al-Marzooqi, W., Al-Lawatia, S.A. and Al-Abri, A.S. (2012) Performance of Omani camels raised under stall-feeding management. *East African Agricultural and Forestry Journal* (in press).

Manefield, G.W. and Tinson, A.H. (1997) Camels, a compendium. Postgraduate Foundation, University of Sydney, Australia.

Megersa, B., Regassa, A., Kumsa, B. and Abunna, F. (2008) Performance of camels (*Camelus dromedarius*) kept by pastoralists with different degrees of experience in camel keeping in Borana, Southern Ethiopia. *Animal Science Journal* 79, 534–541.

Musa, B.E. (1984) Summary of studies on reproduction in the dromedary. In: Cockrill, W.R. (ed.) *The Camel: an All-purpose Animal.* Scandinavian Institute of Africa Studies, Uppsala, Sweden, p. 360–363.

Mutairi, S.E. (1999) Evaluation of Saudi camel calves' performance under an improved management system. *Proceedings of the International Workshop on the Camel Calf.* Ouarzazate, Morocco, 24–26 October, 1999, pp. 219–222.

Mutairi, S.E. (2009) Camel development in the Kingdom of Saudi Arabia. *The 2nd Conference of the International Society of Camelid Research and Development.* Djerba, Tunisia, 12–14 March 2009, pp. 71–72.

Nasser, A.K. (1999) Determination of live weight of one hump camel (*Camelus dromedarius*) at different ages and studying carcass chemical, physical and organoleptical properties. PhD thesis, Department of Animal Resources, Basrah University, Basrah, Iraq.

NRCC (1990). Annual Report, National Research Centre on Camel, Bikaner, India.

Ouda, J.O., Abui, J.O., and Woie, B.M. (1992) Production performance of Somali and Rendille camels and their crossbreds in Northern Kenya. *All Africa Conference on Animal Agriculture*, November, Nairobi, Kenya.

Ouda, J.O. (1995) Camel calf survival and performance under varying amounts of milk intake. *Camel Newsletter* 11, 42–45.

Pacholek, X., Lancelot, R., Lesnoff, M. and Messad, S. (1999) Growth performance of camel calves raised in the pasterol zone of Niger. *Proceedings of the International Workshop on the Camel Calf.* Quarzazate, Morocco, 24–26 October, 1999, pp.189–197.

Patel, M.K., Parnerkar, S., Wadhwani, K.N., Patel, K.S., Solanki, J.V. and Patel, A.M. (2007) Breed characteristics of Kachchhi camel. In: Gahlot, T.K. (ed.) *Proceedings of the International Camel* Conference *Recent Trends in Camelids Research and Future Strategies for Saving Camels*, 16–17 February 2007, Rajasthan, India, pp. 51–54.

Russel, W.S. (1975) The growth of Ayrshire cattle: an analysis of linear body measurements. *Animal Production* 21, 217–226.

Saini, N. and Singh, G.P. (2006) Effect of weaning on growth performance of camel calves. *Indian Journal of Dairy Science* 59, 344–348.

Schwartz, H.J., Dolan, R.B. and Wilson A.J. (1983) Camel production in Kenya and its constraints: productivity. *Tropical Animal Health Production* 15, 169–178.

Sena, D.S., Gorakh Mal, Sharma, N. and Sahani, M.S. (2006) Calf mortality in camels: a report. *Journal of Camel Practice and Research* 13, 171–172.

Shalash, M.R. (1983) The role of camels in overcoming world meat shortage. *Egypt Journal of Veterinary Science* 20, 101–110.

Simpkin, S.P. (1985) The importance of camels to subsistence pastoralists in Kenya. In: *Camel Disease and Productivity in Arid Lands of North Kenya*. Technical report E7, IPAL, Germany, pp. 163–192.

Simpkin, S.P. (1983) The effects of disease as constraints to camel production in Northern Kenya. In: *The Integrated Project in Arid Lands Technical Report E-7. Camel Diseases and Productivity in the Arid Lands of Northern Kenya*. United Nations Educational, Scientific and Cultural Organization, Nairobi, Kenya.

Simpkin, S.P. (1998) The effects of breed and management on milk yield of camels in Kenya. PhD Thesis, Department of Agriculture, University of Newcastle, UK.

Tandon, S.N., Bissa, U.K. and Khanna, N.D. (1988) Camel meat: present status and future prospects. *Annals of Arid Zone* 27, 23–28.

Taylor, S.C.S. (1980a) Genetic size-scaling rules in animal growth. *Animal Production* 30, 161–165.

Taylor, S.C.S. (1980b) Genetically standardized growth equations. *Animal Production* 30, 167–175.

Taylor, S.C.S. and Fitzhugh, H.A. (1971) Genetic relationship between mature weight and time taken to mature within a breed. *Journal of Animal Science* 33, 726–731.

Trenkle, A.H. and Marple, D.N. (1983) Growth and development of meat animals. *Journal of Animal Science* 57 (Suppl. 2), 273–283.

Wilson, R.T. (1978) Studies on the livestock of Southern Darfur, Sudan. V. Notes on camels. *Tropical Animal Health and Production* 10, 19–25.

Wilson, R.T. (1984) *The Camel*. Longman Group Limited, Essex, UK.

Wilson, R.T. (1998) Camel. In: Costa, R. (ed.) *The Tropical Agricultural Series*, Centre for Tropical Veterinary Medicine, University of Edinburgh, UK.

Yagil, R. (1985) *The Desert Camel Comparative Physiological Adaptation*. Vol. 5. Verlag Karger, Basel.

Zeleke, M. and Bekele, T. (1999) Camel herd health and productivity in Eastern Ethiopia selected semi-nomadic households. *Proceedings of the International Workshop on the Camel Calf*. 24–26 October, 1999, Ouarzazate, Morocco, pp. 213–217.

Zhao, X.X. (1995) *Ecophysiology and Reproduction in the Camelidae*. Gansu Science and Technology Press, Lanzhou, Gansu, China, pp. 1–28.

Zhao, X.X., Xue, H.W., Nie, Q.C. and Zhang, Y.K. (1999) Postnatal growth and development pattern of camel calves. In: *Proceedings of the International Workshop on the Camel Calf*. 24–26 October 1999, Ouarzazate, Morocco, pp. 201–208.

5 Slaughtering and Processing of Camels

Isam T. Kadim,[1] Mustafa M. Farouk,[2] Osman Mahgoub[1] and Alaa El-Din Bekhit[3]
[1]*Department of Animal and Veterinary Sciences, College of Agricultural and Marine Sciences, Sultan Qaboos University, Muscat, Sultanate of Oman;* [2]*AgResearch Limited, Ruakura Research Centre, Hamilton, New Zealand;* [3]*Department of Food Science, Division of Sciences, University of Otago, Dunedin, New Zealand*

5.1 Introduction

Dromedary camel slaughtering is rather difficult compared with that of other livestock because of the size of the animal and the amount of manual work involved. There are few specialized dromedary camel slaughtering plants in the world because of the limited production and the low per capita consumption of camel meat. Both male and female dromedary camels are slaughtered for meat, with more males slaughtered than females.

Although most dromedary camels are accustomed to being handled, the slaughtering procedure requires experience in all aspects including the loading and transporting of animals and other pre- and post-slaughter processes. Camels should be carefully unloaded on arrival at the abattoir to avoid stress. The most common method of slaughtering dromedary camels is by tying both the front legs at the knee joint to force the animals into a crouching position, then the head is pushed to one side and the jugular vein severed. In rural areas, dromedary camels are slaughtered on the ground in the open air which exposes the meat to contamination from dust and dirt. In some of the public slaughterhouses, there are also no facilities for carrying out the hygienic processing of meat, including a clean water supply and an adequate cooling system.

The majority of camels slaughtered today are not stunned pre-slaughter. For camel meat to become a mainstream source of animal protein and to be traded across all boundaries there is, however, a need to develop a pre-slaughter stunning method specific for camels to harmonize the religious as well as the modern animal slaughter welfare requirements.

This chapter describes the most important steps in the traditional and modern methods of camel slaughtering and processing with an emphasis on the handling of camels pre- and post-slaughter.

5.2 Pre-Slaughter Handling of Camels

Dromedaries are less susceptible to mishandling practices than other livestock. There are, however, specific mishandling practices that can have a considerable impact on the acceptability of the camel meat (Cortesi, 1994). Pre-slaughter handling of camels has a significant effect on meat quality characteristics. For instance, rough handling of dromedary camels before slaughter results in an abnormal appearance in the hump (Fig. 5.1). This condition results from the increased flow of blood into the peripheral capillaries and inadequate drainage after slaughter. Salmonellae in livestock have been shown to increase with animal stress (Corrier *et al.*, 1990; Galland, 1997).

Dromedary owners usually live outside cities and need to transport their animals to slaughterhouses over considerable distances. If long distances have to be covered, cranes and trucks should be used to load or unload and transport dromedary camels to slaughterhouses (Fig. 5.2).

Camels should be unloaded as soon as they arrive and should be calmly guided into the slaughterhouse. Ante-mortem examination on arrival is essential and resting animals prior to slaughter is highly recommended. Resting prior to slaughter reduces stress and improves meat quality characteristics. Camels that have been transported for long distances or excessively worked should be held in the lairage for 12–24 h before slaughter without feed but with access to water.

Withholding feed results in easier evisceration and minimizes the migration of ingested bacteria from the gastrointestinal tract into the blood stream. Withholding feed immediately before transportation affects the growth of potential pathogens in the rumen (Galland, 1997) and of faecal bacteria (Grau *et al.*, 1968). Access to water enhances complete bleeding, results in a brighter coloured lean carcass and facilitates skin removal.

All aspects of camel handling should be carried out by experienced personnel aware of domestic and international legislations on animal welfare.

Pre-slaughter stress can be reduced by preventing mixing of different groups of dromedary camels, keeping them cool with adequate ventilation and avoiding overcrowding.

5.3 Pre-slaughter Stunning of Camels

Traditionally dromedary camel stunning is not practised at slaughter. Stunning of a camel can, however, be carried out using a captive bolt pistol on the intersection of the medial corner of the eye and upper ear attachment (Herrmann and Fischer, 2004). The pistol needs to be moved slightly to the left or right of the parietal bone because the top of the camel's skull has a prominent apex in the middle.

The purpose of stunning is to render the animal insensible (Gregory, 1998). EFSA (2004) explained the purpose for stunning as follows: most animals that are slaughtered for human consumption are killed by cutting the major blood vessels in the neck or thorax so that rapid blood loss occurs. If not stunned, the animal becomes unconscious only after a certain degree of blood loss has occurred. The animals that are slaughtered have systems for detecting and feeling pain and, as a result of the cut and

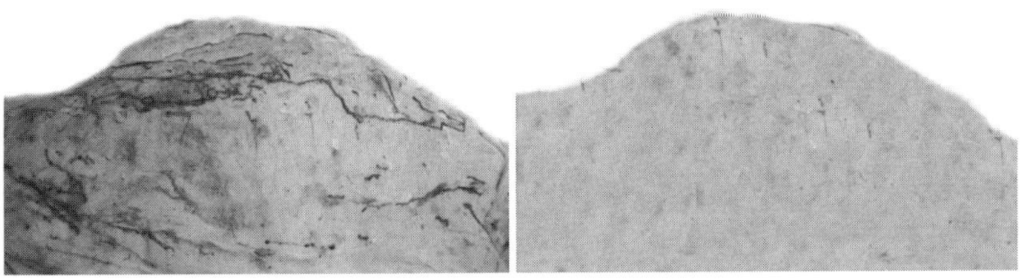

Fig. 5.1. Hump on the left collected from a roughly handled dromedary camel (increased flow of blood into the peripheral capillaries and inadequate drainage) and on the right from a normally handled camel before slaughter.

Fig. 5.2. Transporting and unloading dromedary camels using truck and crane.

the blood loss, if not stunned they will experience pain, fear, panic and other adverse effects such as the inhalation of blood because of bleeding into the trachea. The mechanical and electrical methods of stunning will be briefly discussed because of their relevance in the pre-slaughter stunning of large animals such as camels.

5.3.1 Mechanical stunning

According to Blackmore and Delaney (1988), mechanical stunning of animals for slaughter is achieved by using penetrative captive bolt or non-penetrative percussion stunning. The basic principles are the same and involve the transfer of kinetic energy from a moving object to the brain, which results in neuronal dysfunction and/or destruction and subsequent insensibility. Early studies on mechanical stunning (Blackmore, 1979; Lambooy, 1981; Lambooy and Spanjaard, 1981; Daly *et al.*, 1985, 1986; Daly and Whittington, 1986) were reviewed by Bager (1987), who concluded that captive bolt stunning of domestic animals, except for very large animals, is humane, provided the captive bolt penetrates the skull of the animal at the correct site. The correct site in a large animal is in the frontal position at the point where imaginary lines from the eye to horn cross (Lambooy, 1981). Gregory (1998) reported that concussion – such as caused by captive bolt stunning – is one of the most effective ways of disrupting brain function and stunning an animal; it is instantaneous and can be permanent, as evidenced by the use of evoked potentials (electrical potentials in the brain that occur in response to an external stimulus). Animals that are correctly stunned using captive bolt lost their evoked potentials immediately and they do not return.

With non-penetrative captive bolt (percussion or mushroom stunning), the percussion bolt has a blunt end that looks like a mushroom, designed to concuss without penetrating the brain. This stunning method is essentially similar in its effect to the use of penetrative captive bolt stunning (Gregory, 1998). Anil *et al.* (2002) compared penetrating captive bolt, non-penetrating captive bolt and electrical stunning of different species and found that there was a risk of haematogenous dissemination of central nervous system tissue with the use of pneumatically or cartridge-operated penetrating captive bolt. The dissemination of central nervous system tissue poses a threat to public health in relation to possible slaughter of animals with preclinical bovine spongiform encephalopathy (BSE; Anil *et al.*, 2002).

5.3.2 Electrical stunning

Electrical stunning is the most common method prior to slaughter (Gregory, 1998). It is attractive because it is cheap, suited to high throughputs of animals and can be automated (Bager, 1987). It is also humane from the stand point of animal welfare (Daly and Simmons, 1994). The objective of electrical stunning is to pass sufficient current through the brain to depolarize neurons, which subsequently develop uncoordinated activity; during this period animals are insensible (Blackmore and Delaney, 1988). The stunning can be reversible (head-only) or irreversible (head-to-body) by inducing cardiac arrest (Gilbert, 1993; Grandin, 2003). Electrical stunning results in unconsciousness by producing an epileptic seizure in the brain (Simmons and Daly, 2004).

The two types of electrical stunning, head-only and head-to-body (head-to-back, head-to-forelegs and split current), differ in their effect on the stunned animal. Head-only electrical stunning causes the animal to be unconscious and insensible to pain, yet the animal can fully recover if the slaughter cut is not made; the head-to-body stunning when correctly applied stops the animal's heart resulting in death. Immediately following a head-only stun, noxious stimuli applied to the animal do not elicit movement or autonomic responses. The animals return to normal behaviour within 20–40 min, show no evidence of pain and show no aversion to returning to the stun situation (Cook *et al.*, 1993). According to Gilbert (1993), the head-only electrical stunning is accepted as humane to the animal, safe for the workers and virtuous. There is enough evidence to conclude that head-only electrical stunning does not kill the animal before the animal is slaughtered and the procedure is painless to the animal both at its initiation and while the animal is unconscious before slaughter. Therefore, it is the opinion of the authors that pre-slaughter stunning using head-only electrical stunning is an acceptable method to meet the requirements of industrial processing for camel slaughter and to meet Halal slaughter requirements.

5.4 Slaughtering Procedures

Traditionally, dromedaries are slaughtered in the crouching position; the head is secured in a caudal position (i.e. turned towards the tail; Figs 5.3 and 5.4). A quick cut is made with a very sharp knife at the base of the neck between the neck and the thorax to bleed the animal fast. This is because major blood vessels are mostly

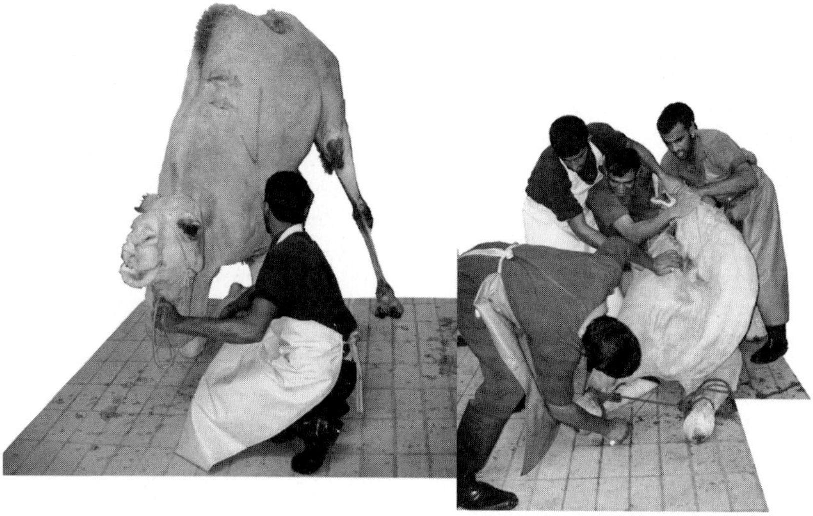

Fig. 5.3. The dromedary camel kneeling down with front legs tied at the knee joints, head turned towards the tail to limit physical movement.

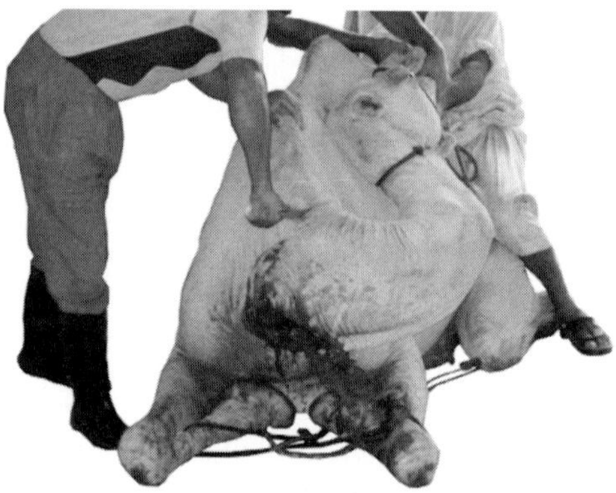

Fig. 5.4. Restraining dromedary camels in the crouching position during slaughtering with the maximum extension of the neck to allow maximum bleeding.

exposed at this point rather than up the neck where the transverse processes of the cervical vertebrae conceal the carotid arteries.

Camels should be allowed to completely bleed out to eliminate meat contamination (Fig. 5.4). The amount of blood at slaughter is estimated at 9% of the body weight (Wilson, 1978; Hill *et al.*, 1993) or 31–53 l, depending on the weight and size of the camel (Al-Ani, 2004). Approximately 40–60% of the total blood is lost at slaughter, which is highly desirable to ensure early brain death and prevent bacterial growth.

5.5 Dressing Procedures of Dromedary Camel Carcasses

5.5.1 Traditional method of dressing

During bleeding, the ligament of the neck is contracted away from the cut site; therefore, the neck is bent towards the back. The bending of the neck causes difficulties in dressing of the carcass, hence the neck is cut off close to the body base between the 6th and 7th cervical vertebrae after bleeding (Fig. 5.5). The oesophagus is separated from the trachea and tied to make sure that the contents of the rumen do not contaminate the carcass.

Dromedary camel carcasses are traditionally skinned (flayed) on the floor in the crouching position (Fig. 5.6). However, because the carcass quality would be affected, some meat packers use the cradle system prior to skinning. The cradle system decreases the fatigue experience from extensive bending-over by the workers and thus substantially increases efficiency. Traditionally the skin is opened from the dorsal instead of

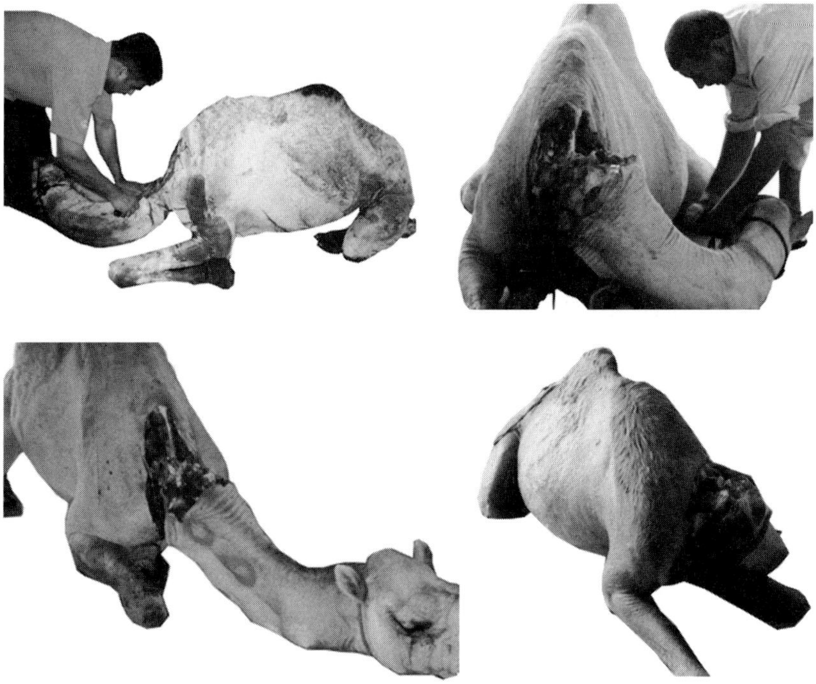

Fig. 5.5. The neck is cut off close to the body base between the 6th and 7th cervical vertebrae after bleeding.

ventral aspect, and dorsal cuts of the carcass are removed before evisceration. It is skinned by cutting along the back from the roots of the tail across the hump (Fig. 5.6). The skin on the legs is then cut and the four legs are removed at the knee joint. The skinning procedure of camel carcass is usually carried out manually using knives and starting from the backbone making the way down both sides of the carcass to the belly.

The traditional method of dressing a slaughtered camel in Kenya and Nigeria has been described by Ulmer and Fischer (2004) and Muhammad and Akpan (2008) to involve the following: (i) hide removal or flaying starts from the backbone or the posterior part of the crouched slaughtered camel and down both sides of the carcass to the belly or the anterior part of the camel; (ii) the flayed skin is laid on the ground with the flesh side uppermost; (iii) the muscles at the posterior cervical vertebrae are cut to relax the neck for ease of flaying; (iv) the hump is split lengthwise and removed; (v) the shoulders are separated and ribs are cut away from the vertebrae; (vi) the gastrointestinal tract is removed; (vii) the backbone is cut out; and (viii) the hind legs are split in the pelvis and divided up into smaller cuts at the joints.

It is important to pull away the skin from the meat to avoid contamination with dirty objects such as dust, faeces and hair. This is done in the following way:

- While the body is maintained in sternal recumbence, the opening incision is made along the midline of the back, and the skin is freed down the sides to reduce contamination of the carcass with dust and dirt.
- The fore and hind legs are severed at the knee joints.
- The skin is slit along the belly and along the insides of the legs and it is important to move the knife from inside to outside.
- The skin is detached by pulling it away from the carcass with the free hand.

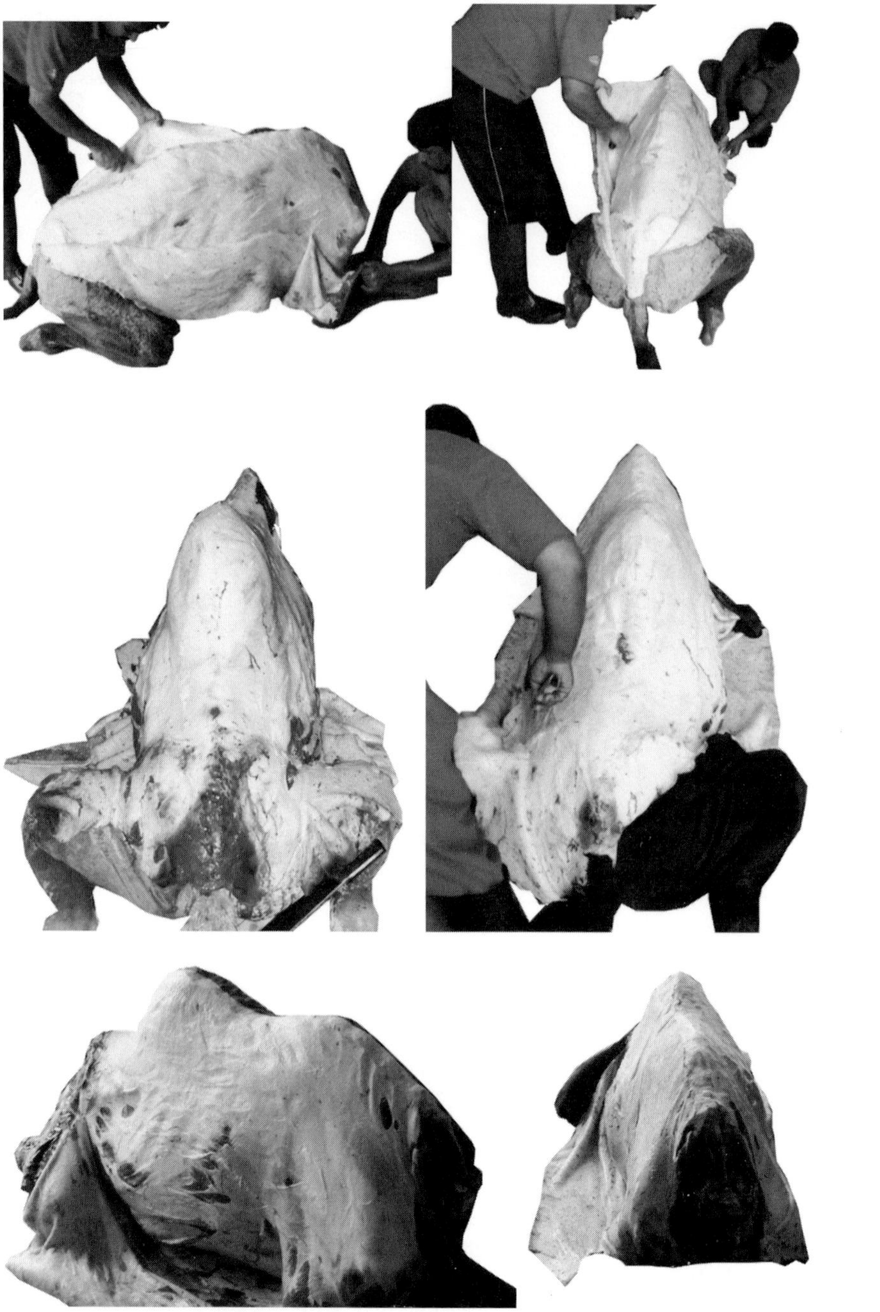

Fig. 5.6. Skinning procedure of the dromedary camel in the crouching position on the floor.

- The skin should be detached from the carcass without damaging the muscles by stretching it fully with the free hand to avoid cutting into the skin or into the meat.
- When the leg, breast and belly areas have been completely skinned, the breast bone is opened with a saw; the carcass can be hung up on the hanging rack by the Achilles tendon.

- A circular cut is made around the anus which should be tied to prevent contamination of the carcass with gastrointestinal contents.
- Skinning of the back and hump is continued in a downward direction until the animal is completely skinned.
- Camel skins can flay and flesh easily depending on the level of hydration. At this stage, the hump is removed from the back in a layer containing connective tissue and hung up separately if it is large, or left attached to the carcass if small.
- The subcutaneous fat is localized in the hump and the absence of a continuous subcutaneous fat layer assists body cooling.
- The shoulder is separated from the thorax working dorso-ventrally and the ribs are removed, followed by the flank.
- The carcass is then eviscerated.

5.5.2 Modern method of dressing

In a modern slaughterhouse, the dressing is carried out by suspending the camel carcasses from the Achilles tendon throughout the slaughtering process (Figs 5.7 and 5.8). The rail procedures are based on gravity systems of flow whereby the carcass is pushed by workers or mechanical power is applied to facilitate movement between stations. The dressing operation involves the removal of the following dress-off items: neck including the head, hide, and fore and hind feet. Herrmann and Fischer (2004) provided a flow chart of a modern method of slaughter and dressing practised in a slaughter house in Kenya. In the camel abattoir in Australia, the camels are stunned, slaughtered and dressed from the caudal to cranial parts of the body and the neck and fore legs are removed before suspension to ensure clearance above the floor. The flesh immediately beneath the skin appears whitish because the subcutaneous fat covers some parts of the carcass (Fig. 5.9). The fat is usually white and soft, whereas the flesh of the camel is brownish red and darker than other livestock meat because of high myoglobin and glycogen contents. The whole carcass is then hung from its hind legs by hooks connected to overhead rails. Post-mortem inspection is usually carried out on the

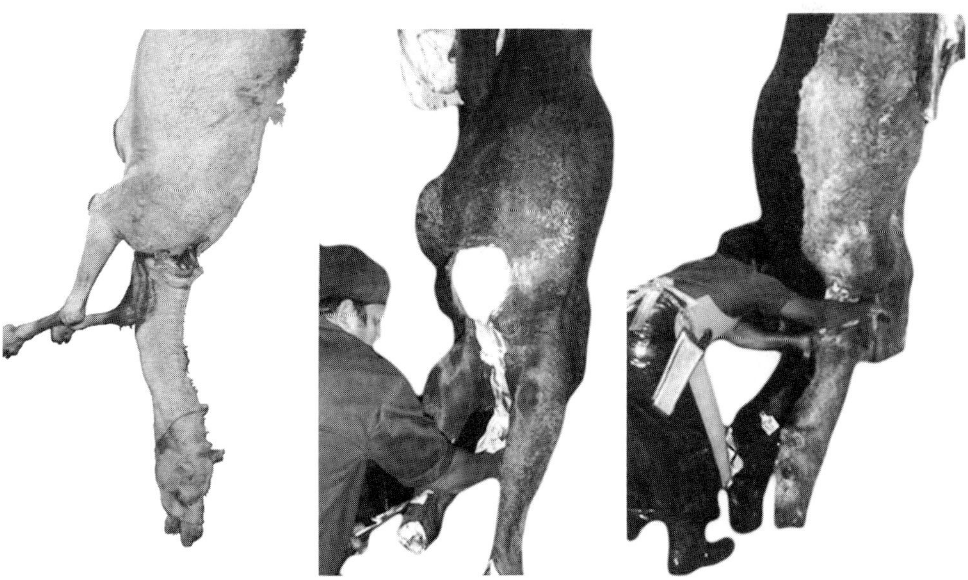

Fig. 5.7. Suspending dromedary camel carcasses from the Achilles tendon before skinning. Removing head and neck (left), removing shank and skinning the thoracic part of camel in a hanging position.

Fig. 5.8. Camel carcasses at different stages of hide removal including the mechanical pulling of the hide from the camel carcass.

camel carcass and offal before its products are released for human consumption.

5.6 Evisceration Procedures

The evisceration process entails the removal of the abdominal and thoracic viscera. It should be performed by a well-trained and skilled worker in either the crouching or hanging position to avoid cutting into the gastrointestinal tract. Both ends of the gastrointestinal tract must be secured to prevent the contents from leaking on or in the carcass. The evisceration process starts by cutting through the middle to lower abdomen making a vertical cut from the thorax to the navel. The abdominal cavity is opened with the carcass hanging by inserting the knife into the abdominal cavity with the blade protruding out of the carcass and the abdominal wall is cut open by pressing in the direction of the chest. The rectum separates from the anus and the gastrointestinal tract is carefully cut free and pulled out of the abdominal cavity. The liver is first detached from the diaphragm and taken out of the carcass. The diaphragm is then cut through and the lungs, heart, trachea and oesophagus are taken out. The internal organs represent approximately 16% of live weight (Kamoun, 1995).

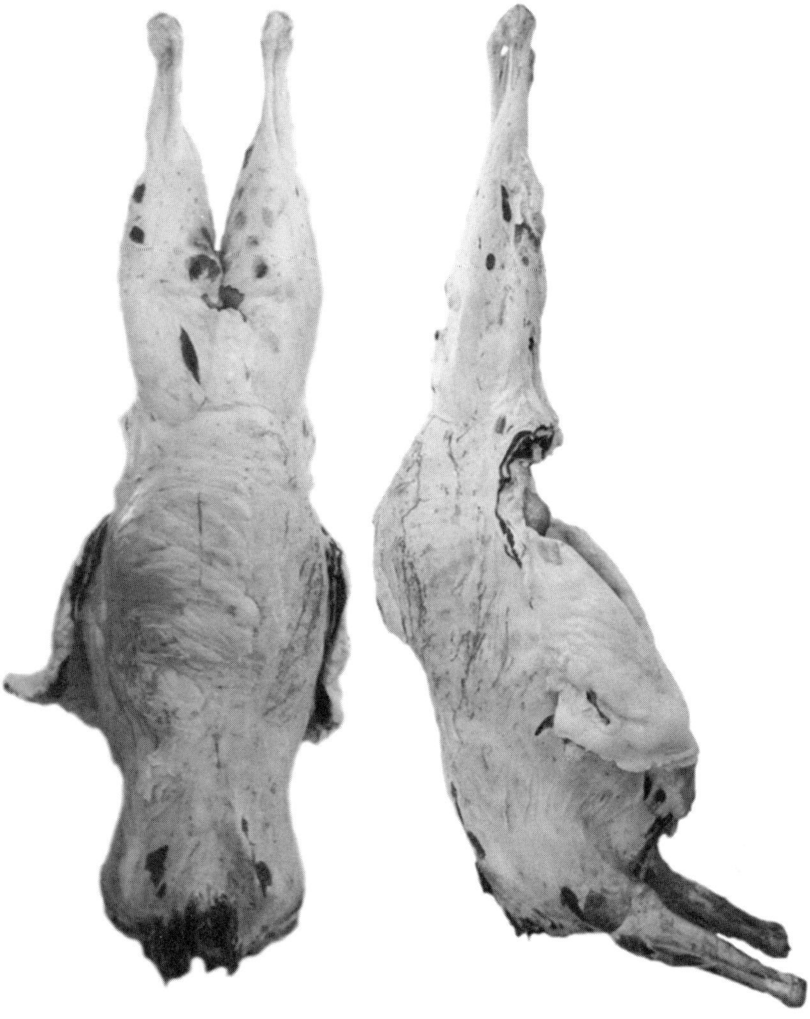

Fig. 5.9. Whitish subcutaneous fat covers dromedary camel carcasses.

In Australia, after bleeding, skinning and the evisceration of all internal digestive, respiratory, excretory, reproductive and circulatory organs, the carcass is minimally trimmed to meet inspection requirements.

5.7 Dromedary Camel Carcass Splitting

Within the modern slaughtering techniques and facilities, because of their large size the dromedary camel carcass is usually longitudinally split (Fig. 5.10). This is done with a handsaw, a cleaver or an electric saw. It is important to cut the spinal column in the middle along the spinal canal. Depending on the age of the animal, splitting the camel carcass is hard, because the bone get harder as the animal gets older. The split carcasses are washed with clean water and stored in a cold room (chiller) for up to 48 h or until rigor mortis is completed.

5.8 Preparation for Chilling

The most important factors in handling fresh camel meat are the control of temperature

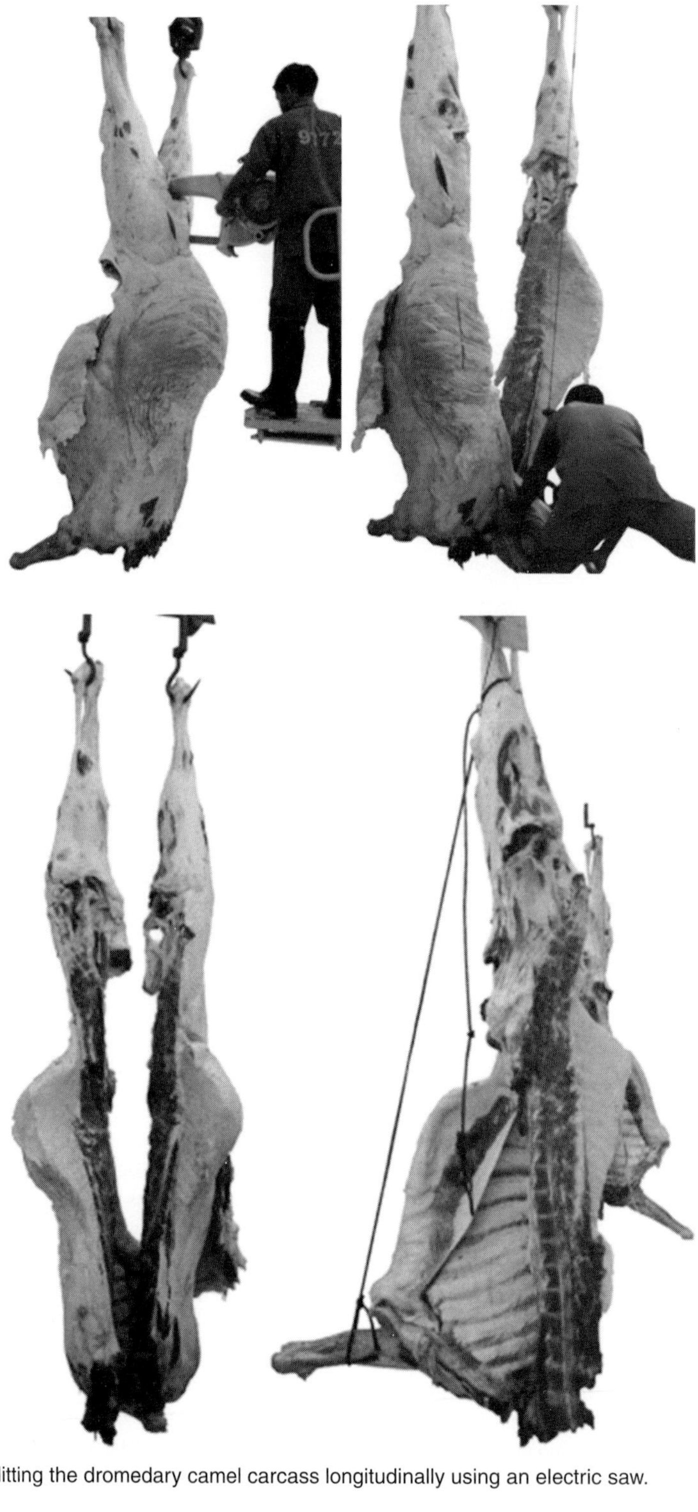

Fig. 5.10. Splitting the dromedary camel carcass longitudinally using an electric saw.

and good hygiene conditions. Slaughtering techniques are required to minimize both physical and microbiological contamination of the carcass. Integrated hygiene control is the most effective approach to increase the storage life of meat and meat products (Smulders, 1995).

Following evisceration, any pieces of adhering skin, bruises and wool or dung spots are carefully removed from the carcass. In modern slaughterhouses, the camel carcass is usually longitudinally split down the centre of the backbone to facilitate rapid cooling, and the carcass is thoroughly washed with high-pressure water. The carcass is then weighed, tagged and placed in a 2–4°C initial chill cooler to remove body heat. After 12–24 h, the carcass is moved to a processing unit or for subsequent storage.

Camel slaughter storage conditions are important factors affecting carcass contamination. Camel meat, like others, is subjected to contamination from a variety of sources within and outside the body, which can occur during slaughter and processing. The level of contamination of a freshly dressed carcass and the composition of the flora depend on the technical structure of the abattoir and the hygienic conditions during the slaughter dressing procedures (Sierra *et al.*, 1995).

5.9 Dromedary Camel Carcass Jointing

Because of the dromedary camel size, the two sides of the carcass are very difficult to handle so each side is subsequently divided into forequarters and hindquarters (Figs 5.11 and 5.12). Each half is separated between the 12th rib and 1st lumbar vertebrae and the forequarter can be hung up with a hook between the ribs or from the shank. The hindquarters constitute an important part of the carcass because they contain large and tender muscle groups that determine the overall profit for butchers.

The forequarter and hindquarter are further fabricated into primal and subprimal cuts mostly by seam boning. The camel forequarter is fabricated into the following cuts:

- **Boneless blade**: consists of a large group of muscles that lie outside of the blade bone and extend from the humerus to the tip of the scapular cartilage.
- **Boneless bolar**: prepared from a blade by the removal of all muscles surrounding and attached to the bolar (*Triceps brachii*) muscle.
- **Boneless cube roll**: made up of a portion of the *Longissimus dorsi* muscle and associated muscles from the 6th to the 12th rib.
- **Boneless chuck**: prepared from the full chuck by the removal of the rib meat/sticking by a straight cut at 25 mm from the chuck eye at the 5th rib and parallel to the shin edge.
- **Boneless chuck tender**: round shape muscle lying lateral to the blade bone on the cranial side of the blade edge.
- **Boneless brisket**: prepared from a 5th rib point (1st to 5th rib) inclusive.
- **Boneless shin/shank**: derived from the shins of the fore and hind legs skinned and tipped.
- **Boneless flank steak**: prepared from the thin flank, it a fan shape muscle located in the leg end of the flank. The muscle is removed along the natural seams; heavy connective tissue and membrane is removed.
- **Boneless thin and thick skirts**: the thin skirt is the costal muscle portion of the diaphragm and is located on the inner cavity of the rib cage; the thick skirt is the thickest portion of the diaphragm located adjacent to spinal column. All fat loose tissue and connective tissue membranes are removed.

The camel hindquarter is fabricated into the following cuts:

- **Topside**: removed from the butt of the hindquarter along the natural seam division separating the silverside and thick flank.
- **Outside**: removed from the butt of the hindquarter along the natural seams between the topside and thick flank. The outside is prepared by the removal of the heel muscle (*Gastrocnemius*) following the natural seam and all associated gland fat.

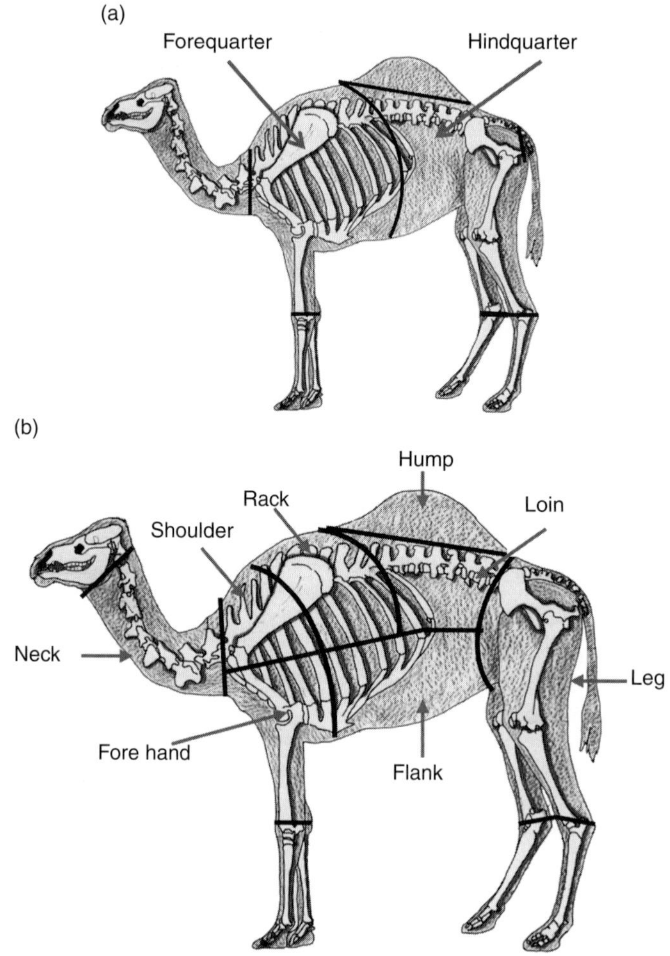

Fig. 5.11. Carcass jointing. (a) Carcasses are divided into forequarters and hindquarters between the 8th and 9th rib. (b) The carcass cuts.

- **Outside flat**: prepared from the outside by the separation along the natural seam of the eye round muscle and the outside flat muscle.
- **Eye of round**: portion of the outside remaining after the removal of the outside flat along the natural seam.
- **Knuckle**: portion of the hindquarter attached to the femur bone and removed from its attachment to the silverside and topside along the natural seam.
- **Rump**: prepared from a full rump removed from the hindquarter. The flank (tail of the rump; the tensor fasciae latae muscle) is removed on a line halfway between the large eye muscle of the rump and the outer flank tip. Fat pocket on the tail of the rump is removed.
- **Striploin**: prepared from the hindquarter and is that portion of the *Longissimus dorsi* muscle attached to and along the edge of lumbar vertebrae.
- **Tenderloin**: removed from the hindquarter in one complete piece.

5.10 Dromedary Camel Fat

Dromedary camels manage their body fat depots in a way that helps them respond to variations in the quality and accessibility of

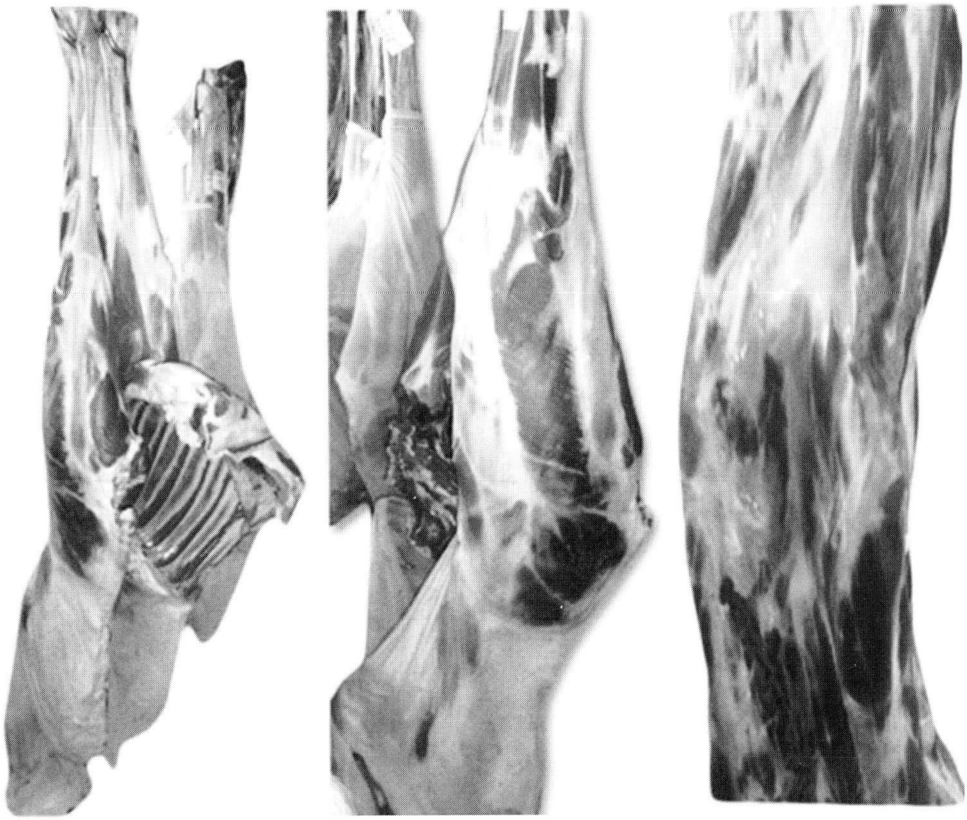

Fig. 5.12. Left: dressed camel neck; middle: dressed forequarter; and right: dressed hindquarter of a camel carcass.

feed resources (Faye *et al.*, 2002). The camel stores its energy reserves in the form of fat in various depots in the body such as the hump, the kidney, and in subcutaneous, intermuscular, abdominal, omental and mesenteric depots. Dromedary camels use their fat stores to maintain their productivity and/or survive by mobilizing adipose tissue.

The fat derived from the camel is of great nutritional importance in the human diet. The edible fats of the camel are obtained from the hump, and the mesentery and kidney fat depots. The hump and abdominal fat depots are used for culinary purposes. The weight of dromedary camel hump, which is mainly composed of fat, accounts for approximately 8.6% of the carcass weight (Kamoun, 1995) and can affect carcass-out percentage (carcass weight without hump). Hump and other fat depots contain mixtures of fatty acids (Emmanuel and Nahapetian, 1980; Kadim *et al.*, 2002, 2008) and most of these are esterified as triglycerides or phospholipids and vary according to their anatomical location in the body (Duncan and Garton, 1967; Kadim *et al.*, 2002, 2008). Many studies have examined the composition of camel hump fat and abdominal depot fat (Mirgani, 1977; Emmanuel and Nahapetian, 1980; Emmanuel, 1981; Orlov *et al.*, 1985; Rawdah *et al.*, 1994; Kadim *et al.*, 2002). Kadim *et al.* (2002), using thin-layer chromatography, found dromedary hump contained more saturated than unsaturated fatty acids. Emmanuel (1981), Orlov *et al.* (1985), Rawdah *et al.* (1994) and Kadim *et al.* (2002) showed that the saturated fatty acid contents in dromedary hump fats were 64.9, 60.2, 60.5 and 63.0% of total fatty acids, respectively, whereas the abdominal fats were 64.9, 60.2, 60.5 and 68.3%, respectively. In the abdominal fat of the single-humped camels, saturated fatty acids accounted for 36.6% of the total fatty acids (Emmanuel and Nahapetian, 1980).

5.11 Dromedary Camel Carcass Grading

Currently, there is no international standard system for camel carcass grading. The camel industry needs to set up a camel carcass evaluation system. The basic intent of carcass evaluation is to provide as much information as possible about the camel carcass. Two basic factors determine carcass merit – the proportion of the carcass that is edible and the indicators of quality and palatability of the edible portion. Thus, the ultimate value of an individual camel carcass could be deduced from two characteristics: (i) quality characteristics of the lean meat (as a measure of expected palatability); and (ii) the combined yield of boneless, closely trimmed cuts from the round, loin, rib and chuck.

5.12 Microbiology of the Camel Carcass

High incidence of contamination and numbers of *Citrobacter freundii, Enterobacter aerogenes, Enterobacter cloacae, Enterobacter sakazakii, Escherichia coli* (O26; K60 (B6), O55; K59 (B5), O111, K58 (B4), O119; K69; (B14) *Serratia liquefaciens, Klebsiella pneumoniae, Proteus mirabilis, Proteus vulgaris, Morganella morganii, Salmonella enteritidis, Salmonella typhimurium* and *Staphylococcus aureus* have been found on the surface of camel carcasses slaughtered in abattoirs of different countries (Hamdy, 1989; Al-Dughaym and Yassien, 2001, Kalalou *et al.*, 2004).

5.12.1 Irradiation

Gamma irradiation can improve the camel meat microbial quality. The total microbial load decreased by increasing the irradiation dose (2–6 kGy) and the effect was greater during storage at 4°C (Al-Bachir and Zeinou, 2009). Coliforms count (log10 CFU/g) was decreased from 3.15 to <1 using a dose of ≥2 kGy and the coliforms count remained at that level during 6 days of storage (Al-Bachir and Zeinou, 2009). Irradiation at 1.5 and 3.0 kGy increased the microbiological shelf life of fresh camel meat by 6 and 12 days, respectively (Fallah *et al.*, 2008) and had no effect on proximate composition, total volatile nitrogen (TVN), cooking loss and sensory properties (Fallah *et al.*, 2008). This confirms an earlier report indicating that irradiation within a dose range of 2–6 kGy had no effect on the sensory properties of the camel meat (Al-Bachir and Zeinou, 2009). The acceptability period of the meat increased from 7 days to 15 and 21 days after irradiation using 1.5 and 3.0 kGy (Fallah *et al.*, 2008). Although Al-Bachir and Zeinou (2009) found irradiation intensity within the range of 2–6 kGy did not affect lipid oxidation in camel meat, Fallah *et al.* (2008) found that lipid oxidation almost doubled on 3.0 kGy treatment. The antioxidant content in the muscle can play a significant role in the oxidative processes during irradiation and subsequent storage, and the conflicting outcomes reported from the above studies might reflect the differing antioxidant levels in the meat. According to the above information, there is potential to use irradiation as an effective preservation method in addition to good manufacturing practices to extend the shelf life of fresh camel meat.

5.12.2 Use of probiotic microorganisms

Lactobacillus delbrueckii subsp. *delbrueckii* was isolated from camel meat and demonstrated inhibitory activity against *E. coli, Pseudomonas aeruginosa, K. pneumoniae, S. aureus, C. freundii, Bacillus subtilis, Bacillus megaterium* and *Bacillus cereus* (Kalalou *et al.*, 2004). This probiotic reduced the total plate count and eliminated coliforms, staphylococci and enterococci owing to the accumulation of lactic acid and the reduction of pH (Kalalou *et al.*, 2004). Similar effects were observed using other *Lactobacillus* species in fermented camel sausage (Kalalou *et al.*, 2004; El Malti and Amarouch, 2008). *Bifidobacterium breve* had little effect on the microbiology and sensory properties of camel meat (Al-Sheddy *et al.*, 1999).

5.12.3 Other methods to improve the camel meat/products safety

Organic acid salts (sodium acetate [10% w/w], potassium sorbate [1.5% w/w], sodium lactate [5% v/v of 60% solution], and trisodium citrate [1.5% w/w]) have been investigated as means to extend the microbiological shelf life of camel meat (Al-Sheddy *et al.*, 1999). Sodium acetate extended the microbial shelf life (>12 days) and minimized surface discoloration. The effects of sodium acetate were augmented by the addition of bifidobacteria. Natural preservatives such as extracts from vegetables (Al-Delaimy and Barakat, 1971) or possibly materials that showed potential antimicrobial activities such as polyphenols or herbs extracts can potentially improve the microbiological shelf life of camel meat.

5.13 Dromedary Camel Slaughter By-products

There is little information available on slaughter by-products of the camel. In general, the dressing-out percentage in the dromedary camel makes up about 55% of the live weight. Accordingly, the proportion of slaughter by-products and gastrointestinal contents account for 45%. For many years, many of the edible camel slaughter by-products have been consumed fresh or used as ingredients in traditional meat products. The by-products of dromedary slaughter are used in many countries where camels are raised and animal protein is scarce.

The proportions of edible slaughter by-products are high in the camel (Table 5.1), thus constituting a very useful source of protein where the camel is raised for meat production. Breed differences and the nutritional state of the animal could be responsible for any variations between the outcomes of different studies. The weight of feet and skin as proportions of live weight are higher for camels than for cattle, but the head is proportionately lower in camels than in cattle (Mahgoub *et al.*, 1995a,b). According to Herrmann and Fischer (2004), the head, skin and feet contributed 2.4, 7.3 and 3.4%, respectively, of live weight in the dromedary camels. The heaviest slaughter by-product of camels is the skin followed by the intestines, whereas the lightest organ was the spleen followed by the reproductive organs (Yousif and Babiker, 1989). The camel liver is heavier than that of cattle (Congiu, 1953). The dromedary body contained an average of about 4.2% offal (liver, heart and lungs). The slaughter by-products included the head (3.5%), feet (3.6%) and skin (8.6%) (Yousif and Babiker, 1989). Al-Ani (2004) reported

Table 5.1. Weight of the carcass (including hump) and slaughter by-products and the same components expressed as a percentage of empty live weight of the camel (Wilson, 1978).

	Weight (kg)		Percentage of empty body weight	
	Mean	Range	Mean	Range
Carcass weight	208.5 ± 38.7	141.0–310.0	60.7 ± 2.09	55.75–65.11
Hump	4.0 ± 4.3	0.0–20.0	1.1 ± 1.04	0.00–4.45
Heart and lung	8.4 ± 1.13	6.5–10.5	2.5 ± 0.33	1.78–3.36
Liver	7.5 ± 1.45	4.5–11.0	2.2 ± 0.41	1.47–3.45
Head (skinned)	12.1 ± 1.81	8.0–16.5	3.6 ± 0.32	2.80–4.49
Feet	14.6 ± 2.25	10.5–19.5	4.3 ± 0.37	3.31–5.16
Skin	34.8 ± 6.11	22.5–47.0	10.2 ± 0.81	8.5–11.76

Table 5.2. Weights of some of dromedary camel (530–730 kg) by-products (Herrmann and Fischer, 2004).

Edible by-product	Weight (kg)	Purpose
Heart	1.9–3.3	Fresh retail product, faggots, haggis
Liver	7.2–9.7	Fresh retail product, sausage, faggots
Lungs	2.3–4.1	Fresh retail products
Rumen		Fresh retail products, andouillette
Spleen	0.4–0.8	Fresh retail product, haggis
Kidneys	1.5–2.0	Fresh retail product
Trachea and oesophagus	1.8–3.6	Haggis
Tongue		Fresh retail product
Intestines		Natural casings for sausages, andouillette

that camels had proportionately heavier kidneys and lighter digestive tracts and head than cattle, sheep or goats. The larger kidney, which was twice that of cattle and four times that of sheep, was possibly owing to adaptation of the dromedary camel to arid desert life. Camel kidneys have been estimated to be up to 50 cm^3 (Abdalla and Abdalla, 1979).

Some of the dromedary camel edible slaughter by-products have substantial nutritive value because of their low fat content and high content of protein, B-vitamins, iron, zinc and copper. The suitable camel edible by-products for human consumption are: heart, liver, lungs, stomach, spleen, kidneys, tongue and brain. They can be boiled, fried or grilled before eating or included in processed camel meat products. Small and large intestines of the camel can be considered as non-carcass parts suitable for human consumption if they are used as sausage casings after cleaning and processing (Table 5.2). It is important to clean the offal with fresh clean water to remove concealed blood after slaughter and to chill at 4°C. The skins, head, fat, genitals and unusable parts of the intestines are regarded as inedible by-products. The skins are used to make leather products.

5.14 Conclusion

The traditional method of camel meat production remains very common in countries where camels are raised and slaughtered. The slaughtering of the dromedary camel requires a comprehensive amount of manual work because of the large size of the animal unit involved. The development of slaughtering techniques, grading systems and a marketing strategy for camel meat products is necessary owing to an increasing demand for high-quality protein and for trading within the gulf region and other potential international camel meat markets. The main cross-border trade in camels is that of live animals, with hardly any camel carcasses being traded across borders. If camel meat is to become a traded commodity between countries, microbiological safety and tenderness of the meat is likely to constitute the main challenge to a successful camel meat industry. Interventions to improve the microbiological safety and shelf life of camel meat have been summarized in this chapter as potential future trends for improving the safety of the meat.

References

Abdalla, M.A. and Abdalla, O. (1979) Morphometric observations on kidney of the camel, *Camelus dromedarius*. *Journal of Anatomy* 129, 45–50.

Al-Ani, F.K. (2004) *Camel Management and Diseases* (1st edn). Dar Ammar Book Publisher and Al-Shraq Printing Press, Jordan.

Al-Bachir, M. and Zeinou, R. (2009) Effect of gamma irradiation on microbial load and quality characteristics of minced camel meat. *Meat Science* 82, 119–124.

Al–Dughaym, A.M. and Yassien, N.A. (2001) Surface contamination of camel carcasses. *Scientific Journal of King Faisal University* 2, 129–138.

Al-Delaimy, Kh.D. and Barakat, M.M.F. (1971) Antimicrobial and preservative activity of garlic on fresh ground camel meat: 1. Effect of fresh ground garlic segments. *Journal of the Science of Food and Agriculture* 22, 96–98.

Anil, M.H., Love, S., Helps, C.R. and Harbour, D.A. (2002) Potential for carcass contamination with brain tissue following stunning and slaughter in cattle and sheep. *Food Contamination* 13, 431–436.

Al-Sheddy, I., Al-Dagal, M. and Bazaraa, W.A. (1999) Microbial and sensory quality of fresh camel meat treated with organic acid salts and/or bifidobacteria. *Journal of Food Science* 64, 336–339.

Bager, F. (1987) Stunning and slaughter procedures in N.Z. A review of the problems of humaneness. MIRINZ Literature on Electrical Stunning. MIRINZ AgResearch, Hamilton, New Zealand.

Blackmore, D.K. (1979) Non-penetrative percussion stunning of sheep and calves. *Veterinary Record* 105, 372–375.

Blackmore, D.K. and Delaney, M.W. (1988) Slaughter of stock. Publication No. 18. Foundation for Continuing Education of N.Z. Veterinary Association, Massey, New Zealand.

Congiu, S. (1953) Dressing percentage, weight distribution and correlations between different parts of the body in the Somali dromedary. *Zootecnica Veterinaria* 8, 188–191.

Cook, C.J., Maasland, S.A., Devine, C.E. and Gilbert, K.V. (1993) Preslaughter stunning and its implications on assuring humaneness. Australian Veterinary Association Conference, May 1993, Brisbane. Massey University, Palmerston North, New Zealand, pp. 459–462.

Corrier, D.E., Purdy, C.W. and Deloach, J.R. (1990) Effect of marketing stress on fecal excretion of *Salmonella* spp. in feeder calves. *American Journal of Veterinary Research* 51, 866.

Cortesi, M. (1994) Slaughterhouse and human treatment. *Scientific and Technical Review of the Office International des Epizooties* 13, 171.

Daly, C.C. and Simmons, N.J. (1994) The use of high frequency currents for the pre-slaughter stunning of livestock. CAPAA Engineers' Conference, Gunnedah, Australia.

Daly, C.C. and Whittington, P.E. (1986) Concussive methods of pre-slaughter stunning in sheep. *Research Veterinary Science* 41, 349–352.

Daly, C.C., Gregory, N.G. and Wotton, S.B. (1985) The effect of captive bolt stunning on brain function in cattle and sheep. *Proceedings 31st European Meat Research Workers Conference.* pp.85–87.

Daly, C.C., Gregory, N.G., Wotton, S.B. and Whittington, P.E. (1986) Concussive methods of pre-slaughter stunning in sheep: effects of captive bolt stunning in the poll position on brain function. *Research Veterinary Science* 41, 353–355.

Duncan, W.R.H. and Garton, G.A. (1967) The fatty acid composition and intermolecular structure of triglycerides derived from different sites in the body of the sheep. *Journal of the Science of Food and Agriculture* 18, 99–102.

EFSA (2004) European Food Safety Authority Welfare aspects of the main systems of stunning and killing the main commercial species of animals. *The EFSA Journal* 45, 1–29.

El Malti, J. and Amarouch, H. (2008) Microbiological and physicochemical characterization of natural fermented camel meat sausage. *Journal of Food Processing and Preservation* 32, 159–177.

Emmanuel, B. (1981) Fatty acid synthesis in camel (*Camelus dromedarius*) hump and sheep (*Ovis aries*) tail fat. *Comparative Biochemistry Physiology* 68B, 551–554.

Emmanuel, B. and Nahapetian, A. (1980) Fatty acid composition of depot fats and rumen wall of the camel (*Camelus dromedarius*). *Comparative Biochemistry Physiology* 67B, 701–704.

Fallah, A.A., Tajik, H., Rohani, S.M.R. and Rahnama, M. (2008) Microbial and sensory characteristics of camel meat during refrigerated storage as affected by gamma irradiation. *Pakistan Journal of Biological Sciences* 11, 894–899.

Faye, B., Bengoumi, M., Messad, S. and Chilliard, Y. (2002) Assessment of body reserves in camels. *Revue Elev. Medical Veterinary Pays Tropical* 55, 69–78.

Galland, J.C. (1997) Risks and prevention of contamination of beef carcasses during the slaughter process in the United States of America. *Scientific and Technical Review of the Office International des Epizooties* 54, 391.

Gilbert, K.V. (1993) Electrical stunning and slaughter in New Zealand. MIRINZ Technical Report 908, AgResearch Ltd, Hamilton, New Zealand.

Grandin, T. (2003) Stunning options. *Meat and Poultry.* October issue, p. 118.

Grau, F.H., Brownlite, L.E. and Roberts, E.A. (1968) Effect of some Preslaughter treatments on the Salmonella population in the bovine rumen and faces. *Journal of Applied Bacteriology* 31, 157.

Gregory, N.G. (1998) *Animal Welfare and Meat Science.* CAB International, Wallingford, UK. pp. 223–253.

Hamdy, M. (1989) Surface contaminants of slaughtered camels. *Zagazig Veterinary Journal* 17, 291–302.

Herrmann, K. and Fischer, A. (2004) Method for hygienic slaughter of camel. In: Farah, Z. and

Fischer, A. (eds) *Milk and Meat from the Camel; Handbook on Products and Processing.* vdt Hochschulverlag AG an ETH Zurich, Zurich/Singen.

Hill, F.I., Death, A.F. and Wyneth, T.K. (1993) Nematode burned of alpacas shearing grazing with sheep in New Zealand. *New Zealand Veterinary Journal* 41, 205–208.

Kadim, I.T., Mahgoub, O., Al-Maqbaly, R.S., Annamalai, K. and Al-Ajmi, D.S. (2002) Effects of age on fatty acid composition of the hump and abdomen depot fats of the Arabian camel (*Camelus dromedarius*). *Meat Science* 62, 245–251.

Kadim, I.T., Mahgoub, O. and Purchase, R.W. (2008) A review of the growth and of the carcass and meat quality characteristics of the one-humped camel (*Camelus dromedarius*). *Meat Science* 80, 555–569.

Kalalou, I., Faid, M., and Ahami, A.T. (2004) Improving the quality of fermented camel sausage by controlling undesirable microorganisms with selected lactic acid bacteria. *International Journal of Agriculture and Biology* 6, 447–451.

Kamoun, M. (1995) Dromedary meat: production, qualitative aspects and acpability for transformation. *Option Mediterraneennes. Series B, Etudes et Recherche* 13, 105–130.

Lambooy, E. (1981) Concussion stunning of veal calves. *Fleischwirtsch* 61, 98–100.

Lambooy, E. and Spanjaard, W. (1981) Effect of the shooting position on the stunning of calves by captive bolt. *Veterinary Record* 109, 359–361.

Mahgoub, O., Olvey, F.H. and Jeffrey, D.C. (1995a) Growth and composition of the Omani Dhofari cattle. I: Body weight growth and carcass composition. *Asian-Australian Journal of Animal Science* 8, 611–616.

Mahgoub, O., Olvey, F.H. and Jeffrey, D.C. (1995b) Growth and composition of the Omani Dhofari cattle. II: Distribution of carcass tissue. *Asia-Australian Journal of Animal Science* 8, 611-616.

Mirgani, T. (1977) Fatty acid composition of hump triglycerides of the camel (Camelus dromedarius). *Comparative Biochemistry Physiology* 58B, 211–213.

Muhammad, B.F. and Akpan, I.N. (2008) Camel (*Camelus dromedarius*) meat utilization in Kano-Nigeria (II): Post-slaughter handling and marketing wholesale cuts. *Research Journal of Animal Science* 2, 113–117.

Orlov, V.K., Servetnik-Chalaia, G.K. and Zagibailova, N.B. (1985) Functional and fatty acid composition of the lipids of horse and camel meat. *Vopr Pitan* 4, 71–76 (abstract).

Rawdah, T.N., El-Faer, M.Z. and Koreish, S.A. (1994) Fatty acid composition of the meat and fat of the one-hump camel (Camelus dromedarius). *Meat Science* 37, 149–155.

Sierra, M., Gonzalez-Fandos, E., Garcia-Lopez, M., Fernandez, M.C.G. and Moreno, B. (1995) Contamination of lamb carcasses at the abattoir. Micro-flora of freshly dressed lamb carcasses: indicators and spoilage organisms. *Arch. Lebenmittelhyg* 46, 135.

Smulders, F.I.M. (1995) Preservation by microbial decontamination: the surface treatment of meats by organic acids. In: Gould, G.W. (ed.) *New Methods of Food Preservation.* Blackie Academic and Professional, Glasgow, UK, p. 253.

Simmons, N.J. and Daly, C.C. (2004) Electrical stunning of sheep: principles and auditing methods. Meat New Zealand Milestone Report E24. Meat New Zealand, Wellington, New Zealand.

Ulmer, K. and Fischer, A. (2004) Traditional slaughter, carcass dressing and processing of camels. In: Farah, Z. and Fischer, A. (eds) *Milk and Meat from the Camel. Handbook on Products and Processing.* Vdf Hochschulsverlag AG an der ETH Zurich, Zurich/Singen, pp. 85–88.

Wilson, R.T. (1978) Studies on the livestock of Southern Darfur, Sudan. V. Notes on camels. *Tropical Animal Health and Production* 10, 19–25.

Yousif, O.K. and Babiker, S.A. (1989) The desert camel as a meat animal. *Meat Science* 26, 245–254.

6 Inspection of Slaughtered Dromedary Camels

Mohammed H. Tageldin,[1] Shawqi Al-Zadjali,[2] Bernard Faye[3] and Salim Al-Mugheiry[2]

[1]*Department of Animal and Veterinary Sciences, College of Agricultural and Marine Sciences, Sultan Qaboos University, Muscat, Sultanate of Oman;* [2]*Muscat Municipality, Madinat Qaboos, Sultanate of Oman;* [3]*FAO Consultant, Camel and Range Research Centre, Al-Jouf-Sakaka, Saudi Arabia*

6.1 Introduction

The abattoir (slaughterhouse) is a valuable source of information on the incidence of animal diseases and conditions, some of which may be zoonotic (Dakkak, 2010; Mellau *et al.*, 2011). Meat is valued as a complete food containing essential amino acids necessary for the human body. The professional examination and judgement of meat and other organs is essential to determine the fitness of the meat for human consumption (Wilson, 2005). Not only does it assure and reinforce consumer confidence in the wholesome meat they are buying, it is mandatory for meat being slaughtered and processed for sale.

Meat inspection offers an effective means of monitoring the level of diseases, particularly zoonotic diseases (Ndukum *et al.*, 2010). Meat inspection data have an important role to play in epidemiology and preventive veterinary medicine (Dakkak, 2010; Mellau *et al.*, 2011). The inspection is an integral part of both quality assurance and the quality control system and gross inspection of carcasses (Ndukum *et al.*, 2010). It should be effectively carried out for the protection of consumers (Wilson, 2005).

In order to prevent cross contamination of meat, it is necessary to adopt a well-planned, well-executed and controlled cleaning and sanitation programme for rooms, machines and equipment (Wilson, 2005). This should also include the personal hygiene of all personnel working in the slaughterhouse. An abundant supply of water, as well as adequate facilities, is important for treatment and disposal. Climatic conditions influence hygiene and processing. In a hot climate it is often necessary to start slaughtering during the night hours, particularly in tropical climates, and to distribute the meat for sale in the morning. Unless meat is preserved by cooling it soon putrefies (Wilson, 2005).

Meat sold to consumers should comply with required standards for cleanliness, purity, safety and wholesomeness (Wilson, 2005). This involves the inspection of live animals (ante-mortem or pre-mortem) and carcass (post-mortem).

6.2 Ante-mortem Inspection

All camels are examined to determine whether they have any disease or condition that would make them unfit for human consumption (Demelash *et al.*, 2009). Camels that show signs of disease, e.g. rabies or tetanus, should not be slaughtered. The animals should be observed at rest and moving, noting any abnormality in health, behaviour, gait posture, discharge or protrusions, and structure conformation. If an animal is not in good health, it will not be slaughtered on that day and it should be examined again if slaughter has been delayed for more than a day (Wilson, 2005).

The main objectives of ante-mortem are to screen all animals destined for slaughter to ensure that they are properly rested and do not show clinical signs of disease. Any animal showing signs of illness should undergo a special examination (emergency slaughter). No dead or dying animal should be brought into the slaughterhouse. After slaughter each carcass and its internal organs are examined for lesions of disease. The ante-mortem should be carried out in adequate lighting where animals can be observed both collectively and individually at rest and in motion. The general behaviour should be observed, as well as the nutritional status.

6.3 Post-mortem Inspection

Post-mortem inspection plays a vital role in safeguarding the health of the public (Mellau *et al.*, 2011). Animals brought for slaughter at the abattoir might harbour chronic or subclinical infections that are rarely detected during ante-mortem (Njoroge *et al.*, 2002; Ibrahim, 2010; Mellau *et al.*, 2011). Post-mortem inspection is carried out through visual examination, palpation and incision of visceral organs (lung, liver, heart, kidney and spleen). Careful examination and inspection of the camel carcasses with all parts of the animal's body should be conducted. The internal organs are examined for lesions of disease that would make all or part of the carcass unfit as human food (Wilson, 2005). Inspection involves visual examination, palpation and prescribed cuts of the carcass and offal with detailed examination of certain lymph nodes by multiple incisions. Upon completion of the post-mortem inspection, a decision should be made to approve the entire camel carcass for human consumption, or condemn organs and/or portions of the carcass that have the abnormal conditions. Organ/offal diseases and lesions are grossly detected on the basis of pathological changes, i.e. colour, size, morphology, consistence, presence of lesions or parasites.

Before describing specific diseases or conditions it is important to discuss certain general physiological and pathological changes.

6.4 Abnormal and General Pathological Conditions

6.4.1 Poor condition

Poor condition is a physiological condition that occurs in very young or older animals. The condition is characterized by a marked scarcity of fat, which is usually of the normal firm consistency. The flesh is usually darker in colour; the cut surface is firm and dry. If the carcass is hanged for some time, the cut surface becomes very dry and dark (Wilson, 2005).

Judgement: Such carcasses are fit for human consumption. If the case is borderline, the carcass can be hanged for 12–24 h before making the final judgement (Wilson, 2005).

6.4.2 Emaciation

This is a pathological condition caused by some chronic diseases such as Johne's disease, chronic trypanosomiasis or parasitic infestation. It is characterized by wasting of muscular tissue and by a reduction in the amount of fat, which becomes soft and gelatinous in advanced stages. The flesh is wet, soft and flabby. The carcass does not set and the outside is wet. When doubt exists, the carcass should be hanged for 12–24 h before final judgement (Wilson, 2005).

Judgement: Total rejection (Wilson, 2005).

6.4.3 Oedema

Oedema is an excessive accumulation of a clear fluid in tissues or serous sacs of the body, e.g. anasarca (generalized oedema), or hydropericardium, hydrothorax and ascites, when there is accumulation of fluid in the pericardium (Awol *et al.*, 2011), pleural and peritoneal cavities, respectively. Generalized oedema occurs in cardiac and renal tissues and in chronic wasting diseases (tuberculosis, Johne's disease, parasitic diseases, etc.).

Judgement: If generalized, then total rejection (Wilson, 2005).

6.4.4 Imperfect bleeding (bleeding insufficiency)

This occurs when a moribund (dying) animal is slaughtered. The flesh and the internal organs including the lungs, liver and kidneys are dark and congested. The intercostal veins are full of blood and clearly visible. Such a carcass sets badly and decomposes rapidly (Wilson, 2005).

Judgement: Such carcasses are unfit for human consumption and should be rejected (Wilson, 2005).

6.4.5 Feverish flesh

Feverish flesh arises through the action of bacteria or viruses and their circulating toxins. The flesh is darker in colour with small scattered petechial haemorrhages. The organs and lymph nodes are congested, and the pleura, peritoneum and fat show a diffuse redness (Wilson, 2005).

Judgement: Total rejection.

6.4.6 Abnormal odours

Abnormal odours of the dromedary camel especially male sexual odour are most apparent immediately after slaughter (Wilson, 2005). They can result from: (i) drugs administered shortly before slaughter, e.g. turpentine oil or chloroform; (ii) sexual odour, which is principally noticed in mature non-castrated males, especially during the rutting period. This odour is due to high androgen hormone levels.

Judgement: All carcasses with pronounced odours are totally rejected (Wilson, 2005).

6.4.7 Jaundice or icterus

The yellow staining of the tissue by bilirubin is the result of an imbalance between production and clearance of bilirubin because there is either excess production or reduced clearance of bilirubin such that it accumulates in the plasma (Myers *et al.*, 2012). Mechanisms leading to icterus can involve one or more of the following:

- Excess production of bilirubin – as in haemolytic disease such as anaplasmosis – or breakdown of erythrocytes in a large haemorrhage such as a haematoma.
- Extensive hepatic necrosis can cause icterus.

Icterus can be classified into:

- Prehepatic, which occurs in haemolytic crisis; high plasma concentrations of unconjugated bilirubin are produced that exceed the uptake capacity of the hepatocytes.
- Hepatic icterus, which is caused by hepatocellular damage, resulting in the release of both conjugated and unconjugated bilirubin into the blood.
- Posthepatic icterus, which is secondary to obstruction of biliary system, either intrahepatic or extrahepatic with an influx of conjugated bilirubin into the blood.

Icteric tissues are coloured yellow. Icterus is detected in the mucous membrane of the oral cavity, urogenital systems and alimentary system, in the omentum, mesentery, adipose tissue and sclera of the eyes (Myers *et al.*, 2012). The condition varies from slight to very severe.

Judgement: The carcass and offal should be rejected if the condition is very severe (Wilson, 2005).

6.4.8 Echinococcosis (cystic hydatid disease)

Cystic echinococcosis (hydatidosis) is one of the most important parasitic zoonotic diseases in the world (M'rad *et al.*, 2005; Borji *et al.*, 2011). Both cystic hydatidosis

(CE) caused by *Echinococcus granulosus* and alveolar echinococcosis (AE) caused by *Echinococcus multilocularis* have been reported in several countries (Ibrahima and Craiga, 1998; Njoroge *et al.*, 2002; Lahmar *et al.*, 2004; M'rad *et al.*, 2005; Azlaf and Dakkak, 2006; Almeida *et al.*, 2007; Borji and Parandeh, 2010; Ibrahim, 2010; Latif *et al.*, 2010; Omer *et al.*, 2010; Borji *et al.*, 2011). All agents of CE fall under the name *E. granulosus* (Almeida *et al.*, 2007). Hydatidosis is of considerable economic and public health importance (Almeida *et al.*, 2007; Dakkak, 2010; Ibrahim, 2010; Borji *et al.*, 2011).

Cystic echinococcosis affects various species of livestock and humans (Dakkak, 2010). It is caused by the larval stages of the tapeworm *E. granulosus* (Mellau *et al.*, 2011). It is one of the most important parasitic infections in livestock, particularly dromedary camels (Ibrahima and Craiga, 1998; Njoroge *et al.*, 2002; Mellau *et al.*, 2011). The life cycle of *E. granulosus* involves domestic carnivores (dogs) and wild carnivores (jackals, hyenas, foxes and wolves) as definitive hosts. They are infested by the ingestion of offal containing the larval forms (hydatid cysts) with viable protoscoleces producing the adult stage in the intestine (Fathi *et al.*, 2011). Stray dogs in urban areas and free or roaming dogs in rural areas are the main definitive host (Azlaf and Dakkak, 2006; Romig *et al.*, 2011). Extensive livestock production provides suitable conditions for the cycle transmission between dogs and livestock animals (Dakkak, 2010). Three distinct cycles of *E. granulosus* have been suggested: a domestic cycle between dogs and livestock; a desert cycle between dogs and camels; and a sylvatic cycle between wild carnivores and wild ruminants (Fathi *et al.*, 2011). Domestic animals as intermediate hosts (cattle, sheep, goats and camels) are a major reservoir for the disease in humans (Njoroge *et al*, 2002; M'rad *et al.*, 2005; Fernanda *et al.*, 2007; Dakkak, 2010; Ibrahim, 2010; Romig *et al.*, 2011). The camel has attracted much interest as an intermediate host (Derbala and El-Massry, 1999; Fathi *et al.*, 2011; Ibrahim, 2010) because they seem to be an important reservoir for human infection (Eckert *et al.*, 1989). The larval stages develop in these animals after oral infection by the ingestion of eggs (Latif *et al.*, 2010). Cystic hydatid disease is prevalent in most parts of Africa, Asia and South America (Njoroge *et al.*, 2002; Lahmar *et al.*, 2004; Fernanda *et al.*, 2007; Borji and Parandeh, 2010; Dakkak, 2010; Latif *et al.*, 2010; Borji *et al.*, 2011; Mellau *et al.*, 2011; Romig *et al.*, 2011).

The prevalence of the disease in camels is higher than in other animals, including cattle, sheep and goats (Njoroge *et al.*, 2002; Dakkak, 2010; Ibrahim, 2010; Mellau *et al.*, 2011). Older camels have a relatively higher rate of infection (Lahmar *et al.*, 2004; Azlaf and Dakkak, 2006; Ibrahim, 2010; Romig *et al.*, 2011). There is no evidence of parasite-induced immunity in camels (Lahmar *et al.*, 2004; Azlaf and Dakkak, 2006). Assessment for this disease is vital in meat inspection because the intermediate stage cysts commonly occur in food animals (Fernanda *et al.*, 2007). Camel lungs are the most frequently infected (Njoroge *et al.*, 2002; Fathi *et al.*, 2011) (Fig. 6.1), followed by the liver (Ibrahim, 2010), spleen (Fathi *et al.*, 2011) and kidneys (Fathi *et al.*, 2011). Few CE cysts are detected in the heart, spleen and kidney (Njoroge *et al.*, 2002). The camel form of *E. granulosus* may vary in morphological and biological features (Derbala and El-Massry, 1999). The inspection of the lung and liver is usually carried out through visual inspection, palpation and incision of the organs where *Echinococcus* nodules can be detected embedded in the tissue. Most hydatid cysts reside in the lung parenchyma but they are also found in the liver parenchyma just below the capsule. Lungs with hydatid cysts have associated multifocal diffused interstitial pneumonia, bronchopneumonia and emphysema (Bekele, 2008). Larger numbers of calcified cysts are detected in the liver (Fathi *et al.*, 2011). Because each cyst might contain hundreds of scolices, each of which is capable of developing into an adult worm, dogs can have a very heavy infestation (Romig *et al.*, 2011). Multiple nodules of various sizes are distributed all over the surface and cut surfaces of the lung (Fig. 6.1) and liver.

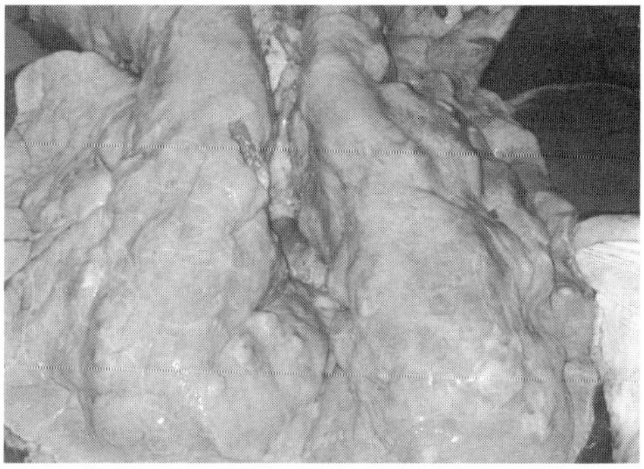

Fig. 6.1. Hydatid cysts in the lung of a camel.

The metacestodes usually form fluid cysts (hydatids) located in liver, lungs and other organs (Dakkak, 2010). The lesions of cystic echinococcosis usually remain for the life of the animal and therefore at post-mortem it is possible to tell whether or not an animal is infected (Njoroge *et al.*, 2002; Ibrahim, 2010). Effort should be made to control the transmission of cystic echinococcosis from slaughter slabs and butcheries by the safe disposal of *Echinococcus* cysts so dogs cannot have access to the cysts (Njoroge *et al.*, 2002; Borji *et al.*, 2011). It is important to enforce legislation that will strictly prevent backyard and roadside slaughtering practices (Bekele, 2008; Dakkak, 2010; Fathi *et al.*, 2011).

Judgement: When just a few cysts are present they can be removed with the surrounding tissue. In a heavy infection, the whole organ should be rejected. The hydatid cysts can be in either one organ or multiple organs. The size of the cyst varies from 2 cm to 8 cm; a number of small-sized and calcified cysts can be seen in the liver sinusoid. Camels more than 10 years old can be highly infected. The greater number of calcified cysts in the liver could be attributed to the abundance of connective tissue in the liver. Because of the greater prevalence of the disease in camels, efforts should be made to control the transmission of cystic echinococcosis from slaughter slabs and butcheries by disposing of echinococcus cysts safely to prevent dogs having access to the cysts (Njoroge *et al.*, 2002; Borji *et al.*, 2011).

6.4.9 Lymph nodes

Lymph nodes are distributed throughout the body. They reflect the health status of region they drain and they are examined during the meat inspection (Wilson, 2005). Camel meat is regularly consumed in many parts of the world and the major lymph nodes are regularly examined in slaughterhouses (Abdel-Magied *et al.*, 2001). The camel lymph nodes differ from those of other mammals in being lobulated and containing blood sinuses. The mandibular and retropharyngeal lymph nodes have lymphatic nodules and diffuse lymphoid tissue dispersed throughout the parenchyma instead of the cortex and medulla (Abdel-Magied *et al.*, 2001).

6.4.10 Caseous lymphadenitis (pseudotuberculosis)

Caseous lymphadenitis is a chronic disease in adult sheep and is worldwide in distribution (Hawari, 2008). The disease is also of economic concern in adult camels (Hawari, 2008). Caseous lymphadenitis in dromedary camels is caused by *Corynebacterium pseudotuberculosis* (Hawari, 2008) and *Corynebacterium ulcerans* (Tejedor *et al.*, 2004). Other bacteria, including *Corynebacterium renal, Corynebacterium equi* and *Staphylococcus aureus*, are isolated from abscesses in the lungs, liver, joints, muscular and subcutaneous tissues of the thigh, axilla, base of the tail, shoulder, elbow, base of the neck and under the jaw (Hawari, 2008). The disease is characterized by the formation of external and internal abscesses, which affect adult animals more than 5 years old (Hawari, 2008). The majority of cases are confined to the externally placed carcass lymph nodes. Affected lymph nodes are the ventral cervical lymph node, the cranial cervical lymph node, the superficial inguinal lymph node, the mammary, ileofemoral and axillary lymph nodes (Tejedor *et al.*, 2004). Multiple large abscesses are also found in the internal organs, particularly the lungs (Hawari, 2008). Occasionally, abscesses of the lymph nodes and the internal organs of various sizes from barely visible to as large as an orange, associated with severe emaciation, are detected. The abscesses are encapsulated by a relatively thick layer of necrotic and fibrous tissues. They contain odourless, non-granular, non-calcified thin homogenous creamy yellowish white pus, sometimes tinged with blood. In some camels, the peripheral lymph nodes are slightly enlarged but without abscess formation (Hawari, 2008).

Judgement: Partial condemnation of the affected parts; total rejection of the carcass if generalized and associated with emaciation.

6.4.11 Tuberculosis

Tuberculosis is a serious chronic infectious disease of humans and animals worldwide (Thoen *et al.*, 2006, 2009; Ndukum *et al.*, 2010). Tuberculosis is rare among camels kept under nomadic conditions (Kinne *et al.*, 2006). It manifests itself particularly in the lungs and lymph nodes or other organs, with granulomas known as tubercles (Kinne *et al.*, 2006). The two most important members of the genus *Mycobacterium* are *Mycobacterium tuberculosis* and *Mycobacterium bovis* (Kinne *et al.*, 2006). *Mycobacterium tuberculosis*, *Mycobacterium bovis*, *Mycobacterium caprae*, *Mycobacterium pinnipedii* and atypical mycobacteria including *Mycobacterium kansasii*, *Mycobacterium aquae*, *Mycobacterium fortuitum* and *Mycobacterium smegmatis* have been isolated in the dromedary camel as causative agents of camel tuberculosis (Kinne *et al.*, 2006).

Post-mortem inspection involves visual examination, palpation and systematic incision of carcasses and visceral organs, particularly the lungs, liver, kidney, heart, spleen and lymph nodes (Demelash *et al.*, 2009; Ndukum *et al.*, 2010). Tubercle bacilli when they enter the body produce a primary lesion generally in the respiratory or digestive tract and their associated lymph nodes (Wilson, 2005). The organs most frequently affected in dromedary camels are the lungs, the bronchial and mediastinal lymph nodes (Kinne *et al.*, 2006), the pleura and the liver (Kinne *et al.*, 2006; Pate *et al.*, 2006). The lungs and mediastinal lymph nodes are the main target (Kinne *et al.*, 2006; Wernery *et al.*, 2007). One or both lungs are consolidated with solid abscesses of different shapes and sizes (Kinne *et al.*, 2006). The lesions appear more frequently in the apical and cardiac lobes of both lungs than in the diaphragmatic lobes (Mamo *et al.*, 2011). Areas of mineralization, solid abscesses and granulomatous lesions have been observed in the lungs (Alvarez *et al.*, 2012). Caseous foci in lung lymph nodes have also been described in camels infected with *Mycobacterium pinnipedii* (Huard *et al.*, 2006). The retropharyngeal, mandibular, parotid, sub-maxillary, mesenteric and portal lymph nodes are those most frequently affected. The mesenteric lymph nodes constitute the most severely affected lymph nodes followed by the mediastinal lymph nodes (Mamo *et al.*, 2011). Other organs including intestine, kidney, spleen and heart are also affected (Pate *et al.*, 2006). Excessive granulomatous lesions can be detected in several internal organs, the costal pleura and the pericardium (Pate *et al.*, 2006).

Judgement: If the lesions are confined to the lungs and associated lymph nodes, the partially affected regions are partially condemned (Wilson, 2005). If the pleura are involved, the whole thoracic cavity should be condemned. Whenever the disease is generalized or disseminated throughout the systems and associated with emaciation, the total carcass should be rejected (Asseged *et al.*, 2004; Wilson, 2005).

6.4.12 Johne's disease (paratuberculosis)

Johne's disease is a chronic infectious disease caused by *Mycobacterium avium* subsp. *paratuberculosis*. It is serious and fatal in the dromedary camel (Alharbi *et al.*, 2012). Affected camels showed severe emaciation, enlargement of mesenteric lymph nodes and thickness and redness of ileum

(Alhebabi and Alluwaimi, 2010). Lesions of paratuberculosis are detected in the ileum, colon, rectum, liver and spleen (Alharbi, 2012). The camel is unique in its high incidence of liver involvement with the disease and generalized nature of the disease. The lesions are granulomas with a highly thickened corrugated mucous membrane that can be seen from the serosal surface (Alharbi *et al.*, 2012). The mesenteric lymph nodes are enlarged and contain greyish granulomas (Alharbi *et al.*, 2012). Generalized lymph node infection has also been reported (Alharbi *et al.*, 2012). There is a wasting of the hindquarter and loss of condition.

Judgement: Dependant upon the degree of emaciation and oedema. It is advisable to retain the carcass for 12–24 h before making a judgement. If the carcass remains wet and does not set, reject the intestines. If there is severe emaciation then total rejection is recommended.

6.4.13 Respiratory system

The lungs of slaughtered camels are examined visually and through palpation for lesions. Incisions are made into the lesion for further observation (Bekele, 2008).

6.4.14 Pneumonia

Pneumonia or inflammation of the lungs can be caused by bacteria, viruses, foreign bodies or parasites. The bacteria that can cause pneumonia in the camel include coagulase negative staphylococci, *Streptococcus* species, *Escherichia coli*, *Francisella tularensis*, *Flavobacterium* species, *Bordetella bronchiseptica*, *Aeromonas hydrophila*, *Neisseria* species, *Streptococcus agalactiae*, *S. aureus*, *Pasteurella trehalosi*, *Pasteurella anatipestifer*, *Pseudomonas aeruginosa* and *Micrococcus* species (Al-Tarazi, 2001; Awol *et al.*, 2011). *Rhodococcus equi* (formerly known as *Corynebacterium equi*) causes suppurative, necrotizing pneumonia with multiple encapsulated abscesses and large caseous nodules distributed throughout the lungs (Awol *et al.*, 2011). It also causes massive pneumonia involving the lungs. Besides large caseous areas, small, tan nodules are distributed throughout the rest of the lung (Kinne *et al.*, 2011). Pulmonary fibrosis, chronic interstitial pneumonia and pulmonary abscesses are the most common lesions recorded in the lungs of dromedary camels (Bekele, 2008). Chronic pleuropneumonia, pleuritis and multiple abscesses in the lungs are frequently detected in older camels. Lungs with hydatid cysts are usually associated with multi-focal or diffuse interstitial pneumonia, bronchopneumonia and emphysema (Bekele, 2008).

Judgement: Consolidation of either lungs or large parts of them and evidence of systemic infection implies total rejection. Otherwise, affected parts should be rejected. Pneumonia associated with pleurisy is a reason for total rejection. Inhalation of regurgitated ruminal contents or deposition of medicine into the trachea and lungs can cause severe pneumonia (Lopez, 2012). During slaughter, blood or ingesta might be aspirated into the trachea and lungs (Wilson, 2005).

6.4.15 Kidneys

The camel kidney is bean shaped and nonlobulated. The kidneys should be longitudinally bisected after the removal of the capsule, exposing the cortex and medulla. A variation in shape, colour, size or the presence of any lesion should lead to the total rejection of the kidney (Figs 6.2 and 6.3).

6.4.16 Liver

The liver of the dromedary camel has a peculiar shape, differing from that of other domestic animals (Abdalla *et al.*, 1971). The organ is irregular in shape and has four lobes, namely, cranial, quadrate, caudate and caudal lobes (Abdalla *et al.*, 1971). It has a large and small lobe, which in turn are divided by fissures into very small lobes (Abdalla *et al.*, 1971). The hepatic lobules are surrounded by fibrous connective tissue that renders the liver a hard consistency. The gall bladder is absent (Abdalla *et al.*, 1971). Liver diseases of camels such as

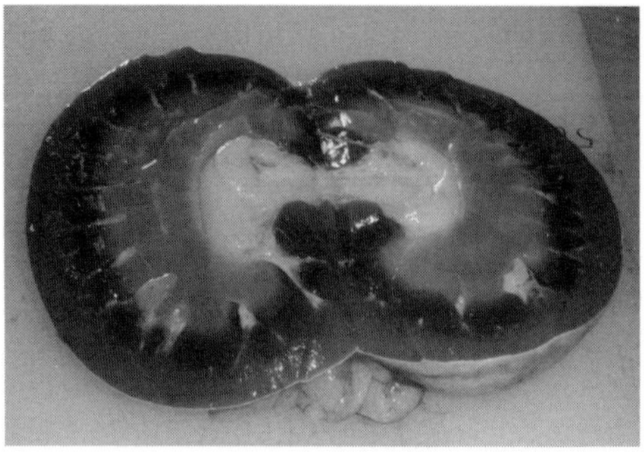

Fig. 6.2. Normal appearance of the camel kidney.

Fig. 6.3. Discoloration of a camel kidney.

hepatic lipidiosis and liver cirrhosis are frequently detected during meat inspection (Tejsingh *et al.*, 2006). Liver inspection is carried out by visual examination, palpation and incision of the organ (Mellau *et al.*, 2011). The liver should be carefully inspected on both the parietal and visceral surfaces by several incisions for the presence of the following conditions.

Hepatic lipidiosis

Lipids are normally transported to the liver from the adipose tissue and gastrointestinal tract in the form of either free fatty acids or chylomicrons, respectively. The presence of excessive lipids within the liver is termed lipidiosis or steatosis (fatty liver or fatty change) and occurs when the rate of triglyceride accumulation within hepatocytes exceeds either their rate of metabolic degradation or their release as lipoproteins (Cullen and Brown, 2012). It arises from conditions that cause increased mobilization of the body fat stores as in late pregnancy, early lactation and starvation (increased mobilization of triglycerides). The liver is enlarged, yellow, soft and friable, and the edges of the lobes are rounded and broad instead of being sharp and flat. When incised, the hepatic parenchyma is soft and friable and has a greasy texture attributable to lipids within hepatocytes (Myers *et al.*, 2012).

Judgement: The liver should be rejected in severe cases.

Tumours

Tumours are abnormal growths of new tissue of unknown cause and having no purposeful function (Wilson, 2005). The liver is a common site of tumours.

Judgement: If the tumour is localized the tumour can be removed with its surroundings but a large tumour occupying most of the liver implies total rejection (Wilson, 2005).

Necrosis

Necrosis can be in various forms, e.g. focus or foci (Fig. 6.4) and they are more or less circular in outline. In the early stage, the peripheries of the lesions are hyperemic but, at a later stage, the necrotic areas become encapsulated.

Judgement: Reject the affected parts and in severe cases total rejection is recommended.

Abscesses

Liver abscesses are very frequent in the liver of the camel and if found the liver has to be rejected.

Cirrhosis

Cirrhosis is fairly common in the liver of camel. It is either focal or diffuse and the liver should be rejected.

Calcified cysts

Calcified cysts are commonly found in the liver of camel (Fig. 6.5). If few cysts are seen, they can be removed. If there are many, however, then total rejection of the liver is recommended.

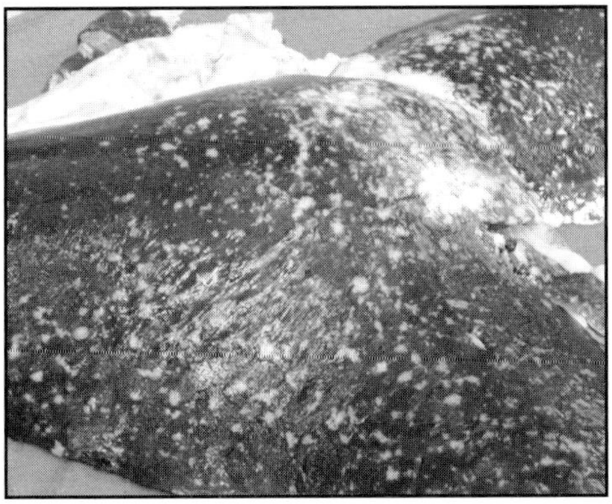

Fig. 6.4. Random foci of necrosis in the liver of the camel.

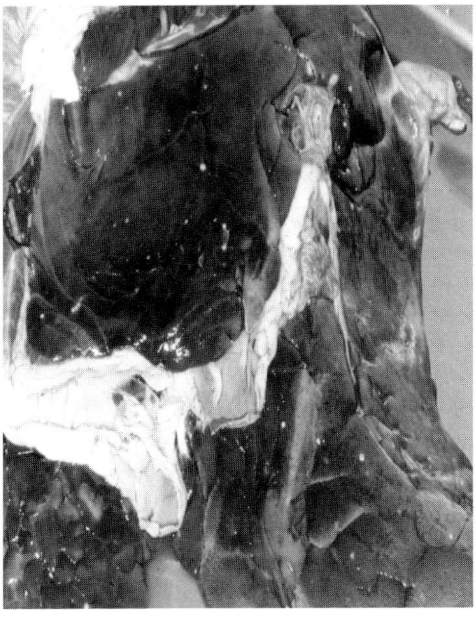

Fig. 6.5. Minute calcified cysts in the liver of the camel.

6.4.17 Male genital organs

These comprise the two testicles, ductus deferens, seminal vesicles, prostate, bulbourethral glands and penis; they are rejected.

6.4.18 Female genital organs

These comprise ovaries, uterus, vagina, vulva, mammary glands with their associated lymph nodes; total rejection.

Fig. 6.6. Degenerative myocarditis lesions in the heart of a 1-year-old camel.

6.4.19 Muscle degeneration

The main cause of muscle degeneration is linked to selenium deficiency (white muscle disease) including the heart (white degeneration of the heart; Fig. 6.6), which is common in the Arabian Peninsula (Faye and Seboussi, 2009). There is no risk for consumers, but it is commercially not appreciated and could be rejected.

6.5 Conclusion

Meat inspection not only protects human health against zoonotic diseases but also plays a pivotal role in monitoring the level of diseases. To prevent cross contamination of meat, it is obligatory to adopt well-planned, well-executed and controlled cleaning and sanitation of the slaughterhouse, machines and equipment to ensure purity, safety and health.

Ante-mortem examination is imperative for camels destined for slaughter to ensure that they do not show signs of disease. Post-mortem inspection is conveyed through careful visual examination, palpation and incision of visceral organs including the lungs, liver, heart, spleen and kidneys. A decision should be made to approve the entire carcass for human consumption or condemn organs and/or part of the carcass.

Cystic echinococcosis is of considerable economic and public health importance and is one of the most important parasitic infections in the dromedary camel. Safe disposal of *Echinococcus* cysts so that dogs cannot have access to the cyst is crucial for the interruption of the life cycle.

References

Abdalla, O., Arnautovic, I. and Fahmy, M.F.A. (1971) Anatomical study of the liver of the camel (*Camelus dromedarius*).1. *Topography and Morphology, Acta Morphologica Neerland-Scandinavica* 9, 85–100.

Abdel-Magied, E.M., Taha, A.A.M., Al-Qarawi, A.A. and Elfaki, M.G. (2001) The parotid, mandibular and lateral retropharyngeal lymph nodes of the camel (*Camelus dromedarius*). *Anatomy, Histology and Embryology* 30, 199–203.

Alharbi, K.B., Al-Swailem, A., Al-Dubaib, M.A., Al-Yamani, E., Al-Naeem, A., Shehata, M., Hashad, M.E., Albusadah, K.A. and Mahmoud, O.M. (2012) Pathology and molecular diagnosis of paratuberculosis of camels. *Tropical Animal Health and Production* 44, 173–177.

Alhebabi, A.M. and Alluwaimi, A.M. (2010) Paratuberculosis in camel (*Camelus dromedarius*): The diagnostic efficiency of ELISA and PCR. *The Open Veterinary Science Journal* 4, 41–44.

Almeida, F.B., Rodrigues-Silva, R., Neves, R.H., Romani, E.L.S. and Machado-Silva, J.R. (2007) Intraspecific variation of *Echinococcus granulosus* in livestock from Peru. *Veterinary Parasitology* 143, 50–58.

Al-Tarazi, Y.H. (2001) Bacteriological and pathological study on pneumonia in the one-humped camel (*Camelus dromedarius*) in Jordan. *Revue Elevage Medicine Vetrinair Pays Tropeau* 54, 93–97.

Alvarez, J., Bezos, J., de Juan, L., Vordermeier, M., Rodriguez, S., Fernandez-de-Mera, I.G., Mateos, A. and Dominguez, L. (2012) Diagnosis of tuberculosis in camelids: Old problems, current solutions and future challenges. *Transboundary and Emerging Diseases* 59, 1–10.

Asseged, B., Woldesenbet, Z., Yimer, E. and Lemma, E. (2004) Evlauation of abattoir inspection for the diagnosis of Mycobacterium bovis infection in cattle at Addis Ababa abattoir. *Tropical Animal Health and Production* 36, 537–546.

Awol, N., Ayelet, G., Jenberie, S., Gelaye, E., Sisay, T. and Nigussie, H. (2011) Bacteriological studies on pulmonary lesions of camel (*Camelus dromedarius*) slaughtered at Addis Ababa abattoir, Ethiopia. *African Journal of Microbiology Research* 5, 522–527.

Azlaf, R. and Dakkak, A. (2006) Epidemiological study of the cystic echinococcosis in Morocco. *Veterinary Parasitology* 137, 83–93.

Bekele, S.T. (2008) Gross and microscopic pulmonary lesions of camels from Eastern Ethiopia. *Tropical Animal Health and Production* 40, 25–28.

Borji, H. and Parandeh, S. (2010) The abattoir condemnation of meat because of parasitic infection, and its economic importance: results of a retrospective study in north-eastern Iran. *Annals of Tropical Medicine and Parasitology* 104, 641–647.

Borji, H., Azizzadeh, M. and Afsai, A. (2011) An abattoir-based study of Hydatidosis in the dromedary (*Camelus dromedarius*) in Mashhad, Iranian. *Journal of Helminthology* doi: 10.1017/S0022149X10000933.

Cullen, J.M. and Brown, D.L. (2012) Hepatobiliary system and exocrine pancreas. In: Zachary, J.F. and McGavin, M.D. (eds) *Pathologic Basis of Veterinary Disease*, 5th edn, Mosby–Elsevier, St Louis, Missouri, Chapter 8.

Dakkak, A. (2010) Echinococcosis/hydatidosis: A severe threat in Mediterranean countries. *Veterinary Parasitology* 174, 2–11.

Demelash, B., Inangolet, F., Oloya, J., Asseged, B., Badaso, M., Yilkal, A. and Skjerve, E. (2009) Prevalence of bovine tuberculosis in Ethiopian slaughter cattle based on postmortem examination. *Tropical Animal Health and Production* 41, 755–765.

Derbala, A.A. and El-Massry, A. (1999) Some studies on the growth and development of *Echinococcus granulosus*, camel origin in expermintally infected dogs. *Veterinary Parasitology* 83, 25–36.

Eckert, J., Thompson, R.C.A., Michael, S.A., Kumaratilake, L.M. and El-Sawah, H.M. (1989) *Echinococcus granulosus* of camel origin: development in dogs and parasite morphology. *Parasitology Research* 75, 536–544.

Fathi, S., Dehaghi, M.M. and Radfar, M.H. (2011) Occurrence of hydatidosis in camels (*Camelus dromedarius*) and their potential role in the epidemiology of *Echinococcus granulosus* in Kerman area, southeast of Iran. *Comparative Clinical Pathology* doi 10.1007/s00580-011-1200-0.

Faye, B. and Seboussi R. (2009) Selenium in camel – A review. *Nutrients* 1, 30–49.

Fernanda, B.A., Rosangela-Selva, R., Renata, H.N., Elizabeth, L.S.R. and Machado-Silva, J.R. (2007). Intraspecific variation of Echinococcus granulosus in livestock from Peru. *Veterinary Parasitology* 143, 50–58.

Hawari, A.D. (2008) *Corynebacterium pseudotuberculosis* infection (Caseous Lymphadenitis) in camels (*Camelus dromedarius*) in Jordan. *American Journal of Animal and Veterinary Sciences* 3, 68–72.

Huard, R.C., Fabre, M. de H.P., Lazzarini, L.C. van, Soolingen, D., Cousins, D. and Ho, J.L. (2006) Novel genetic polymorphisms that further delineate the phylogeny of the *Mycobacterium tuberculosis* complex. *Journal of Bacteriology* 188, 4271–4287.

Ibrahim, M.M. (2010) Study of cystic echinococcosis in slaughtered animals in Al Baha region, Saudi Arabia: Interaction between some biotic and abiotic factors. *Acta Tropica* 113, 26–33.

Ibrahima, M.M. and Craiga, P.S. (1998) Prevalence of cystic echinococcosis in camels (*Camelus dromedarius*) in Libya. *Journal of Helminthology* 72, 27–31.

Kinne, J., Johnson, B., Jahans, K.L., Smith, N.H., Ul-Haq, A. and Wernery, U. (2006) Camel tuberculosis- a case report. *Tropical Animal Health and Production* 38, 207–213.

Kinne, J., Madarame, H., Takai, S., Jose, S. and Werney, U. (2011) Disseminated *Rhdodococus equi* infection in dromedary camels (*Camelus dromedarius*). *Veterinary Microbiology* 149, 269–272.

Lahmar, S., Debbek, H., Zhang, L.H., McManus, D.P., Souissi, A., Chelly, S. and Torgerson, P.R. (2004) Transmission dynamics of the *Echinoccocus granulosus* sheep-dog strain (G1 genotype) in camels in Tunisia. *Veterinary Parasitology* 121, 151–156.

Latif, A.A., Tanveer, A., Maqbool, A., Siddiqi, N., Kyaw-Tanner, M. and Traub, R.J. (2010) Morphological and molecular characterization of *Echinococcus granulosus* in livestock and humans in Punjab, Pakistan. *Veterinary Parasitology* 170, 44–49.

Lopez, A. (2012) Respiratory system, mediastinum and pleurae. In: Zachary, J.F. and McGavin, M.D. (eds) *Pathologic Basis of Veterinary Disease*, 5th edn, Mosby–Elsevier, St Louis, Missouri, pp. 516–517.

Mamo, G., Bayleyegn, G., Tessema, T.S., Legess, M., Medhin, G., Bjune, G., Abebe, F. and Ameni, G. (2011) Pathology of camel tuberculosis and molecular characterization of its causative agents in pastrol region of Ethiopia. *PLoS ONE* 6(1). Available at: www.plosone.org volume 6 Issue 1 e15862 (Accessed January 2011).

Mellau, B.L., Nonga, H.E. and Karimuribo, E.D. (2011) Slaughter stock abattoir survey of carcasses and organ/offal condemnations in Arusha region, northern Tanzania. *Tropical Animal Health and Production* 43, 857–864.

M'rad, S., Filisetti, D., Oudni, M., Mekki, M., Belguith, M., Nouri, A., Sayadi, T., Lahmar, S., Candofi, E., Azaiez, R., Mezhoud, H. and Babba, H. (2005) Molecular evidence of ovine (G1) and camel (G60 strains of *Echinococcus granulosus* in Tunisia and putative role of cattle in human contamination. *Veterinary Parasitology* 129, 267–272.

Myers, R.K., McGavin, M.D. and Zachary, J. (2012) Cellular adaptation, injury, and death: morphologic, biochemical, and genetic bases. In: Zachary, J.F. and McGavin, M.D. (eds) *Pathologic Basis of Veterinary Disease*, 5th edn, Mosby–Elsevier, St Louis, Missouri, p. 32.

Ndukum, A., CalebKudi, A., Bradley, G., Ane-Anyangwe, I.N., Fon-Tebug, S. and Tchoumboue, J. (2010) Prevalence of bovine tuberculosis in abattoirs of the littoral and western highland regions of Cameroon: a cause for public health concern, SAGE-Hindawi Access to Research. *Veterinary Medicine International* Article ID 495015, doi: 10.4061/2010/495015.

Njoroge, E.M., Mbithi, P.M.F., Gathuma, J.M., Wachira, T.M., Gathura, P.B., Magambo, J.K. and Zeyhle, E. (2002) A study of cystic echinococcosis in slaughter animals in three selected areas of northern Turkana, Kenya. *Veterinary Parasitology* 104, 85–91.

Omer, R.A., Dinkel, A., Romig, T., Mackenstedt, U., Elnahas, A.A., Aradaib, I.E., Ahmed, M.E., Elmalik, K.H. and Adam, A. (2010) A molecular survey of cystic echinococcosis in the Sudan. *Veterinary Parasitology* 169, 340–346.

Pate, M., Svara, T., Gombac, M., Paller, T., Zolnir-Dovc, M., Emersic, I., Prodinger, W.M., Bartos, M., Zdovc, I., Krt, B., Pavlik, I., Cvetnic, Z., Pogacnik, M. and Ocepek, M. (2006) Outbreak of tuberculosis caused by Mycobacterium caprae in a zoological garden. *Journal of Veterinary Medicine* B 53, 387–392.

Romig, T., Omer, R.A., Zeyhle, E., Huttner, M., Dinkel, A., Siefert, L., Elmahdi, I.E., Magambo, J., Ocaido, M., Menezes, C.N., Ahmed, M.E., Mbae, C., Grobusch, M.P. and Kern, P. (2011) Echinociccosis in sub-Saharan Africa: emerging complexity. *Veterinary Parasitology* 181, 43–47.

Tejedor, M.T., Martin, J.L., Corbera, J.A., Shulz, U. and Gutierrez, C. (2004) Pseudotuberculosis Dromedary camels in the Canary Islands. *Tropical Animal Health and Production* 36, 459–462.

Tejsingh, G.D., Sharms, A., Singh, R.D. and Surender, S. (2006) Incidence and pathology of degenerative changes in the liver of camels. *Veterinary Practitioner* 7, 35–36.

Thoen, C.O., Steel, J.H., and Gilsdorf, M.J. (2006) Mycobacterium bovis *Infection in Animals and Humans*, 2nd edn. Blackwell Publishing, Ames, Iowa, p. 329.

Thoen, C.O., LoBue, P., Ernason, D., Kaneen, J.B. and de Kantor, N. (2009) Tuberculosis: a re-emerging disease in animals and humans. *Veterinaria Italiana* 45, 135–181.

Wernery, U., Kinne, J., Jahans, K.L., Vordermeier, H.M., Esfandiari, J., Greenwald, R., Johnson, B., Ul-Haq, A. and Lyashchenko, K.P. (2007) Tuberculosis outbreak in a dromedary racing herd and rapid serological detection of infected camels. *Veterinary Microbiology* 122, 108–115.

Wilson, W.G. (2005) *Wilson's Practical Meat Inspection*, 7th edn, Blackwell Publishing Ltd, Oxford, UK.

7 Prospects for Online Grading of Camel Meat Yield and Quality

Howard J. Swatland
University of Guelph, Guelph, Ontario, Canada

7.1 Introduction

Judging from current trends, it seems that the international shipment of camel meat to Europe, North America and Asia will continue to increase. As problems of global warming and the expansion of arid areas become more severe (Lutz, 1998; Makoto *et al.*, 2006), the contribution of camel meat to world protein production might increase (McCloy and Rowe, 2000), thus justifying the development of new technology for quality control and grading of premium exports. Camels allow meat production on arid land where not much else is possible, and, on the other hand, international markets for Halal products continue to increase. At present, shippers might claim their meat is from young camels (<36 months) with pale, soft meat – but how can buyers be sure of this? If the trading of established products such as beef, pork and lamb is any guide, there will be a commercial incentive to develop reliable criteria for grading camel meat yield and quality. The grading of established products is moving away from subjective judgement, and moving towards greater objectivity, on the basis of instrumentation. Subjective evaluation works best when retailers can personally examine carcasses in a wholesale market, or when a premium is paid for commercial integration from source to retailer (branded products). If heterogenous carcasses from various sources come into a wholesale market and are purchased remotely, how are they to be evaluated? A lack of objective criteria for meat quality also hinders the feedback of information to breeders and feeders, as well as the feed-forward of information to retailers and consumers. Thus, with camel meat at an early stage of market development, there might be an opportunity to bypass subjective grading and proceed directly to objective grading. This calls for educated guesses. What is required to facilitate commerce? What existing methods might be adapted? What new methods are needed?

Some ideas for the prospects of online grading of camel meat yield and quality are outlined in this chapter, building on our established knowledge of meat animal growth and structure (Swatland, 1994), online measurement of meat quality (Swatland, 1995) and novel meat products (Swatland, 2004). These books provide the academic references for statements made in this chapter, leaving us free to focus on the main topic: camel meat.

7.2 Camel Meat

Camelus dromedarius, the Arabian camel or dromedary, has one hump, long slender limbs, runs swiftly and can be used for racing.

Camelus bactrianus, the Bactrian camel of the cold deserts of Asia, has two humps and thick limbs. Camel meat tastes rather like beef (some say mutton) and has a similar nutrient value, although it is usually lower in fat content (Kadim *et al.*, 2008a) and vitamin E (Soltanizadeh *et al.*, 2010). Prime meat from young camels can be cooked rapidly with dry heat, whereas meat from the extremities of young animals and all the meat from older animals requires cooking with moist heat. Thus, both the toughness and fat content of camel meat increase with age (Kadim and Mahgoub, 2008). In a zoological context, camels are intermediate between pigs (which have upper incisors) and cattle and sheep (which have a horny pad in place of upper incisors). Camels have vestigial upper incisors, plus well developed canine teeth that may be tusk-like (canine teeth are present in pigs but not in cattle and sheep).

The shape of a camel carcass (Fig. 7.1) differs radically from the shape of a beef, lamb or pork carcass. Apart from the obvious shape of the dorsal hump, the most notable feature is the restriction of hindlimb muscles near the pelvis: they do not overlap with the abdominal muscles. Thus, when the hindlimb is stretched back in a hanging carcass, it forms an indentation ventral to the ilium. This is because the camel has long limbs capable of considerable rotation relative to the vertebral axis. In a sitting camel, the distal end of the femur projects downwards towards the ground, whereas in sitting cattle and sheep the distal end of the femur projects upwards. The camel also has a broad cutaneous pad on each foot instead of two hooves. Camels are ruminants with four stomach compartments, but the omasum is indistinct, not hard and round as in cattle (Clutton-Brock, 1987; Elgasim and Alkanhal, 1992). The names used in international meat cutting form a complex linguistic labyrinth (Swatland, 2004), but a basic international vocabulary seems to be developing for international trading in camel meat (Fig. 7.1).

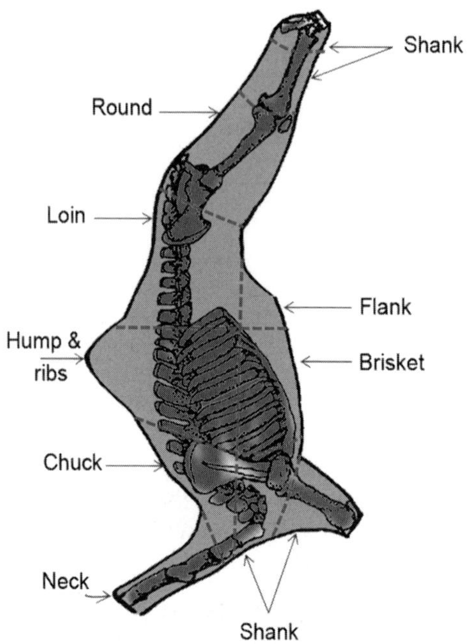

Fig. 7.1. Camel cuts in the United Arab Emirates (from a personal communication, G. Alhadrami, 1999, United Arab Emirates University) with typical international names.

The llama (*Lama glama*) is a multipurpose domesticated camelid of the Andes. Most meat is used domestically, although in Bolivia and Peru there are commercial markets. Llama meat is similar in taste to mutton and has technological characteristics similar to other meats (Salva *et al.*, 2009). A llama produces about 12–15 kg of charqui (Calle Escobar, 1984; Iniguez *et al.*, 1998). Charqui is an intermediate moisture (45%) meat product with a high (15%) sodium chloride content. Typically it has some protein denaturation in the A-band and M-line, together with empty fluid channels created during dehydration. For alpaca (*Lama pacos*) charqui, the meat is cut into slices 0.5–1 cm thick, treated with salt, and then soaked in brine for 2 or 3 days. Open-air drying for 2–3 weeks in the Andes resembles freeze-drying. Final drying is done under a roof (Calle Escobar, 1984; Biscontini *et al.*, 1996). The alpaca is famous for its long, fine wool. The meat is consumed domestically when only a few animals are kept, but the surplus from larger ranches may be used for charqui.

7.3 Meat Yield

Estimation of meat yield from pork and beef carcasses is based on a number of assumptions. The first assumption is that the bone content of a carcass is approximately constant, both in volume and density. In most cases this is acceptable for a population of animals of similar age and body type, but there must be numerous exceptions that elude detection. The next assumption is that, of the remaining carcass volume (total minus bone), there are only two compartments – saleable meat and fat. This ignores a small volume attributable to tendons, ligaments and other major connective tissues such as fasciae (sheets of collagenous tissue binding major muscles). In carcass dissection reports, this compartment is usually called trim. Camel carcasses may have a composition of 56% meat, 19% bone and 14% fat (Yousif and Babiker, 1989). In estimating the meat yield of pork carcasses, the sometimes obvious differences in subcutaneous fat distribution over the anterior thoracic and posterior lumbar regions are ignored, and a single fat depth measurement is made near the thoraco–lumbar junction. This single linear measurement is usually made with an automated fat depth probe using light-emitting diodes and applied to a regression equation to predict meat yield (saleable meat as a fraction of carcass weight). It is not difficult to see all the problems involved, such as predicting a three-dimensional parameter (mass of meat) from a linear measurement (subcutaneous fat depth), hot versus cold carcass weight (shrink loss), pattern of meat cutting for saleable meat, etc. A more reliable input can be obtained ultrasonically by scanning along much of the vertebral axis of the carcass. This multiple measurement encompasses antero-posterior differences in subcutaneous fat depth, but still suffers from the other problems. Pork carcasses are seldom ribbed to expose major muscles such as the *Longissimus thoracis* and it is usually only in research methodology that there is any reliable input on muscle mass. Thus, in many countries, the meat yield of a pork carcass is estimated as carcass weight minus an assumed constant bone volume minus a predicted volume of fat to be trimmed from saleable meat. True, a lot of assumptions, but this approach is widely used and accepted as a basis for incentive payments to producers of lean, high-yielding pork carcasses.

Yield grading for beef follows similar assumptions as for measuring fat depth in pork, but most beef carcass are ribbed (separating the forequarter from the hindquarter near the posterior part of the rib cage) and this gives extra information on the exposed rib-eye (*Longissimus thoracis*) to be incorporated into a regression equation to predict meat yield. Several measurements of subcutaneous fat depth over the rib-eye may be used in some countries, as well as multiple measurements of the rib-eye (width × depth or overall area). Although two-dimensional measurements are an improvement over a linear measurement in predicting meat yield, there remains a serious problem – cattle are not all equal in antero-posterior length. Potentially useful scientific information is also lost for feedback to producers. The *Longissimus thoracis* is a complex muscle with many parts, and its fibres pass through the rib-eye area on a ribbed carcass at an angle. Rib-eye width is a function of muscle fibre number × diameter (genetics versus nutrition, respectively), whereas rib-eye depth is a function of muscle fibre length (nutrition).

There is considerable international variation in methods and technology for yield grading. Pork grading gives the least attention to meat quality but, so far, has the best developed technology. Optical fat depth probes are widely used. They detect the fat to muscle boundary by a change in reflectance detected by diodes in the probe shaft, with boundary depth being found relative to a plate remaining on the surface of the carcass. Ultrasonic systems detect the same boundary but at multiple sites along the back of the carcass, as well as providing some information on muscle depth. For beef, where a rib-eye area is traditionally exposed for grading, and where meat quality information is essential, the best technology is video image analysis (VIA). VIA might therefore be the first choice as a potential tool for yield grading of camel carcasses but not without many concerns.

The scientific dissection of beef carcasses to understand meat production has a long history, starting in the 1800s with Lawes (1814–1900) and Gilbert (1817–1901) when the primary objective was to explain how animals deposited fat. Veterinary anatomists soon provided detailed information on bovine myology and this gave agricultural scientists a reliable basis for extensive carcass dissection studies. This was a major field of academic activity in meat research laboratories from the 1940s to the 1970s, and survives today in government laboratories responsible for national meat grading systems. But carcass dissection is very expensive in terms of both facilities and labour. Will the commercial advantages of yield grading for camel carcasses ever justify such a massive investment of research resources? Are there less expensive alternatives? Within a commercial environment, with its ever increasing use of computer-based tracking and inventory control, will it be possible to use the monetary value of products as the target for prediction? For example, the price of steak meat is currently three times that of stewing meat from the shank (Fig. 7.1). Thus, to establish the pricing for the various cuts it is necessary to know their weight as a fraction of cold carcass weight so commercial companies might already be using the carcass cut-out data we would like to predict with yield grading. Continuing advances in optoelectronics are almost certain to increase the sophistication of instrumentation, enabling agricultural scientists to work in a realm once the exclusive domain of astronomers and space scientists. Thus, we might be able to bypass the traditional carcass dissection approach altogether.

Let us assume that the rib-eye area of a camel carcass is a useful indicator of total muscle mass. It will be essential to check this assumption – perhaps another muscle area will more reliable? Whatever muscle area is to be used, the major technical problem in measuring muscle areas automatically is caused by the difficulty of separating adjacent muscles that touch the area of interest, sometimes without any intermuscular fat separation. Several muscles may have a muscle to muscle boundary with the *Longissimus thoracis*, such as the *Multifidus dorsi*, *Longissimus costarum* and *Spinalis dorsi*. In camel steaks it will be necessary to detect and ignore the dorsal extensions of abdominal muscles (Fig. 7.2). When two separate muscle areas touch each other to produce an apparently continuous structure, they can often be resolved separately with two VIA operations: erosion and dilation.

In the erosion operation, pixels are turned off all around the borders of the major areas, thus reducing the area. After several erosions, a crack separating two adjacent areas that touch may be completely opened. Once this is detected computationally, the area is dilated by turning on the pixels around its edges, except where they would cause it to rejoin the area from which it has just been separated. Dilation is repeated until the areas are back to their original size, except in the seam where adjacent areas would touch. Thus, erosion and dilation can be used to create an imaginary seam of intermuscular fat between two adjacent muscles, and this enables key parameters such as area, width and depth of the *Longissimus thoracis* to be found quite easily.

Small specks of intermuscular fat lined up along a true intermuscular separation might facilitate finding a seam between the two adjacent areas in the video image. Thus, not only must this information be preserved if the image is smoothed (using a VIA kernal to remove noise), but colour information such as a high G value for a pixel (green colour, lost by myoglobin absorbance in muscle) can be used to enhance specks of fat, both for marbling determinations and opening seams. Camel carcasses often lack marbling fat (and even intermuscular fat in certain areas) so this will be a challenge with VIA.

In finding seams between contiguous muscle areas using VIA, it may be useful to have a probability model of the likely radius of curvature of the separation, by reference to the outline of the main muscle area that has already been established or by referring to a library of previous operator-guided separations. These two reference sources (the outline established so far and a library of reliable separations) could be useful for identifying separations that should not be

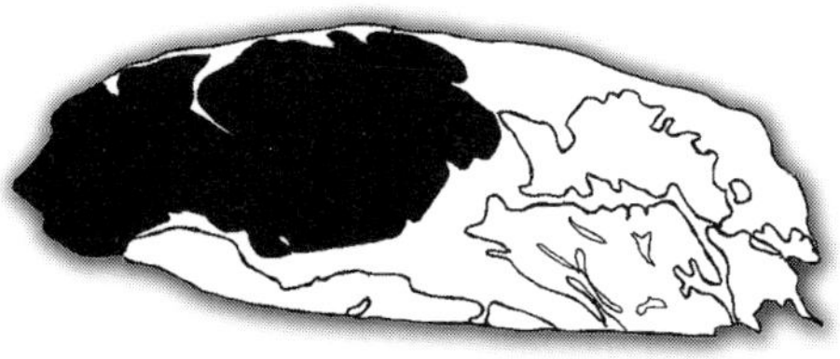

Fig. 7.2. *Longissimus thoracis* or rib-eye area (black) in a camel steak.

pursued. With compound muscles such as the *Longissimus thoracis*, a major crack through the muscle area may be caused by the partial separation of one of the subunits (Fig. 7.2), and knowing the likely angle and position of subunit separation might be required to prevent the erroneous subdivision of the whole compound muscle.

Erosion and dilation operations are available in many general purpose software packages for VIA. In a typical application, the operator writes a macro that goes through a series of erosions and dilations for the most difficult separation in the set of images to be processed, and then the remainder of the images are processed automatically. But there is still considerable subjectivity in the operation; a macro that works satisfactorily for a series of fat carcasses with easy separations between muscles because of abundant intermuscular fat might fail if applied to images from leaner carcasses. For rib-eye areas, interactive decision making by the operator may be essential for erosion and dilation methods, and complete automation of the rib-eye area measurement solely by erosion and dilation could be unreliable.

Manufacturers of VIA systems for meat yield keep their algorithms a secret, but artificial intelligence to recognize specific muscles of the carcass is bound to happen, if it is not already being used. Some of the systems currently available still rely on the knowledge and hand–eye coordination of a human operator for a critical operation: locating the video camera at a specific point and orientation relative to the rib-eye area. This is facilitated by mounting the video camera on a rigid frame that the operator places against the rib and split vertebrae of a ribbed forequarter. Thus, the software may be written with the knowledge that the rib-eye area is somewhere within the video image at a constant position relative to the skeletal reference points. If a specified muscle area is contained totally within the video frame, and if the light intensity and camera response are regulated so that a constant threshold level can be used to distinguish muscle from fat, then this will allow several methods to be used to locate the approximate centre point of the specified muscle. The first step may involve a Boolean rejection of pixels below the threshold, regarding all the survivors as equal. The largest areas of contiguous surviving pixels then may be skeletonized (progressively eroding the outer pixels of the area to converge on the centre of the area). Then the muscle area may be reconstructed by flooding (turning on peripheral pixels around the skeletonized areas, but not if this would involve linking up with another rebuilt area).

Another method of finding a muscle area, if the operator has placed it within the video frame, is to drop a plumb line down the *y*-axis of the image after setting an acceptance threshold. The plumb line is lowered until it contacts an *x*-axis vector of on-pixels greater than a minimum value known for the muscle. In other words, this requires some prior knowledge of the minimum *x*-axis width of the specified muscle. Once inside the specified area, an edge-finding algorithm can be used to delineate the *x*:*y* coordinates of the muscle perimeter.

VIA is not the only method for yield grading to be considered by researchers pursuing yield grading in the camel carcass. Dual-emission X-ray absorptiometry (DXA or DEXA) is another possibility (Ribeiro *et al.*, 1998). Its original biomedical application was for measuring bone mineral density in patients at risk from osteoporosis. Two or more X-ray beams differing in energy levels are used, one that is absorbed by bone and the other that has less absorbance (providing a background level for a ratiometric determination). This allows hard tissue (bone) to be separated from soft tissues such as muscle and fat. Applied to complex models, such as a camel carcass, there might be problems with beam hardening (Gotfredsen *et al.*, 1997). This occurs when low energy photons are easily absorbed, leaving high energy photons that then cause problems in ratiometric determinations. The medical literature contains methods to control this error using samarium (an element in the lanthanide series), which has many interesting properties and is used as a neutron absorber in control rods for reactors; it is not dangerous and is used in alloys such as those found in domestic ignition systems. Another concern is the effect of animal shape. X-ray methods have been used for many years with boxed beef (for example, the MeatMaster developed in the 1990s). Their high reliability is dependent upon the fixed photon pathway through meat filling a standard-sized box. Thus, to be applied to the irregular shape of carcass meat, preliminary studies will be required to control this source of error. In medical applications, the state of hydration of tissues might be a problem (Pietrobelli *et al.*, 1998), but this might be used to advantage if it allows some assessment of water content in meat (relative to fluid losses and the effect of pH, as explained later in this chapter).

In conclusion, it is important to see yield grading of carcasses for what it is – an imperfect process but one that has become essential for remote purchasing by buyers who cannot directly choose from carcasses on a rail. Perhaps VIA grading methods for beef yield can be adapted for camel carcasses; only research will prove this one way or another. Possibly an entirely different approach will be needed, such as using VIA for overall carcass conformation. For beef carcasses, if differences in bone length are taken into account, carcasses with bulging muscles have a higher relative meat yield than carcasses with sunken muscles. An increase in the girth of a deep muscle resulting from the radial growth of muscle fibres causes overlying superficial muscles to bulge outwards. Thus, the length of the superficial muscle is increased in a curvilinear manner as the deep muscle grows in size. When both fatness and carcass length are taken into account, an appraisal of muscle conformation may be a useful guide to the anticipated lean yield of a carcass. Is this true for camel carcasses as well? Can this by detected by VIA of the carcass, thus removing the need to rib a carcass?

Finally, as scientists, we must consider the null hypothesis – that camel carcasses are so uniform in meat yield that no yield grading is required. From a commercial perspective, yield grading might not be needed in a vertically integrated commercial system if camels are farmed to reach a constant end point in body composition. But when camels are culled from wild populations, as in Australia (McCloy and Rowe, 2000), constant body composition seems unlikely. If major shippers of vacuum-packed primal cuts currently source their own raw material, yield grading could be useful internally within the company, and a publically known yield grade would only be useful when carcasses rather than primal cuts are traded.

7.4 Meat Quality

Table 7.1 shows a summary of the methods available to assess meat quality online (Swatland, 1995). The methods are based on the basic biophysical properties of common meats such as beef and pork. They may be applicable to camel meat but research is required to prove this. As a preliminary study, results are presented here from a small number of previously frozen samples of Australian camel meat purchased in Canada. Future studies are required to characterize biological variation in fresh meat from known sources. The preliminary results reported here merely show what measurements are possible.

7.4.1 Muscle reflectance

Reflectance from meat is far more complex than is realized by routine researchers using a commercial colorimeter on a meat surface to record its chromaticity coordinates. First we need to know the emission spectrum of the illuminator; it might look like bright, white light to the casual observer but every

Table 7.1. Summary of existing online methods.

Basis	Methods	Prediction
Subcutaneous fat depth and muscle cross-sectional area (usually *Longissimus thoracis*)	Optical probes using diodes, or ultrasonics, or video image analysis (VIA) of cut surfaces	Meat yield. Assuming bone content is constant (which is not always true), subtract an estimate of fat content from total mass and the remainder is an estimate of the meat content.
Acidity, pH	Glass (calomel half-cell) or ion-sensitive field-effect transistor (ISFET) electrode	Paleness–darkness, fluid exudation, softness
Electrical impedance	Two or four electrodes, conductivity, capacitance, phase angle	Paleness–darkness, fluid exudation, softness
Muscle internal reflectance	Fibre-optic spectrophotometry	Myoglobin concentration, paleness–darkness
Fat internal reflectance	Fibre-optic spectrophotometry	Carotene yellowness, short-chain triglyceride translucency
Connective tissue	Subcutaneous fat-depth probe adapted for UV fluorescence	Amount and distribution of collagen and elastin, and pyridinoline cross-linking of collagen
Rheology	Electromechanical probes using compression or rotation, and elastic deformation detected ultrasonically	Toughness
Surface appearance	Video image analysis	Carcass shape (muscularity), rib-eye area and marbling, subcutaneous fat colour
Near-infrared reflectance	Fibre-optic and surface reflectometers	Triglyceride content, collagen content

light source is different, largely depending on its temperature. Some light is reflected directly from the meat surface without entering the meat (specular or mirror-like reflectance following Fresnel equations). Specular reflectance is polarized, and may be partly extinguished by rotating a polarizer (as in Polaroid sunglasses) to the appropriate angle. Thus, a polarizer should be used in any system attempting to quantify flecks of marbling using VIA, otherwise it is difficult to separate the fat flecks from the bright flecks of specular reflectance (which are a function of surface irregularity). Light entering the meat is scattered. Some of it scatters back to the meat surface to appear as diffuse or Lambertian reflectance. Lambertian reflectance appears similar at all angles, whereas specular reflectance, like a mirror, has a strong angular effect. Fortunately, randomization of numerous, small reflective surfaces on a typical meat sample tends to obscure angular effects. Far more important for meat reflectance is the intrinsic anisotropy of meat.

Meat is composed of microscopic muscle fibres. Their size depends on the size to which an animal has grown, but typically they are about 0.1 mm in diameter and hundreds of millimetres in length. Muscle fibres conduct light by a series of total internal reflections, just like optical fibres, because their myofibrillar cores have a higher refractive index than the surrounding intercellular fluid. So, if muscle fibres are cut perpendicularly to the measured surface, they will conduct light deep into the meat and the meat will appear dark. But if the muscle fibres are parallel to the measured surface they will scatter rather than conduct light and the meat will appear pale. Measuring meat reflectance with an uncontrolled angle of muscle fibres to the measured surface may be a noticeable source of error, as may bulging of a meat surface into the measuring aperture of the apparatus. To make measurements that are repeatable by other researchers, the optical geometry of the apparatus must be known, particularly the angles of the illuminator and the photometer relative to the sample.

How is light scattered in meat? At extremely low pH values, sarcoplasmic proteins may be precipitated between myofibrils to create the extremely high scattering observed in severely pale, soft, exudative (PSE) meat. Kadim *et al.* (2008a) found Arabian camel meat (*Longissimus thoracis*) to have a mean pH of 5.89, with a range from 5.56 to 6.61. Rates of post-mortem glycolysis may be relatively slow in camel meat (Soltanizadeh *et al.*, 2008). Low ultimate pH values have been reported in some camel muscles (Gheisari *et al.*, 2009) and in fermented products (El Malti and Amarouch, 2009). Post-mortem electrical stimulation of camel meat also might lower pH levels (Kadim *et al.*, 2009) but sarcoplasmic proteins are unlikely to be a noticeable source of light scattering in this pH range, thus leaving a clear optical pathway between the myofibrils. Myoglobin is dissolved in this clear pathway, creating the redness of camel meat by strongly absorbing violet and green light before it can escape from the meat surface as diffuse reflectance (Fig. 7.3).

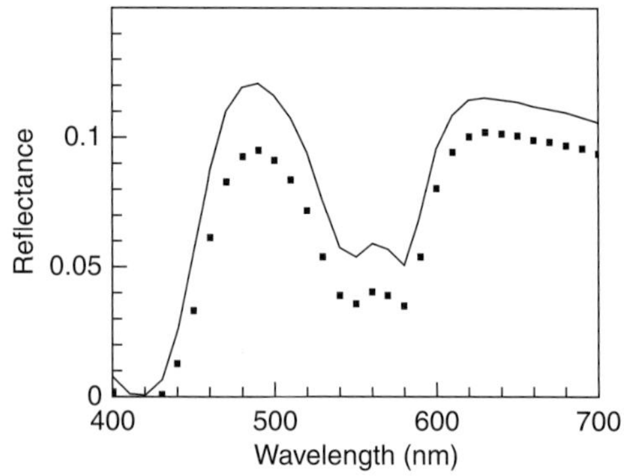

Fig. 7.3. Fibre-optic reflectance of camel meat (*Longissimus thoracis*) showing mean (solid line) and standard deviation ($n = 5$) subtracted from mean (solid squares). This is the reflectance spectrum of myoglobin modified by the state of the myoglobin and light scattering in the meat. It also depends on the optical apparatus used to measure it.

The features seen in Fig. 7.3 are similar to those of beef, and indicate strong absorbance by haemoproteins in the Soret band (420 nm) and a secondary absorbance band with a dimple at 560 nm. The presence of the dimple shows oxidation of myoglobin to metmyoglobin, as would be expected in a frozen sample slowly thawed. Metmyoglobin formation might be responsible for the subjective appearance of camel meat being described as from raspberry red to dark brown (Kadim *et al.*, 2008b).

7.4.2 Myofibrillar refraction

Although camel meat has not yet been widely reported as suffering from pH-related problems such as PSE and dark, firm, dry (DFD) meat, the basic sources of light scattering are still important because they underlie every aspect of meat reflectance and colour. For example, reflectance is becoming a popular method in attempts to predict meat toughness, and toughness is important, especially in meat from older camels (Kadim *et al.*, 2008b). Without a scientific understanding of light scattering in meat, the best that can be achieved is an empirical correlation highly dependent on the apparatus and a particular set of samples and, hence, not easily transferable or repeatable for online use.

The only experimentally verified source of light scattering in meat is myofibrillar refraction. Other sources of scattering are possible, such as reflectance from cell membranes forming refractive index boundaries with intracellular and extracellular fluids, but they require verification. Beef, pork and chicken all exhibit myofibrillar refraction (Swatland, 2008) and the preliminary data reported here show the existence of myofibrillar refraction in camel meat. To explain how refraction affects reflectance, consider the example of white paint. White paint contains refractive granules and very white, expensive paints have granules with the highest refractive index (Williamson and Cummins, 1983). Refractive inclusions in paint cause light scattering to return light to the observer, making the paint appear white when illuminated by white light. The dominant refractive components of meat are the contractile myofibrils that almost fill every muscle fibre. Myofibrils are sensitive to pH. When the pH is relatively high, as in living muscle and in meat with minimal post-mortem glycolysis, negative electrostatic repulsion between myofilaments keeps the myofilaments relatively far apart – separated by water. When the pH decreases in muscle post-mortem, the myofilaments move closer together with two important consequences: water is released to contribute to drip losses from meat (purge in packaged meat) and the refractive index of myofibrils is increased. Increases in refractive index increase refractive scattering, so that meat becomes paler as pH decreases. To measure these refractive changes, we exploit the fact that myofibrils are birefringent: they have two refractive indices. The familiar names of myofibrillar striations – A-band (anisotropic band) and I-band (isotropic band) – are derived from this feature discovered many years ago.

Interferometry is the measurement of the interference between waves. One set is superimposed on a second set, either to reduce wave amplitude (destructive interference) or to increase the amplitude (constructive interference). Myofibrils are birefringent, so that polarized transmitted light splits into two pathways – the ordinary and extraordinary ray paths. When recombined after being transmitted through the myofibrils, the rays combine to produce the first-order white interference commonly seen when a muscle fibre is examined with a polarizing microscope. However, it is possible to increase the interference order by using a compensator in the polarizing microscope. A fixed compensator such as a first-order (λ) red plate produces a visible background interference colour (red) against which a muscle fibre can be viewed and measured. In one orientation the muscle fibre adds to the background interference, whereas perpendicular to the first orientation it subtracts from the background interference. In other words, when the slow axis (γ) of the compensator is parallel to the slow axis of the muscle fibre, the interference order is advanced, and vice versa. A rotary compensator produces the whole range of background interferences so that, in the optical axis, measurements can be made at any selected background interference. This greatly simplifies interferometry when the interference colour is measured with a monochromator (Swatland, 2009). So how do camel muscle fibres behave – can we detect any effect of pH on myofibrillar refraction? As shown in Fig. 7.4, adjusting the pH of camel muscle fibres using 0.2 M phosphate buffer causes a strong change in refraction. Thus, we can expect that, as the pH declines as a consequence of post-mortem glycolysis, myofibrillar refraction will be increased to scatter more light and increase the paleness of the meat.

7.4.3 Connective tissue fluorescence

Connective tissue contains two predominant proteins, collagen and elastin, and both proteins are well-known sources of meat toughness. The amount, tensile strength and heat stability of collagen increase as animals grow older, especially if they use their muscles for locomotion. Elastin is completely resistant to cooking and tends to follow the physiological pattern of muscle activity, being particularly abundant in postural

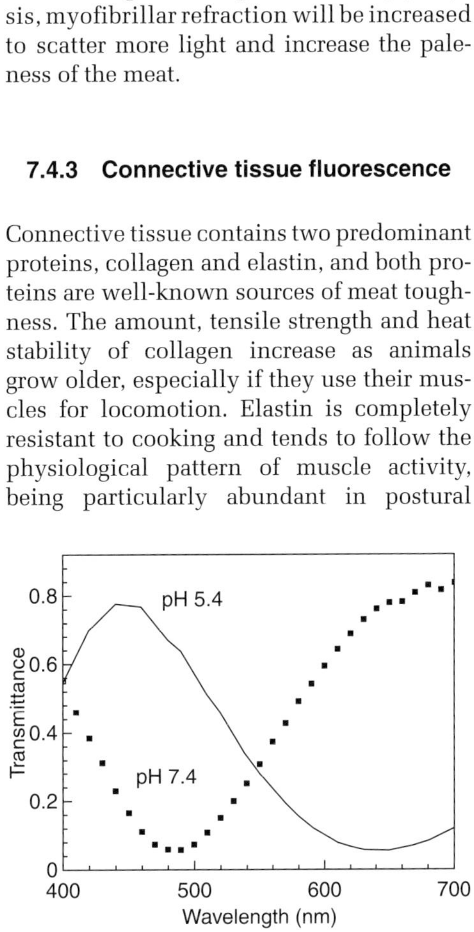

Fig. 7.4. Polarized-light interferometry of two camel muscle fibre fragments of equal diameter orientated with their slow axis parallel to that of a Nikitin–Berek compensator at tilt 5.5°. This shows that the lateral negative electrostatic repulsion between myofilaments in camel meat changes with pH and causes changes in refractive index.

muscles. Both collagen and elastin exhibit a blue–white fluorescence when illuminated by UV light. Pyridinoline linkages in collagen increase both tensile strength and fluorescence. Thus, UV fluorescence may be used to identify meat from older animals that is likely to be tough when cooked. The method even detects seasonal trends in beef probably associated with differential growth rates (Swatland, 2003). Seasonal changes in camel meat also have been reported and should be taken into account (Abdelhadi *et al.*, 2012). Many of the studies published on meat yield and quality fail to take seasonal variation into account, a mistake that future researchers working on camel meat might easily avoid. The strong fluorescence of major seams of connective tissue in meat may be detected using a fibre-optic probe. The main question is do seams of connective tissue in camel meat fluoresce as strongly as they do in beef? Figure 7.5 shows the fluorescence of the aponeurosis (epimysium) over the *Longissimus thoracis* in a camel steak. The fluorescence is as strong as that of beef and, hence, we can expect that technology developed for the detection of connective toughness in beef also will work for camel meat.

7.4.4 Adipose tissue reflectance

Reflectance of adipose tissue is important if there are quality concerns with the nature of the fat. For example, pork fat is sometimes dark and oily, whereas beef fat may be coloured yellow by dietary carotene. Both traits are judged as undesirable in many countries. With optical fat-depth probes, if the fat to muscle boundary is not distinct, a probe may give erroneous predictions of fatness. This might be a serious commercial problem if a producer receives a premium payment for a lean carcass and a penalty for a fat carcass. If this approach is to be used for camel carcasses, then Fig. 7.6 is of interest. It shows, as expected, that fat has a much higher reflectance than lean muscle (Fig. 7.3) but, in the sample shown in Fig. 7.6, there is a problem. The reflectance spectrum of adipose tissue is normally flat, with little indication of selective absorbance by haemoproteins. But, in Fig. 7.6, there is evidence of selective absorbance, probably from haemoglobin in adipose tissue capillaries rather than from myoglobin as in Fig. 7.3. This could be due to chance, but might also be an indication that camel adipose tissue may retain erythrocytes more readily than the fat of other species. Further study is required to test this possibility.

7.5 Instrumentation

Very few of the methods used routinely in a meat science laboratory can be applied online in the meat industry, and this calls for

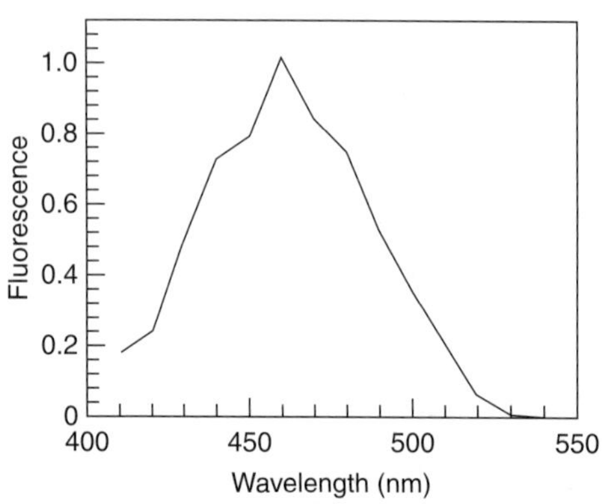

Fig. 7.5. Relative fluorescence of the aponeurosis over camel *Longissimus thoracis.*

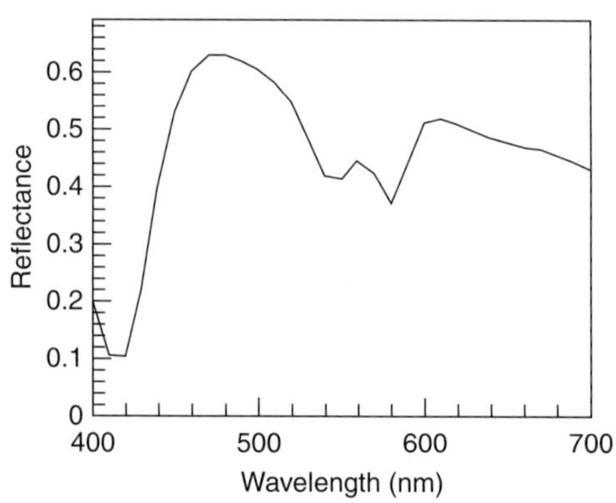

Fig. 7.6. Fibre-optic reflectance spectrum of camel subcutaneous adipose tissue.

a serious attempt to develop new technology. The apparatus must be rugged, non-destructive and easy to operate in the difficult working conditions found in the meat industry. Although this is a great challenge, the advantages to be gained from successful methods are enormous. Biophysical methods offer the greatest opportunities for online delivery but meat biophysics is a poorly developed subject. Another problem is that commercial production of new technology requires either a high profit margin from manufacturing a small number of instruments or a low margin profit from a large number.

Methods are already available for estimating quality parameters such meat yield, muscle colour, marbling, pH, fat colour and translucency (Table 7.1). But meat toughness is the most important and most difficult to detect online and is likely to be a primary requirement in quality control of camel meat. There are many causes of toughness in meat, but the dominant three are: (i) inadequate ageing; (ii) connective tissue; and (iii) short sarcomeres. The first can be solved by ensuring adequate post-mortem ageing, although this might fail if endogenous autolytic enzymes are inactive. Possible detection methods include measurements of electrical impedance (capacitance and resistance) and light scattering as an indicator of pH affecting autolytic enzymes. Sarcomere length is the final obstacle. Meat from any source might be made intolerably tough if muscles contract as rigor mortis develops. A typical cause is rapid, pre-rigor refrigeration (cold shortening). The effects are visible under the microscope where the contractile units (sarcomeres) along muscle fibres are unusually short. But the methods we use at present, such as phase contrast, differential interference contrast or polarized light, cannot withstand the scattering of light that occurs in bulk meat. Polarized light offers some interesting possibilities. Also, we must not prejudge the importance of a signal versus its background noise. Thus, when scattering is the background noise, the scattering also contains information on pH (which affects autolytic enzymes) and sarcomere length (Swatland, 2005). There are many possibilities for future research in this area, exploring the biophysical properties of camel meat and adapting them for commercial use.

From the preliminary results on camel meat given in this chapter, we can see the importance of three parameters: (i) imaging, as in quantification of meat yield; (ii) wavelength, as in evaluating meat colour; and (iii) polarization, as in evaluating fluid loss and sarcomere length. These are the types of information seen in Figs 7.2 to 7.5, and they have traditionally been studied by themselves. But new technology allows us to study them all at once. For each of the pixels forming an image, it is possible to gather information on wavelength and plane of polarization. This is often called hyperspectral or multispectral polarization imaging. The front runner in this technology is the acousto-optical tuneable filter (AOTF), which has already been packaged for use in the meat industry (Fig. 7.7). The AOTF is a piezoelectric transducer coupled with a crystal (tellurium or silicon dioxide) to create acoustical waves that alter the refractive index of the crystal at a high frequency. This creates something like a diffraction grating (as used in many spectrophotometers) with the advantage that, instead of rotating the grating mechanically (which is a slow operation with mechanical problems), the wavelength is changed by the electrical frequency applied to the piezoelectric transducer. Thus, we have a fast system without mechanical constraints. Using traditional methods (Swatland, 1995), we can detect almost everything we need to know about meat quality if we encompass an image of a muscle area, wavelength, polarization, angle of light path through the meat, fluorescence, etc. The challenge for researchers in the field will be to do this in real time, online and cost-effectively. The AOTF might enable this. It is important to be aware, however, that camel meat has great optical complexity, comparable to the optical complexity of the AOTF. Muscle fibres act as pH-sensitive optical fibres (Fig. 7.4), their myofibrillar cores may act as diffraction gratings (Huxley, 1990), and multilayer interference effects may occur longitudinally, causing phenomena such as iridescence (Swatland, 2011).

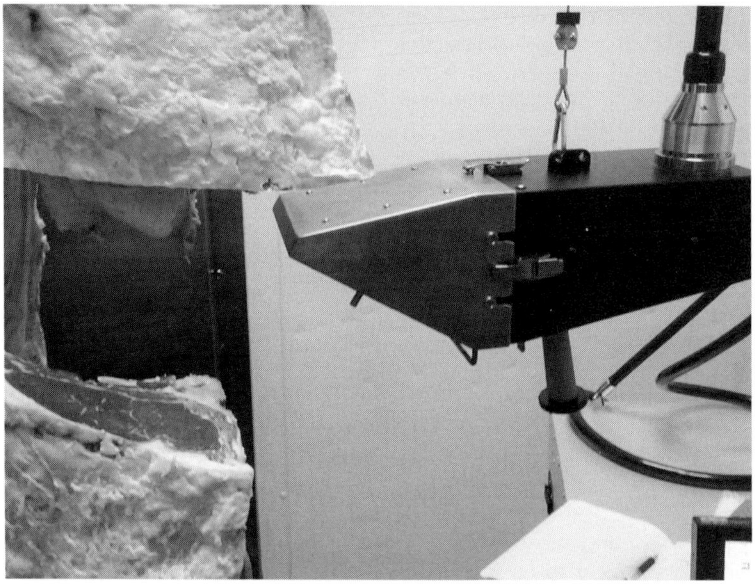

Fig. 7.7. A hyperspectral imaging system for beef carcasses (courtesy of A.Y. Fong, Gooch & Housego, Orlando, Florida).

7.6 Conclusion

International trading in camel meat is still a novelty, but one that might increase in importance. Predicting camel meat yield and quality with online measurements looks possible but the commercial advantage will depend on how international trading develops. The most cost-effective approach might be to combine product tracking and computer-based inventories with predictive methods such as hyperspectral polarization imaging.

References

Abdelhadi, O.M.A., Babiker, S.A., Picard, B., Jurie, C., Jailler, R., Hocquette, J.F. and Faye, B. (2012) Effect of season on contractile and metabolic properties of desert camel muscle (*Camelus dromedarius*). *Meat Science* 90, 139–144.

Biscontini, T.M.B., Shimokomaki, M., Oliveira, S.F. and Zorn, T.M.T. (1996) An ultrastructural observation on charquis, salted and intermediate moisture products. *Meat Science* 43, 351–358.

Calle Escobar, R. (1984) *Animal Breeding and Production of American Camels.* Ron Hennig, Patience, Lima, Peru.

Clutton-Brock, J. (1987) *A Natural History of Domesticated Mammals.* Cambridge University Press, and the British Museum (Natural History), London.

Elgasim, E.A. and Alkanhal, M.A. (1992) Proximate composition, amino acids and inorganic mineral content of Arabian camel meat. *Food Chemistry* 45, 1–4.

El Malti, J. and Amarouch, H. (2009) Microbiological and physicochemical characterization of the natural fermented camel meat sausage. *African Journal of Biotechnology* 8, 4199–4206.

Gheisari, H.R., Aminlari, M. and Shekarforoush, S.S. (2009) A comparative study of the biochemical and functional properties of camel and cattle meat during frozen storage. *Veterinarski Arhiv* 79, 51–68.

Gotfredsen, A., Bæksgaard, L., and Hilsted, L. (1997) Body composition analysis by DEXA by using dynamically changing samarium filtration. *Journal of Applied Physiology* 82, 1200–1209.

Huxley, A.F. (1990) A theoretical treatment of diffraction of light by a striated muscle fibre. *Proceedings of the Royal Society – Biological Sciences (Series B)* 241, 65–71.

Iniguez, L.C., Alem, R., Wauer, A. and Mueller, J. (1998) Fleece types, fiber characteristics and production system of an outstanding llama population from Southern Bolivia. *Small Ruminant Research* 30, 57–65.

Kadim, I.T. and Mahgoub, O. (2008) Effect of age on quality and composition of one-humped camel *Longissimus muscle*. *International Journal of Postharvest Technology and Innovation* 1, 327–336.

Kadim, I.T., Mahgoub, O. and Al-Marzooqi, W. (2008a) Meat quality and composition of *Longissimus thoracis* from Arabian camel (*Camelus dromedarius*) and Omani beef: a comparative study. *Journal of Camelid Science* 1, 37–47.

Kadim, I.T., Mahgoub, O. and Purchas, R.W. (2008b) A review of the growth, and of the carcass and meat quality characteristics of the one-humped camel (*Camelus dromedarius*). *Meat Science* 80, 555–569.

Kadim, I.T., Mahgoub, O., Al-Marzooqi, W., Khalaf, S.K., Mansour, M.H., Al-Sinani, S.S.H. and Al-Amri, I.S. (2009) Effects of electrical stimulation on histochemical muscle fiber staining, quality, and composition of camel and cattle *Longissimus thoracis* muscles. *Journal of Food Science* 74, S44–S52.

Lutz, E. (1998) *Agriculture and the Environment. Perspectives on Sustainable Rural Development.* World Bank, Washington, DC.

Makoto, Y., Hitoshi, M., Kazuki, N., Hiroyuki, A., Toshihiko, S., Masayuki, Y. and Mitsunori, K. (2006) Impact of global warming on broiler meat production estimated from changes of the mean ambient temperature. *Nihon Chikusan Gakkaiho* 77, 231–235.

McCloy, L. and Rowe, P. (2000) *Assessing the Potential for a Commercial Camel Industry in Western Australia.* RIRDC 00/123, Canberra.

Pietrobelli, A., Wang, Z., Formica, C. and Heymsfield, S.B. (1998) Dual-energy X-ray absorptiometry: fat estimation errors due to variation in soft tissue hydration. *American Journal of Physiology, Endocrinology and Metabolism* 274, E808–E816.

Ribeiro, F.R.B., Rhoades, R.D., Tedeschi, L.O., Smith, S.B., Martin, S.E. and Crouse, S.F. (1998) Evaluating the application of dual X-ray energy absorptiometry to assess dissectible and chemical fat and muscle from the 9th-to-11th rib section of beef cattle. *Journal of Animal Science* 87, Supplement 2, p. 55.

Salva, B.K., Zumalacarrequi, J.M., Figueira, A.C. Osorio, M.T. and Mateo, J. (2009) Nutrient composition and technological quality of meat from alpacas reared in Peru. *Meat Science* 82, 450–455.

Soltanizadeh, N., Kadivar, M., Keramat, J. and Fazilati, M. (2008) Comparison of fresh beef and camel meat proteolysis during cold storage. *Meat Science* 80, 892–895.

Soltanizadeh, N., Kadivar, M., Keramat, J., Bahrami, H. and Poorreza, F. (2010) Camel cocktail sausage and its physicochemical and sensory quality. *International Journal of Food Science and Nutrition* 61, 226–243.

Swatland, H.J. (1994) *Structure and Development of Meat Animals and Poultry.* Technomic Publishing, Lancaster, Pennsylvania.

Swatland, H.J. (1995) *On-line Evaluation of Meat.* Technomic Publishing, Lancaster, Pennsylvania.

Swatland, H.J. (2003) Seasonal changes of connective tissue fluorescence in well-aged beef roasts, correlating fat reflectance and signal source. *Canadian Journal of Animal Science* 83, 445–450.

Swatland, H.J. (2004) *Meat Cuts and Muscle Foods.* Nottingham University Press, Nottingham, UK.

Swatland, H.J. (2005) On correlating penetrometry with photometry in beef roasts. *Archives of Animal Breeding* 48, 318–323.

Swatland, H.J. (2008) How pH causes paleness or darkness in chicken breast meat. *Meat Science* 80, 396–400.

Swatland, H.J. (2009) Effect of granule size on the interference colors of starch in polarized light. *Biotechnic and Histochemistry* 84, 329–336.

Swatland, H.J. (2011) Iridescence in beef caused by multilayer interference from sarcomere discs. *Meat Science* 90, 398–401.

Williamson, S.J. and Cummins, H.Z. (1983) *Light and Color in Nature and Art.* John Wiley, New York, pp. 363–375.

Yousif, O.K. and Babiker, S.A. (1989) The desert camel as a meat animal. *Meat Science* 26, 245–254.

8 Camel Carcass Quality

Isam T. Kadim and Osman Mahgoub
Department of Animal and Veterinary Sciences, College of Agricultural and Marine Sciences, Sultan Qaboos University, Muscat, Sultanate of Oman

8.1 Introduction

A camel carcass can provide a considerable amount of meat for human consumption. Although the marketing systems for camel meat are not well organized, there is evidence of a high demand for fresh camel meat as well as for camel blended meat products, even among societies where camel herding does not take place (Kadim *et al.*, 2008). Camel meat is relished equally to beef in some Middle East and African countries, and is highly appreciated in many parts of Arabia, Libya, Algeria and Tunisia. There is, however, often some lack of acceptance of the consumption of camel meat in non-camel-herding societies. The carcass characteristics of camels vary considerably owing to differences in age, sex, breed, type and health status. The characteristics depend mainly on live weight and husbandry practices and the condition of the vegetation. The role of the camel as a meat producer is becoming more evident because of the resourceful role it plays rather than as a symbol of social status. None the less, the camel as a potential meat producer has received little attention (Kadim *et al.*, 2008). This is because camels have been raised in less-developed countries and research for improving their reproductive and productive parameters has been limited (Skidmore, 2005). One of the problems relating to camel meat production is the lack of information on carcass quality. Slaughter, carcass and meat characteristics enable information to be gathered on how camel muscles are transformed into meat. More information on these parameters may be required to compare the meat production potential of camels, using different breeds and under different management conditions.

Camels slaughtered worldwide in 2009 produced around 373,565,000 tonnes of meat, most of which was produced in Somalia, Sudan, Mauritania, Saudi Arabia and Egypt (FAO, 2011). A considerable number of camels are managed and bred for slaughtering in the Near East and northern Africa as well as for export. Both Somalia and Sudan export large numbers of camels to Saudi Arabia, Egypt and the Gulf States, whereas Libya imports camels from Sudan, Mali, Algeria and Mauritania every year for slaughter. In pastoral societies, camels are rarely slaughtered except during ritual ceremonies.

Camel meat is a good source of food to meet the growing needs for meat in developing countries, especially for low-income population groups (El-Mossalami *et al.*, 1996; Saparov and Annageldiyev, 2005). This chapter highlights carcass quality characteristics of the dromedary camel for meat production, with special emphasis on carcass weight, dressing-out percentage and carcass composition.

8.2 Carcass Conformation

Carcass conformation refers to the proportional size of carcass parts and the relationship of the thickness of soft tissues to the skeletal size. Carcass conformation of the dromedary camel is, therefore, an important trait because carcasses with a superior conformation attract higher prices. The importance of conformation as an indicator of commercial values is based on the assumption that carcasses with a better conformation have advantages in terms of lean meat content, proportion of higher priced cuts and possibly greater muscle size or area (Kempster *et al.*, 1982). Camel breeders, meat traders and scientists can use carcass conformation as an indicator of good meat producers. The difference in carcass conformation between the different groups of dromedary camels was attributed to variations in carcass depth from the scapula to sternum, width behind shoulder, maximum shoulder width and gigot width (Table 8.1). Similarly, Nsosoa *et al.* (2000) found large differences in carcass depth measurements between animal breed types and concluded that an increase in carcass weight significantly increased linear measurements in absolute terms, but reduced them relative to weight. A lack of information for camel carcass conformation has led, however, to difficulties in interpreting and understanding camel carcass quality. Therefore, there is a need to adopt a common carcass conformation to enable the easy communication of carcass data, research results and trading terms in the camel-meat industry. The values of assessed carcass conformation are considered as useful indicators of individual camel carcass composition. Visually assessed conformation as a useful indicator of carcass composition in various meat animal species has been the subject of many studies in the literature (Kempster *et al.*, 1982), whereas the camel has not been investigated.

The conformation of the dromedary camel carcass fundamentally differs from that of other meat animal carcasses. Apart from the obvious shape of the dorsal hump, the most notable feature is the restriction of the hindlimb muscles near the pelvis, which do not overlap with the abdominal muscles. When the hindlimb is stretched back in a hanging carcass, it therefore forms a serration ventral to the ileum (Fig. 8.1). This is because the camel has long limbs capable of considerable rotation relative to the vertebral axis. In a sitting-camel position, the distal end of the femur projects downwards towards the ground, whereas, when other meat animals sit, the distal end of the femur projects upwards.

In the meat market, good carcass conformations are valued more highly and receive better prices than those with an average or poor conformation. The steaking method clearly demonstrates the superiority of good conformation carcasses in terms of the higher yield of saleable meat than for poor carcass conformation. Total sellable meat (steaks, breast, scrag, fillet, lean, trim and mince) was 3% higher in good conformation carcasses than in poor ones.

Table 8.1. Effect of three levels of feed intake on camel carcass linear measurements (Al-Kharusi, 2011).

	Group 1	Group 2	Group 3	SEM[a]
Carcass length (mm)	2220.5	2224.5	2217.0	52.05
Leg length (mm)	465.0	498.7	472.5	7.73
Depth from the scapula to sternum (mm)	635.5	637.5	610.0	9.97
Width behind shoulder (mm)	395.0	392.5	315.0	90.38
Maximum shoulder width (mm)	362.5	342.5	330.0	12.44
Gigot width (mm)	391.2	385.0	357.5	12.95

[a]SEM, standard error of mean.

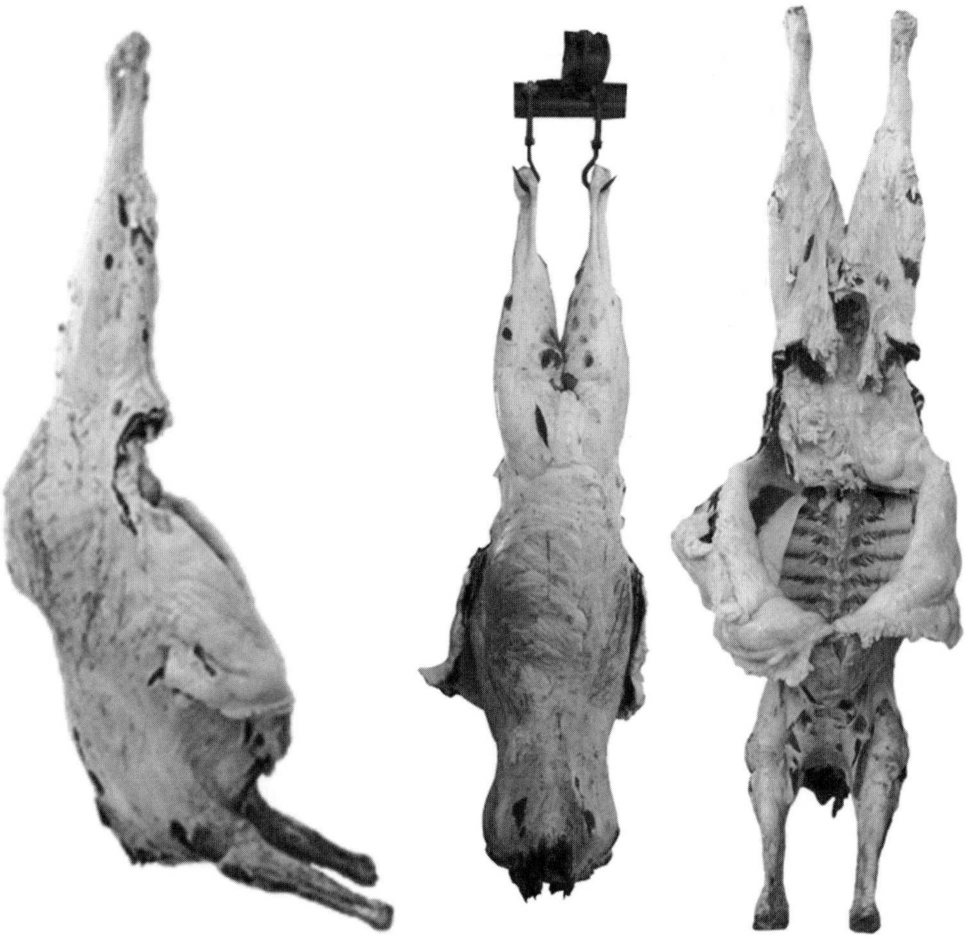

Fig. 8.1. Three positions of hanging camel carcasses.

The differences were evident in different parts of the carcass. The good conformation carcasses had a significantly higher yield in leg (6% in weight of steaks and 5% in area of steaks) and loin (13% in weight of steaks and 17% in area of steaks) than carcasses of poor conformation. These differences mean better financial returns from carcasses of good than those of poor visual conformation. To demonstrate the higher meat yield of good than of poor visual conformation carcasses, new cutting techniques such as steaking might be desirable. It is therefore, important that there should be more research focusing on objective carcass conformation with the view to utilizing results to increase meat production from camels.

8.3 Carcass Weight

Camels are a good potential source of meat because they yield reasonably heavy carcasses under inexpensive management systems. A wide range of carcass weights have been reported for camels, with the variation apparently owing to body condition, sex, breed or type, age at slaughter and depending mainly on husbandry practices and the condition of the grazing pastures. Carcass weight is higher when animals are well managed from weaning to maturity. Camel carcass weight, which generally ranges from 125 kg to 400 kg, increases with increasing body weight (Fig. 8.2). Although recent figures of camel carcass weight are consistent

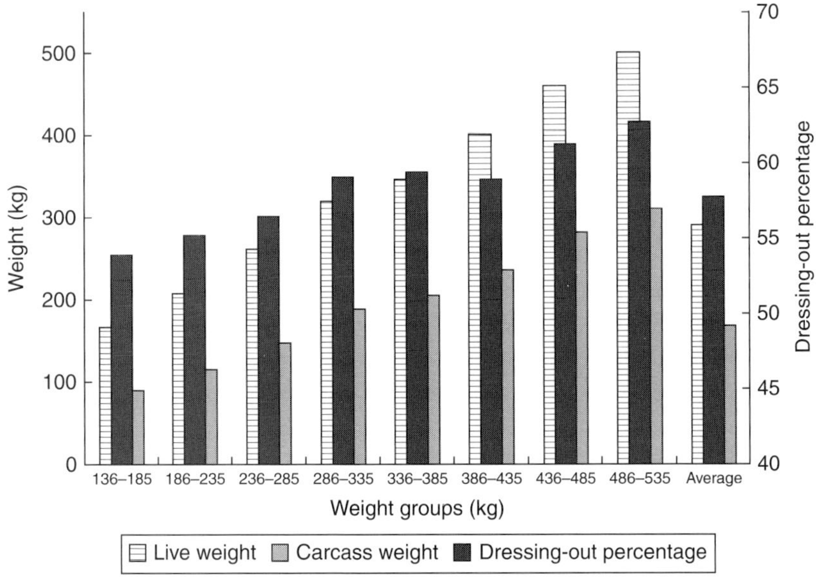

Fig. 8.2. Body weights, carcass weights (±) and dressing-out percentage of Najdi male camels showing that dressing-out percentage generally increases with increasing weight (Abouheif *et al.*, 1986).

with the average dromedary weights of 450–600 kg, the average carcass weight was 168 kg in Saudi camels (Abouheif *et al.*, 1986), but was much higher at 300–400 kg in Iranian camels (Khatami, 1970) and was 194 kg in Nigeria (Muhammad and Akpan, 2008). The Iranian camels must have been well fattened. The average carcass weight of mature Sudanese desert camels was 239.9 kg (Yousif and Babiker, 1989), whereas the mature castrated Algerian camels had heavier carcasses (244.2 kg; Bahhamou and Baylik, 1999). An average Tunisian camel carcass weight is 231–244 kg but there is a great variation in the weight of humps of between 4 kg and 31 kg (Kamoun, 2004). The male camels ranging from 15 to 50 months yielded carcasses of between 150 kg and 343 kg.

Following the trend in camel live weight under the same environmental conditions, Kurtu (2004) reported the range of carcass weight of Ethiopian camels was 225–280 kg and 149–190 kg for males and females, respectively (Table 8.2). The male camel carcasses were greater than that of females by 32–34%. Meat production of the Issa type of camel in Eastern Ethiopia was assessed by Mekonnen (2004). The average carcass weights were 327 kg and 261.5 kg for males and females, respectively. Wilson (1978) reported an average of 209 kg for Sudanese camel carcass weights, with 231 kg for male carcass weight being heavier than 196 kg for that of females (Table 8.2). Higher carcass weights of 240 kg and 232 kg of male and female Sudanese camels, respectively, were reported by Yousif and Babiker (1989). The average carcass weight of the Iranian dromedary was higher for males (300–400 kg) than for females (250–350 kg; Khatami, 1970). Bremaud (1969) noted that the average carcass of Somali camels was 286 kg, whereas a carcass weight of 231 kg was reported for male and 196 kg for female Sudanese camels (Wilson, 1978). In Tunisia, 15 fattened camels had a mean value of carcass hot weight of 231 kg with a range of 150–343 kg (Kamoun, 1995). Hertrampf (2004) reported an average carcass weight for males of 283 kg and 251 kg for female camels. Bakkar *et al.* (1999) found that the average camel carcass weights/kg (dressing-out percentage) of three feeding groups (6–14 months of age) were 180.6 kg (57.3%) for concentrate plus alfalfa hay, 170.7 kg (57.1%) for concentrate

Table 8.2. Carcass weight and dressing-out percentage in dromedary camels.

Number/breed and sex	Carcass weight (kg)	Dressing-out percentage	Reference	Remarks
21 Sudanese males	231.3 ± 49.18	51.4 ± 2.88	Wilson (1978)	Sex effects
39 Sudanese females	196.3 ± 24.94	47.4 ± 3.25		
227 Najdi males and females	88.81 ± 1.4 68.0 ± 4.2	53.8–57.7	Abouheif *et al.* (1986)	Live body weight and sex effect
52 Males	200–288.5	51.1–67.2 52.5–74.2	Yousif and Babiker (1989)	Full and empty body weight effect
21 Najdi males	105.3–273.4	61.5–60.6	Abouheif *et al.* (1990)	Age effects (8–26 months of age)
6 Libyan males	146.8	51.0	Baila *et al.* (1990)	Nutrition effects
15 Males	184–343	60.3–71.4	Kamoun (1995)	Nutrition and age effect
12 Saudi males	119.5–132.5	52.1–56.1	El-Gasim and El-Hag (1992)	Nutrition effects
Male	231.3	51.4	Wilson (1998)	Sex effects
Female	196.3	47.4		
47 Algerian males	244.2	53.3	Bahhamou and Baylik (1999)	Castration
11 Najdi males	148.6 ± 9.1–153.5 ± 8.3	48.7 ± 0.8–49.2 ± 0.73	Al-Owaimer (2000)	Nutrition effects
Majaheem and Harrah	119.5–132.5	52.1–56.1	Al-Ani (2004)	Breed effects
88 Somalian males and 12 females	170.01 ± 20.49–252.27 ± 26.58	50.65 ± 3.7 54.03 ± 5.13	Kurtu (2004)	Sex effects
8 Somali × Rurkana males	302.1–414.8	47.5–58.4	Herrmann and Fischer (2004)	Body weight effects
8 Males	283.2	53.7 ± 3.26	Hertrampf (2004)	Sex and region effect
8 Females	251.1	50.7 ± 4.67		
24 African	231.1	53.7 ± 2.8		
8 Asian	393.7	62.1 ± 12.7		
108 Ethiopian males	261.5–327	59.6–73.5	Mekonnen (2004)	Sex effects
10 Nigerian males	194.4 ± 9.51	49.9	Muhammad and Akpan (2008)	Environments
10 Omani males	104.5–131.5	45.7–48.7	O. Mahgoub *et al.*, unpublished data	Nutrition effects
108 Ethiopian	327.0 (male) 261.5 (female)	73.5 59.6	Mekonnen (2004)	Mature camel (>10 years old)

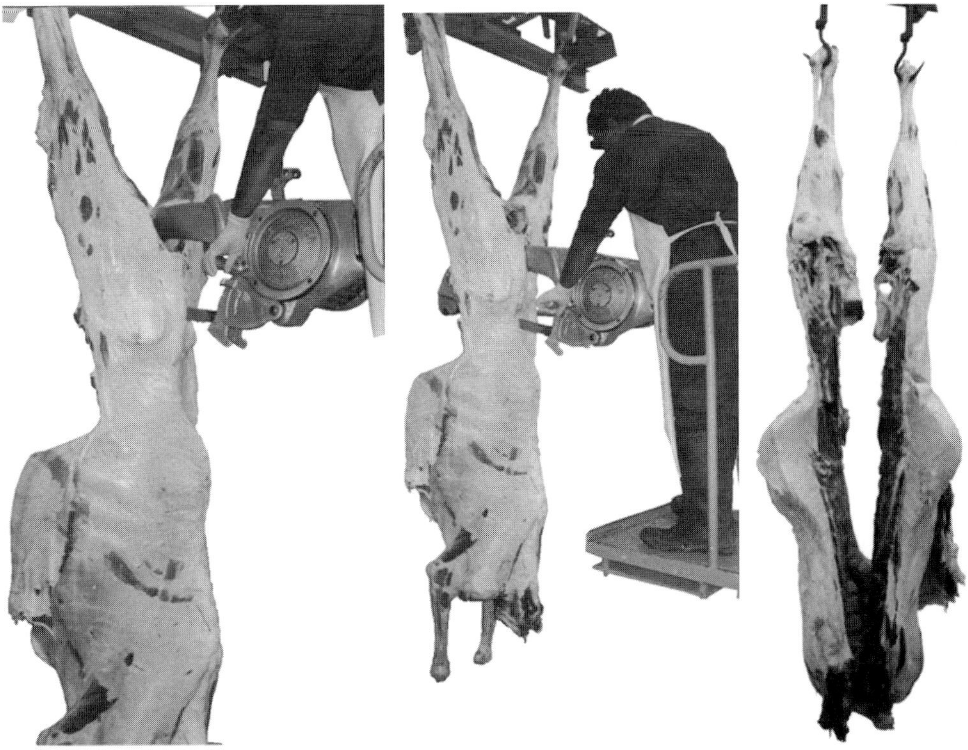

Fig. 8.3. Splitting the camel carcass longitudinally using an electric saw.

plus Rhodes grass hay, and 168.1 kg (57.7%) for concentrate plus wheat straw treated with ammonia gas.

Within the modern slaughtering technique and facilities and because of the large size, the camel carcass is usually split longitudinally using an electric saw for easy handling and storage (Fig. 8.3). For more details see Chapter 5. The carcass is washed with clean water and stored in cold room (chiller) for up to 48 h or until rigor mortis is completed. Prior to storage or 48 h post-mortem, the camel carcass is usually divided into forequarters and hindquarters because the two sides of the carcass are very difficult to distribute whole because of their size. Each half is separated between the 8th and 9th rib and the forequarter can be hung up with a hook between the ribs or by the shank. The right and left forequarters constitute an important meat component that determines the overall profit realized by the butcher. The right and left hindquarters include the femur and tibia bones and are considered the meaty part of the camel carcass and are usually sold intact or cut into small pieces.

8.4 Dressing-out Percentage

Dressing-out percentage is an important measure of carcass quality and yield in meat animals, but it varies owing to factors such as age, sex, slaughter weight, fatness, carcass weight, dressing procedures and degree of gut fill at slaughter (Table 8.2). In the camel, dressing-out percentage varies from 47% to 62% (Kadim *et al.*, 2008) depending on sex, age, body condition, breed or type and the digestive tract content. In general, the dromedary has a higher dressing-out percentage than cattle (Wilson, 1984) but it is not clear whether such a percentage can be sustained under all management and ecological systems.

Limited studies have provided some information on the relationship between slaughter weight, carcass weight and dressing-out percentage calculated either on the basis of live weight or empty live weight of dromedary camels (Wilson, 1978; Yousif and Babiker, 1989; Biala *et al.*, 1990; El-Gasim and El-Hag, 1992; Kamoun, 1995; Bahhamou and Baylik, 1999; Mahgoub *et al.*, unpublished data). The dressing-out percentage on an empty body weight basis is higher than those based on full slaughter weight. Fifteen Tunisian camels were fattened from weaning to slaughtering (Kamoun, 1995). Live weight and carcass weight were 413.8 kg and 231.1 kg, respectively, with a dressing-out percentage of 55.8% and 65.4% on a body weight and an empty body weight basis, respectively (Table 8.3). The slaughter weight and carcass weight of mature Sudanese desert camels in non–fattened males were 306–581 kg and 144–310 kg with a dressing-out percentage based on body weight and on empty body weight of 46.2–55.6% and 55.7–65.1%, respectively. In fattened male camels of 395–512 kg and 208–295 kg body weight, the dressing-out percentages on a body weight and empty body weight basis were 47.2–62.8% and 53.1–74.7%, respectively (Yousif and Babiker, 1989). The dressing-out percentage of the mature Algerian camels was 53.3% (Bahhamou and Baylik, 1999). The discrepancy in dressing-out percentage between different camels within the same breed may be explained by differences in the weight of the digestive tract content, which in turn is influenced by the duration of fasting between last weighing and slaughtering. Kamoun (2004) found that blood and digestive tract content represent 6.8 ± 3.5% and 15.1 ± 5.1% of the live weight, respectively, whereas the dressing-out percentage on a body weight and on an empty body weight basis were 55.67 ± 2.77% (52.3–61.4%) and 65.40 ± 3.74% (60.30–72.12%), respectively. The weight of hump, which is mainly made up of fat, can reach up to 40 kg and account for 8.6% (5–13%) of the carcass weight (Kamoun, 1995), and can affect dressing-out percentage (Table 8.2). Large fat camels had a dressing-out percentage of 56.6%, whereas relatively thin camels had a dressing-out percentage of 51.4% (Wilson, 1978; Yousif and Babiker, 1989). The differences in dressing-out percentage in previous studies might have been due to variations in slaughter weight and fatness because the camels were fed rations that were different in quantity and quality. Carcass weight, yield and hump increased as slaughter weight increased. Camel carcass weight (including the hump) makes up about 55% of the camel live weight (Herrmann and Fischer, 2004). As mentioned, for an average camel carcass weight of 231–244 kg there is a great variation in the weight of humps, 4–31 kg (Kamoun, 2004). The rate of live body growth did cause a change in the camel carcass yield and characteristics and they increased as live weight increased.

Male camels were found to have higher dressing-out percentages than females, varying between 51% and 54% for Ethiopian

Table 8.3. Live weights, carcass weight and dressing-out percentage in camel males.

Number/age	Slaughter weight (kg)	Empty body weight (kg)	Carcass weight (kg)	Dressing-out % (live body weight)	Dressing-out % (empty body weight)	Reference
15 males / 3 years	413.8	351.5	231.1	55.8	64.4	Kamoun (1995)
6 males / 2 years	288.0	241.3	146.8	51.0	60.7	Biala *et al.* (1990)
12 males / 2 years	226.8–271.0	194.0–235.0	119.5–132.5	52.1–56.1	60.0–65.5	El-Gasim and El-Hag (1992)
21 castrated males	459.7		244.2	52.3		Bahhamou and Baylik (1999)
10 males / 2 years	228.0–268.5	192.3–241.9	104.5–131.5	45.7–48.7	54.3–54.4	O. Mahgoub *et al.*, unpublished data

Table 8.4. Live weight, carcass weight and dressing-out percentage of dromedary camels.

		Male	Female	Average
Wilson (1978)	No. of animals	21	39	
	Live weight (kg)	447.9 ± 84.10	414.4 ± 50.83	431.2 ± 65.74
	Range	305.5–581.0	307.5–522.5	305.5–581.0
	Carcass weight (kg)	231.3 ± 49.18	196.3 ± 24.94	213.8 ± 38.78
	Range	144.0–310.0	141.0–248.0	141.0–310.0
	Dressing-out percentage	51.4 ± 2.88	47.4 ± 3.25	49.4 ± 3.65
	Range	46.2–55.6	41.3–53.5	41.3–55.6
Kurtu (2004)	No. of animals	88	12	
	Live weight (kg)	465.8 ± 63.85	335.7 ± 42.20	400.8 ± 56.0
	Range	402–530	293–378	293–530
	Carcass weight (kg)	252.3 ± 26.58	170.0 ± 20.49	211.2 ± 24.5
	Range	225–280	149–191	149–280
	Dressing-out percentage	54.03 ± 5.13	50.7 ± 3.70	52.4 ± 4.68
	Range	49.0–59.0	47.0–54.0	47.0–59.0

camels (Kurtu, 2004). On the other hand, Mekonnen (2004) for the same breed found 72.9% and 59.6% dressing-out percentages for males and females, respectively. Table 8.4 shows that the average dressing-out percentage was 48.8% for Sudanese camels, with 51.4% for males and 47.4% for females (Wilson, 1978). Babiker and Yousif (1987) reported dressing-out percentages of male Sudanese camels of 54.4% for cold carcasses and 55.9% for hot carcasses. Higher dressing-out percentages, (57% for females and 63.8% for males) were reported by Yousif and Babiker (1989). Congiu (1953) reported a 56.1% dressing-out percentage for male and 54.1% female Somali camels. Male camels were also reported to have a higher dressing-out percentage than females by Kuznestov and Tretyakov (1972), varying between 52.8% and 76.6%.

The dressing-out percentage might not always correspond to live weight, however; for instance, a female camel could be slaughtered while pregnant, which will influence the slaughter weight and consequently the dressing-out percentage. In Libya, fattened camels of 2 years old had a dressing-out percentage of 51% and 60.7% calculated on a live weight the empty body weight basis, respectively (Biala *et al.*, 1990). Well fed Sudanese camels had a dressing-out percentage of 56.6% (Yousif and Babiker, 1989), whereas underfed Sudanese camels had a dressing-out percentage of 51.4% (Wilson, 1978) at similar slaughter weights. Table 8.5 shows three groups of Omani male camels received a feed intake equivalent to 1.5%, 2.0% and 2.5% of body weight for 162 days (Mahgoub *et al.*, unpublished data). The dressing-out percentage increased with increasing feed intake (Table 8.5). Age has a significant effect on carcass components with distinct advantages to slaughter camels at an early age. Dressing-out percentage of a 32-month- old camel was 62.8% compared with 55.8% for a 19–20-month-old one (Kulaeva, 1964). In Australian camels, the dressing-out percentage was 53% for 4-year-old male camels and 48% for 7-year-old females (Camel Newsletter, 1997). Herrmann and Fischer (2004) reported an average 53.6% dressing-out percentage for castrated 7–10-year-old Somali × Turkana camels in Kenya. Mukasa-Mugerwa (1981) reported a dressing-out percentage of 57% for a 590 kg live weight camel. However, Abouheif *et al.* (1990) found no significant differences in the dressing-out percentage of 21 Najidi male camels slaughtered at 8, 16 and 26 months of age. The variation in the dressing-out percentages among those reported by various researchers is most probably due to a variation in body weight, feed, age or environmental effects.

Table 8.5. Effect of three feeding levels on slaughter weight, carcass weight and dressing-out percentage of Omani camels (O. Mahgoub *et al.*, unpublished data).

	Feeding level (percentage of body weight)		
Traits	Low (1.5%)	Medium (2.0%)	High (2.5%)
Slaughter weight (kg)	228.0 ± 22.66	259.1 ± 15.98	268.5 ± 15.91
Empty body weight (kg)	192.3 ± 1.67	222.1 ± 1.18	242.0 ± 1.84
Hot carcass weight (kg)	104.5 ± 14.78	118.8 ± 10.45	131.5 ± 10.54
Dressing-out percentage	45.7 ± 1.67	45.8 ± 1.84	48.7 ± 1.38

8.4 Carcass Composition

The camel carcass and meat composition are important characteristics. Information on such parameters is required for the improvement of camel meat production. There is a lack of standard methods for the assessment of camel carcass characteristics including a standard cutting system for camel carcasses compared with other meat animal species, which makes comparisons between different studies difficult. The camel carcass side is usually divided into a forequarter and a hindquarter by cutting between the 11th and 12th ribs (Fig. 8.4). The forequarter is usually divided into wholesale regions (neck, shoulder, rack, brisket and plate; Kadim *et al.*, 2008).

Figure 8.5 shows a cutting procedure for eight wholesale cuts (Kadim *et al.*, 2008). Herrmann and Fischer (2004) and Kamoun (2005) proposed a different method of different cuts presented in Table 8.6, along with those of some other studies. The forequarter is larger than the hindquarter with the latter being about two thirds of the former (Table 8.6); this is mainly due to the presence of the hump, which comprises about 1–5% of live weight (2.0–8.4 kg). Kurtu (2004) reported similar figures for male and female camels (Table 8.6). Using standard cattle butchery procedures, Khan *et al.* (2003) found that the forequarters comprised about 34% of the total carcass, whereas the hindquarters constitute 25%. Excluding the hump (4.6%), the forequarter contributed 23.8%, whereas the hindquarter contributed 21.3% of live weight in Somali × Turkana camels (Herrmann and Fischer, 2004). In the same study, the forequarter, hindquarter, neck and hump constituted 44.3, 39.7, 7.1 and 8.6% of the carcass, respectively. The forequarter, hindquarter, *Longissimus dorsi* muscle, neck and hump constitute the major edible parts of the carcass. The neck, being long, and usually separated from the carcass in the camel, contributed about 4% of live weight (Herrmann and Fischer, 2004). Average forequarters weights (and percentage of the carcass) were 98.9 kg (54.8%), 95.6 kg (56%) and 88.1 kg (52.4%) for the three feeding groups of concentrate plus alfalfa hay, concentrate plus Rhodes grass hay and concentrate plus wheat straw treated with ammonia gas, respectively. The hindquarters average weights (and percentage of the carcass) for the three respective groups were 81.7 kg (45.2%), 75.1 kg (44%) and 80 kg (47.6%) (Bakkar *et al.*, 1999). Mekonnen (2004) found the proportions of forequarter (71.6 ± 11.6 kg; 62.8 ± 9.9 kg) and hindquarters (60.8 ± 8.8 and 54.1 ± 14.9 kg), for male and female camels, respectively.

Carcass components are unevenly distributed within the carcass with variation between the hind and fore halves. Muscle, bone and fat components were reported as 59.3, 4.5 and 36.2% in the fore half and 66.5, 14.9 and 17.3% in the hind half, respectively (Kamoun, 1995). The higher proportions of fat in the forequarter could mainly be attributed to the hump fat. The hump fat accounted for 8.6% of the carcass weight. The back and leg contained 77.6 and 74.1% of muscle, respectively. The proportion of muscle, bone and fat in Sudanese camels was 56, 19 and 13.7%, respectively, with a muscle: bone ratio of 3.0 (Yousif and Babiker, 1989).

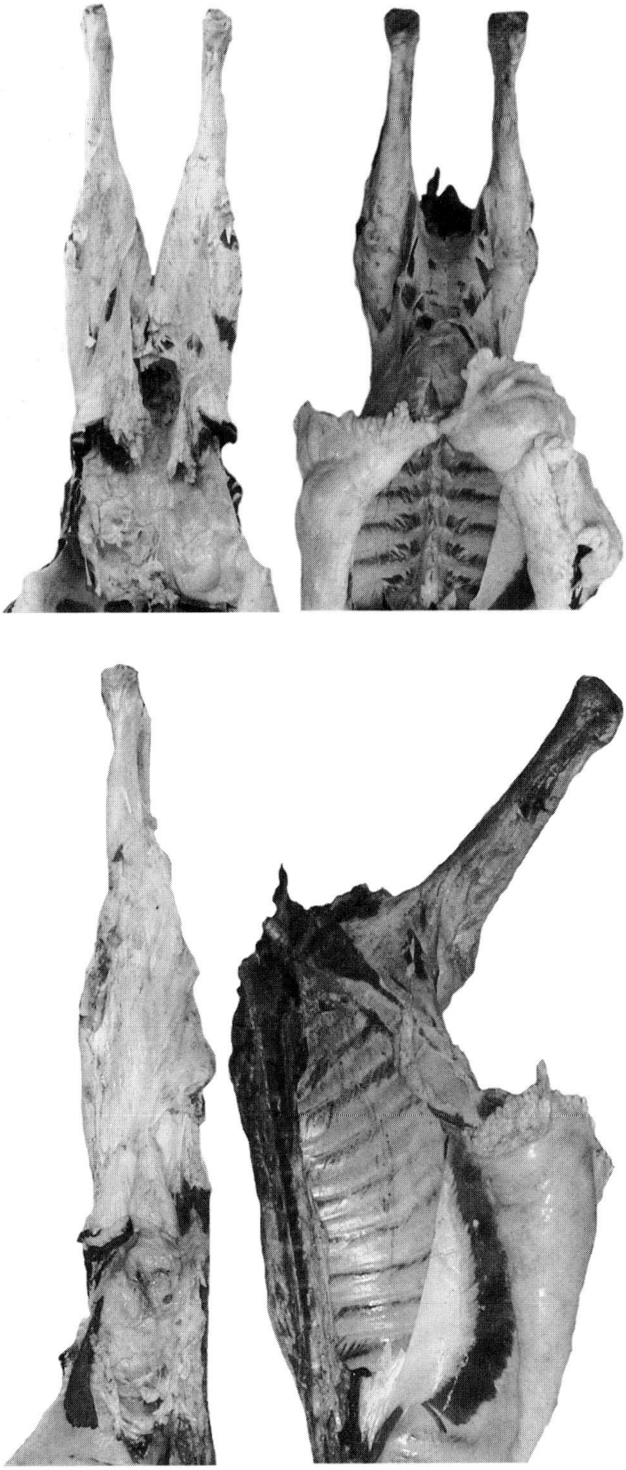

Fig. 8.4. The camel carcass divided into forequarters and hindquarters.

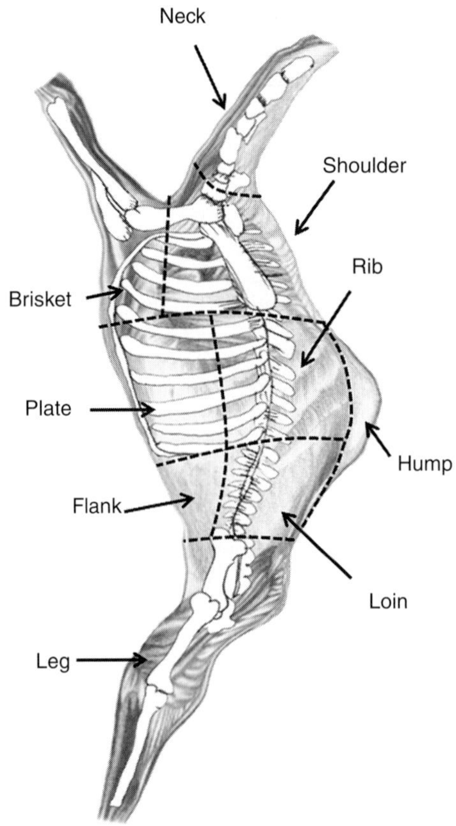

Fig. 8.5. A side of carcass showing a general plan of the cuts using dotted lines.

The leg and shoulder, as a proportion of the carcass weight, were the heaviest cuts in the carcass, followed by the thoracic region (dorsal + flank) and neck, whereas the lumbar and flank were of lighter weight. Wholesale yield cuts have been studied for 15 fattened male camels slaughtered at different body shape (Kamoun, 1995). The neck, the forequarter, the thoracic-back region, the ribs, the loin region, the hindquarter, the flank, the hump and others (kidney fat + tail + diaphragm muscle) were: 9.4, 22.6, 8.1, 10.8, 7.5, 24.5, 5.7, 8.4 and 3.0% of carcass weight, respectively. In general, the forequarters were heavier (51 and 49% of carcass weight) than the hindquarters. The largest cut of the carcass used is the leg followed by the shoulder. Different cuts for mature camel males (>10 years old) were heavier than those from females of a similar age (Mekonnen, 2004). Significant differences were found for the weights of the forequarter, neck, muscles of the back, *Longissimus thoracis*, fascia and associated muscles, and pectoral and ventral abdominal muscles (Mekonnen, 2004).

Males have higher forequarter:hindquarter ratios mainly because of their higher proportions of neck and hump. The forequarter:hindquarter ratio was reported as 1.61% for males and 1.27% for females. Although intact males during the mating season stop growing and might even lose weight, males are known to have more developed heads, necks and shoulders, which are necessary characteristics for competing males during the breeding season. The average weights of the forequarters, hindquarters and the hump were 71.6 ± 11.6 kg, 60.8 ± 8.8 kg and 7.5 ± 4.4 kg for males, and 62.8 ± 9.9 kg, 54.1 ± 14.9 kg and 7.4 ± 2.9 kg for female Ethiopian camels slaughtered at 10 years old (Mekonnen, 2004).

An important characteristic of camel meat is its low fat content compared with many other meat species. There are, however, some reports of higher fat content in camel carcasses, apparently depending on the feeding system. Kamoun (1995) determined the tissue composition by dissecting 12 carcasses and found that the dromedary male camel carcasses (average weight 256.6 kg; range of 181–343 kg) contained, on average, 60.9% muscle (range of 57.3–64.9%), 20.9% bone (range of 16.2–23.7%) and 18.2% fat (range of 12.5–24.0%). The muscle to bone ratio decreased as body weight increased with the ratio ranging from 2.48 to 3.76 with a mean of 2.95 ± 0.39 (Table 8.7).

Wilson (1998) reported a proportion of 57% muscle, 25.5% bone and 16.9% fat in average camel carcasses. The proportion of muscle in nine Sudanese camels was 56% (43.6–67.6%), with 19% (13.4–25.3%) bone, and 13.7% (9.7–18.4%) fat, with a muscle: bone ratio ranging from 2.7 to 3.0 (Yousif and Babiker, 1989). This characteristic of camel meat containing less intermuscular and intramuscular fat than other meat animals could be used in marketing strategies for camel meat (Dawood and Alkanhal,

Table 8.6. Live weight, carcass weight and proportion of carcass components (percentage of carcasses) of dromedary camels.

Items	Kurtu (2004)	Kamoun (1995)	Herrmann and Fischer (2004)	Wilson (1978)	Biala *et al.* (1990)	Bahamou and Baylik (1999)	El-Gasim and El-Hag (1992)
Live weight (kg)	465.8 (M) 335.7 (F)	413.8	530–800	447.9	288	459.7	226.8–271
Carcass weight (kg)	252.3 (M) 170.0 (F)	231.1	309.7–414.8	231.3	146.8	244.2	119.5–132.5
Hindquarter	47.3 (M) 36.0 (F)	49.1	131.0–149.3	40.5	36.8	41.5	
Forequarter	76.0 (M) 45.9 (F)	50.9	123.4–196.8	57.5	63.2	58.5	
Neck	13.5 (M) 10.3 (F)	9.4	22.0–25.0		10.4	10.2	7.5
Shoulder		22.6			22.8	23.6	31.6
Brisket							
Rib		18.9			30.0	24.7	14.6
Plate							
Loin		7.5			16.3	9.3	9.3
Leg		24.5			20.5	28.7	28.8
Remarks	7 years old; 88 male and 12 female	3 years old; 15 well-fattened camels	Eight dromedaries	60 Sudanese dromedaries	2 years old; 6 fattened camels	47 mature castrated non-fattened camels	2 years old; 12 non-fattened camels

Table 8.7. Live weight and carcass characteristics in dromedary males.

	Yousif and Babiker (1989)	Biala *et al.* (1990)	Kamoun (1995)	Kurtu (2004)
Number	9	6	12	88
Age (year)	Mature	2	3	Mature
Live weight (kg)	456.1	288	455	465.8
Carcass weight (kg)	239.9	146.8	256.6	252.3
Hump weight (kg)	30.8	9.1	20.1	33.5
Carcass tissue (%)				
Muscle	57.8	60.5	60.9	68.0
Bone	18.8	28.1	20.9	20.0
Fat	13.7	9.2	18.2	12.0
Muscle:bone ratio	2.95	2.2	3.31	3.4
Remarks	Well fattened	Fattened	Well fattened	Fattened

1995). The proportion of muscle in the camel carcass is comparable to that of cattle (Preston and Willis, 1975; Babiker, 1984; Mahgoub *et al.*, 1995a, b), whereas the proportion of carcass bone is higher and therefore the muscle to bone ratio is lower for camels (Babiker, 1984). This may possibly be attributed to increased bone length in the camel. The muscle:bone ratio was 3.0 in Sudanese camels (Yousif and Babiker, 1989).

Muscle distribution varies according to the anatomical site on the carcass. The highest muscle distribution in the carcass has been reported to be in the ribs and backbone (30%), shoulder (28%), leg (22%) and the neck (8%) (El-Gasim and El-Hag, 1992). Tissue components of certain camel cuts are presented in Table 8.8. The muscle, bone and fat tissues were unevenly distributed in the carcass. Percentages of muscle, bone and fat tissues were 66.9 ± 1.3%, 26.8 ± 3.7% and 6.3 ± 3.6%, respectively, for the forequarter and 54.5 ± 3.2%, 14.8% ± 1.7% and 30.7 ± 3.9%, respectively, for the hindquarter (Kamoun, 1995). Moreover, the same author found that when the hump fat was included, it accounted for 60.0 ± 4.7% of the hindquarter fat. The muscle:bone ratio was, respectively, 3.73, 2.55, 4.30, 3.59, 2.67, 2.49, 1.53 in the hindquarter, forequarter, forelimb, hindlimb, neck, lumbar region, and thoracic back and rib region (Kamoun, 1995).

Table 8.8 shows that the camel shoulder and leg have a muscle proportion of around 75%, whereas the neck and lumbar regions have a muscle proportion of 71 and 60%, respectively. The proportion of bone in carcass cuts was reported to be highest in the thoracic dorsal region and lowest in the flank, whereas the proportion of fat was high in the flank, ranging from 25 to 45%, and low in the neck, shoulder and leg cuts (Table 8.8).

Age, sex, breed and the nutritional state influence camel body composition. Age has a significant effect on carcass components with distinct advantages in slaughtering camels at an early age. Muscle content was highest for 2-year-old castrated camels. Hump fat represented 1.9% of the dressed carcass of a 24-month-old and 5.19% of the carcass of a 44-month-old camel (Kulaeva, 1964). Sex is an important factor in determining carcass yield in the camel. The total meat weight from male camels was higher than from females by 35% (Kurtu, 2004). As in other farm animal species, females are fatter than males, especially at older ages. Congui (1953) reported carcass fat 8.8% for male and 20.5% for female 10–12 year old Somali camels.

The major characteristic of camel meat is the low fat level. Camel meat contains less intermuscular and intramuscular fat than other meat animals, a fact that can be used in marketing strategies of camel meat (Dawood and Alkanhal, 1995). The intramuscular fat content of muscle is of particular importance because it enhances the palatability traits of meat products, such as flavour, juiciness and tenderness. Fat ranged between 0% and 4.8%, whereas bones ranged between 15.9 and 38.1% (Shalash, 1979). There are, however, reports of higher fat contents in camel carcasses. Kamoun (1995) reported that the camel carcass contains 57% muscle, 25.5% bone and 16.9% fat. Kuznestov and Tretyakov (1972) reported that the camel carcass consists of 52.8–76.6% muscle, 0–4.8% fat and 15.9–38.1% bone.

Table 8.8. Muscle, bone and fat percentages in the carcass and in different cuts from the camel carcass.

	Kamoun (1995)			Biala *et al.* (1990)		
	Muscle (%)	Bone (%)	Fat (%)	Muscle (%)	Bone (%)	Fat (%)
Carcass	60.9	20.9	18.2	62.5	28.3	9.2
Forequarter	66.9	26.8	6.3	61.8	29.4	8.8
Neck	71.3	27.5	1.2	70.7	28.1	1.2
Shoulder	76.7	18.2	5.1	69.6	28.9	1.5
Thoracic dorsal	56.0	36.4	7.6	52.1	30.4	17.5
Hindquarter	54.5	14.8	30.7	64.1	27.4	8.5
Loin	60.4	25.6	14.0	53.7	27.0	19.3
Leg	74.5	21.0	4.5	72.1	27.6	0.3

8.5 Conclusion

Only few sources provide data on carcass characteristics such as carcass weight, dressing-out percentage and carcass components. Camels reach live weights of about 650 kg at 7–8 years of age, and produce carcass weights ranging from 125 to 400 kg with dressing-out percentage values from 55% to 70%. Camel carcasses contain about 57% muscle, 26% bone and 17% fat, with the forequarter being significantly heavier than the hindquarter.

References

Abouheif, M.A., Basmaeil, S.M. and Bakkar, M.N. (1986) Estimation of body and carcass weights in Saudi Arabian Najdi male camels. *Arab Gulf Journal of Science Research* 4, 733–743.

Abouheif, M.A., Basmaeil, S.M. and Bakkar, M.N. (1990) A standard method for jointing camel carcasses with reference to the effect of slaughter age on carcass characteristics in Najadi camels. I: wholesale cut weight. *Asian-Australian Journal of Animal Science* 3, 97–102.

Al-Ani, F.K. (2004) Use and production of camels. In: Al-Ani, F.K. (ed.) *Camel Management and Diseases*, 1st edn. Al-Sharq Printing Press, Qatar, pp. 91–114.

Al-Kharusi, A.H.H. (2011) Composition, quality and histochemistry of the Arabian camel (*Camelus dromedarius*) muscles. Msc Thesis, Department of Animal and Veterinary Sciences, Sultan Qaboos University, Oman.

Al-Owaimer, A.N. (2000) Effect of dietary Halophyte *Salicornia bigelovii* Torr on carcass characteristics, minerals, fatty acids and amino acids profile of camel meat. *Journal of Applied Animal Research* 18, 185–192.

Babiker, M.M. (1984) Abundance and economic potentials of camels in the Sudan. *Journal of Arid Environment* 7, 377.

Babiker, S.A. and Yousif, K.H. (1987) Carcass yield and characteristics of mature male camels of the Sudan. *Annual Report, Camel Research Unit*, University of Khartoum, Sudan, 120–124.

Bahhamou, M. and Baylik, R. (1990) Contribution a létude du rendement de la carcasse et du cinquieme quartier du dromadaire (population targui) Algerie. CARDN/ACSAD/CAMEL/P 531999, pp. 102.

Bakkar, M.N., Basmaeil, S.M. and Hamam, A.A. (1999) Meat production and quality: two unique properties of young camels. In: *Proceedings of the International Workshop on the Camel Calf*, Ouarzazate, Morocco, 24-26 October, pp. 218.

Biala, S.A., Shareha, A.M., Hermas, S.A., Khalifa, S. and Abussaod, F. (1990) Dressing percentage proportions of edible and non-edible portions of camel carcasses. In: Wardah, M.F., Wilson, R.T. and Zaied, A.A. (eds) *Proceedings of the International Conference on Camel Production and Improvement*. Tobruk-Libya. ACSAD SYRIA, 187–198.

Bremaud, O. (1969) Notes sur lelevage camelin dans les district Nord del la Republique du Kenya. Avril Maj-Juin, 1969. Rapport LEMVT, Maisons, Alfort. p.105.

Camel Newsletter (1979) The Central Australian Camel Industry. *Camel Newsletter* 13, 52–58.

Congiu, S. (1953) Dressing percentage, weight distribution and correlations between different parts of the body in the Somali dromedary. *Zootecnica Veterinaria* 8, 188–191.

Dawood, A. and Alkanhal, M.A. (1995) Nutrient composition of Najdi-Camel Meat. *Meat Science* 39, 71–78.

El-Gasim, E.A. and El-Hag, G.A. (1992) Carcass characteristics of the Arabian camel. *Camel Newsletter* 9, 20–24.

El-Mossalami, E., Awad, Y., Ibrahim, A. and Diab, O. (1996) Major chemical constituents of beef, buffalo and camel's meat. *Veterinary Medical Journal Giza* 44, 49–53.

FAO (2011) FAOSTAT database. Food and Agricultural Organization, United Nations. Available at: http://faostat.fao.org/site/339/default.aspx (accessed 2 October 2012).

Herrmann, K. and Fischer, A. (2004) Method for hygienic slaughter of camel. In: Farah, Z. and Fischer, A. (eds) *Milk and Meat from the Camel; Handbook on Products and Processing*. vdt Hochschulverlag AG an ETH Zurich, Zurich/Singen.

Hertrampf, J.W. (2004) The "ship of the desert" as a meat supplier. *Fleischwirtschaft* 84, 111–114.

Kadim, I.T., Mahgoub, O. and Purchas, R.W. (2008) A review of the growth and carcass and meat quality characteristics of the one-humped camel (Camelus dromedarius). *Meat Science* 80, 555–569.

Kamoun, M. (1995) Dromedary meat: production, qualitative aspects and capability for transformation. *Option Mediterraneennes. Serie B, Etudes et Recherches* 13, 105–130.

Kamoun, M. (2004) Meat recording systems in camelids. In: Cardellino, R., Rosati, A. and Mosconi, C. (eds) *Current Status of Genetic Resources, Recording and Production Systems in African, Asian and American Camelids*, Sousse, Tunisia, 30 May 2004. *ICAR Technical Series* 11, 105–130.

Kempster, A.J., Cuthbertson, A. and Harrington, G. (1982) *Carcass Evaluation in Livestock Breeding, Production and Marketing.* Granada Publishing Limited, London, pp. 306.

Khan, B. B., Iqbal, A. and Riaz, M. (2003) *Production and Management of Camels.* Pak TM Printers, Pakistan.

Khatami, K. (1970) Camel Meat: A new promising approach to the solution of meat and protein in the arid and semi-arid countries of the world. Ministry of Agriculture, Tehran.

Kulaeva, V. (1964) The production of the Bactrian camel. *Animal Breeding Abstract* 32.

Kurtu, M.Y. (2004) An assessment of the productivity for meat and carcass yield of camel (Camelus dromedarius) and the consumption of camel meat in the Eastern region of Ethiopia. *Tropical Animal Health and Production* 36, 65–76.

Kuznestov, V.A. and Tretyakov, V.N. (1972) Carcass classification in the Turkmen single-humped camel. *Animal Breeding Abstract* 40, 2365.

Mahgoub, O., Olvey, F.H. and Jeffrey, D.C. (1995a) Growth and composition of the Omani Dhofari cattle. I: Body weight growth and carcass composition. *Asian-Australian Journal of Animal Science* 8, 611–616.

Mahgoub, O., Olvey, F.H. and Jeffrey, D.C. (1995b) Growth and composition of the Omani Dhofari cattle. II Distribution of carcass tissue. *Asian-Australian Journal of Animal Science* 8, 617–625.

Mekonnen, G.A. (2004) Meat production potential of Issa type camels under traditional management systems in Eastern Ethiopia. *Bulletin of Animal Health and Production in Africa* 52, 215–220.

Muhammad, B.F. and Akpan, I.N. (2008) Camel (*Camelus dromedarius*) meat utilization in Kano-Nigeria (II): Post-slaughter handling and marketing wholesale meat cuts. *Research Journal of Animal Sciences* 2, 113–117.

Mukasa-Mugerwa, E. (1981). The camel (*Camelus dromedarius*): A biographical review. *International Livestock Centre for Africa, Addis Ababa, Ethiopia.* p.147.

Nsoso, S.J., Young, M.J. and Beatson, P.R. (2000) A review of carcass conformation in sheep: assessment, genetic control and development. *Small Ruminant Research* 35, 89–96.

Preston, T.R. and Willis, M.B. (1975) *Beef Production*, 2nd edn. Pergamon, Oxford, UK.

Saparov, G. and Annageldiyev, O. (2005) Meat Productivity of the camel Arvana breed and ways to increase it. In: Faye, B. and Esenov, P. (eds) *Desertification Combat and Food Safety.* IOS Press, Amsterdam, pp. 211–214.

Shalash, M.R. (1979) Effect of age on quality of camel meat. *First Workshop on Camel.* Khartoum International Foundation for Science, Sudan.

Skidmore, J.A. (2005) Reproduction in dromedary camels: an update. *Animal Reproduction* 2, 161–171.

Wilson, R.T. (1978) Studies on the livestock of Southern Darfur, Sudan. V. Notes on camels. *Tropical Animal Health and Production* 10, 19–25.

Wilson, R.T. (1984) *The Camel.* Longman Group Ltd, Essex, UK.

Wilson, R.T. (1998) Camel. In: Costa, R. (ed.) *The Tropical Agricultural Series*, Centre for Tropical Veterinary Medicine, University of Edinburgh, UK.

Yousif, O.K. and Babiker, S.A. (1989) The desert camel as meat animals. *Meat Science* 26, 245–254.

9 Distribution and Partitioning of Tissues in the Camel Carcass

Osman Mahgoub and Isam T. Kadim
Department of Animal and Veterinary Sciences, College of Agricultural and Marine Sciences, Sultan Qaboos University, Muscat, Sultanate of Oman

9.1 Introduction

The camel can survive and thrive under harsh environmental conditions and produce high-quality animal protein, especially for deprived segments of society. Camels can also be raised under intensive management to produce good-quality carcasses and meat for the modern supermarket industry. Carcass tissue distribution (i.e. muscle, bone and fat) is not well studied in camels. Camel carcasses are unique because of their shape, with a large variable-sized hump, long neck and shallow hind side.

Muscle tissue distribution in the carcass is important in determining various cut values. For instance, some of the cuts that contain higher muscle content are known as expensive cuts, such as the proximal hindlimb and proximal forelimb and loin (Butterfield, 1988).

Fat partitioning differs in rate and intensity in the carcass and varies with species. Some animal species such as goats lay more fat in the interior of the body, whereas sheep lay more fat on the subcutis and intermuscular areas (Mahgoub *et al.*, 2012). Fat distribution and partitioning in the camel is not well studied.

Skeletal growth is essential for body growth and carcass quality. Longitudinal bone growth affects bone and attached muscle growth (Mahgoub, 1988). Proportions of bone in the carcass influence other edible components (muscle and fat). Consequently, cuts with higher bone content are less valuable than those with lower bone and subsequently higher muscle content.

The factors that affect carcass tissue distribution in the camel include: age, body weight, sex, breed and nutrition (Kadim *et al.*, 2008). Muscle content was higher in camels of 2 years of age than in older animals (Kadim *et al.*, 2008). The hump fat content was, however, higher in older Bactrian camels than in young ones (Kulaeva, 1964). Total meat yield was greater from male camel carcasses than from females (Kurtu, 2004). This is mainly because females are usually fatter than males, especially when older. Male camels contain 8.8% fat, whereas females contain 20% fat in their carcasses (Congiu, 1953).

This chapter discusses the distribution of the essential carcass tissues, muscle, bone and fat, in the camel carcass and the practical implications of that on camel carcass quality and marketability.

9.2 Muscle Distribution in the Camel Carcass

Distribution of muscle tissue in the carcass is important from a body development and commercial point of view. Very little is known, however, about muscle distribution in the camel compared with cattle, sheep, pigs or goats. This is because the camel has been slaughtered and its meat consumed mainly within traditional societies where no modern marketing systems are available. Camel meat is usually sold in bulk with no standard cutting system applied. Muscle distribution in other species has been well studied. In several studies, muscle content has been expressed as proportions of individual muscle and muscles were grouped to reflect muscle distribution in various carcass regions (Butterfield, 1983; Mahgoub *et al.*, 1998) but there are not many similar published studies on the camel.

Mahgoub *et al.* (2012) carried out detailed analyses of camel carcass tissue including individual muscle and bone dissections (O. Mahgoub, unpublished data). Figure 9.1 shows the distribution of camel carcass muscle in nine major muscle groups in the nomenclature defined by Butterfield (1988) for sheep. The largest proportion of muscle (about 30% of whole muscle carcass) was in the muscles in the proximal hindlimb (muscle group 1). This muscle group is made up of a large number of muscles including some of the largest muscles in the carcass such as *Gluteoceps*, *Semimembranosus*, *Semitendinosus* and *Abductor*. This is an important muscle group from a commercial point of view because it is regarded as one of the high-priced cuts in beef (topside and silverside). The contribution of this muscle group to total side musculature seems to be lower in camels than in cattle. The total proportion of this muscle group in the total side camel muscle was lower than that published by Butterfield and May (1966) for cattle (29.87 versus 34.05%, respectively). This is logical from a conformational point of view with the proximal limb of camels being slender and shallower, possibly owing to its role in camel movement and in camel squatting. The difference is more pronounced in the Mm. *Biceps* (*Gluteoceps* in camel), *Gluteal* group, *Semimembranosus* and *Semitendinosus* in favour of cattle. However, more studies are needed in large camels, which are more suited to and selected for meat production. This is an area that needs to be improved in the camel carcass because this carcass region is important from an economical point of view. None the less, the proportions of muscles in this muscle group were similar to those in sheep and goats (Table 9.1).

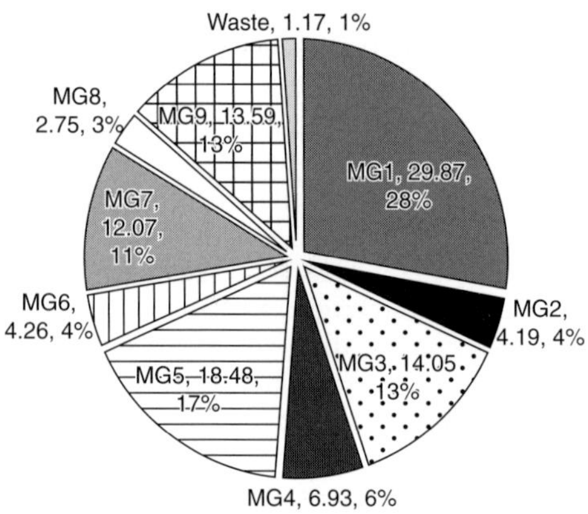

Fig. 9.1. Muscle distribution in nine muscle groups on the camel. MG1, muscles of the proximal hindlimb; MG2, muscles of the distal hindlimb; MG3, muscles surrounding spinal column; MG4, muscles of abdominal wall; MG5, muscles of the proximal forelimb; MG6, muscles of the distal forelimb; MG7, muscles connecting the thorax to the forelimb; MG8, muscles of the shoulder; MG9, intrinsic muscles of neck; expensive muscle group (EMG) = MG1 + MG3 + MG5 = 58%.

Muscle group 2, which contains the lower hindlimb, contributes about 4% of the total muscle with the *Gastrocnemius et soleus* muscle being the largest of the individual muscles. Its proportion in the total carcass muscle seems to be lower than that in sheep and goats (Table 9.1). This muscle group makes a less important leg cut. However, with a careful selection programme for meat production, the conformation of this region could be improved.

Muscle group 3, which contains muscles surrounding spinal column, contributes about 13% of the total side muscle content. This is an important muscle group that contributes to many high-priced muscles including: sirloin, fore rib and rump. Some of its muscles will contribute to the chuck and blade cut in beef. The group contains the largest muscle of the carcass, the *Longissimus thoracis et Lumborum* muscle. The proportion the group forms of the

Table 9.1. Muscle groups (% total side muscle) in the carcass of camel, cattle, sheep and goats.

		Camel[a]		Cattle[b]	Sheep[c]		Goats[c]	
Muscle Group	Description	Mean	SD	Mean	Mean	SE	Mean	SE
Muscle Group 1	Proximal hindlimb	29.87	5.77	32.15	28.78	0.15	27.82	0.16
Muscle Group 2	Distal hindlimb	4.19	0.72	4.84	5.46	0.06	5.20	0.06
Muscle Group 3	Surrounding spinal column	14.05	4.56	12.61	16.34	0.11	14.41	0.12
Muscle Group 4	Abdominal wall	6.93	1.89	10.10	9.55	0.13	11.96	0.14
Muscle Group 5	Proximal forelimb	18.48	2.89	11.13	11.78	0.13	12.28	0.14
Muscle Group 6	Distal forelimb	4.26	0.61	2.69	3.11	0.04	3.27	0.04
Muscle Group 7	Connecting thorax to forelimb	12.07	2.69	9.91	9.96	0.07	9.73	0.07
Muscle Group 8	Connecting neck to forelimb	2.75	1.75	6.95	4.34	0.04	4.57	0.04
Muscle Group 9	Intrinsic neck and thorax	13.59	1.69	9.05	10.67	0.10	11.06	0.11
EMG[d]	Expensive Muscle Group	57.93	1.87	55.59	56.90	0.19	54.51	0.20
Forequarter[e]	Forequarter	46.19	1.90		36.75	0.14	37.34	0.15

[a]Mahgoub *et al.* (2012); [b]Charles and Johnson (1976); [c]Mahgoub and Lodge (1998); [d]Expensive Muscle Group = total of muscle groups 1, 3 and 5; [e]Forequarter = total muscle groups 5, 7, 8 and 9.

muscle carcass was lower than that in sheep and goats (Table 9.1).

The abdominal muscle group (muscle group 4) contributes about 7% of side muscle content in the camel, which was lower than in sheep and goats (Table 9.1). This muscle group, which make the thin and thick flank cuts in beef and camel, is regarded as being among the lower-priced cuts. These cuts do not contain bones but usually have a high-fat content including the abdominal wall fat flap in the camel, which will be discussed later. The tissue content of various cuts is important for commercial value, retail cutting and also for the method of cooking.

The proximal forelimb (muscle group 5) contributes a significant 18.5% of side muscle in the camel. This seems to be a much higher contribution than in cattle, sheep and goats (Table 9.1). It is a very muscular area of the camel body because it needs to support the large neck and strong legs of the camel. It is also important from an economical point of view because it contributes to one of the high-priced cuts of the carcass, the chuck and blade and the thick rib in beef, and the shoulder cut in the camel (Fig. 9.1).

Muscle group 6 includes muscle of the distal forelimb and contributes approximately 4% of the total side muscle content. It forms the less important shank cut. The extensor muscle group is the largest component of this group. The proportion of muscle group 6 in the total muscle is not much different from that of sheep and goats (Table 9.1).

Muscle group 7, i.e. the muscles connecting the thorax to the forelimb, contributes 12% to the total side muscle. This is an important muscle group and it is well developed in the camel because of the large weight load of the heavy forequarter in the camel. It also contributes to the important chuck and blade cut. Its proportion in the camel side total muscle is greater than that in cattle, sheep or goats (Table 9.1).

Muscle group 8, connecting the neck to the forelimb, contributes only 2.8% of the side muscles so is less significant (Fig. 9.2).

Muscle group 9, which contains the intrinsic neck muscles, is very important and contributes about 13.6% of the side muscle, which is greater than that in sheep and goats (10.7 and 11.1%, respectively). However, the neck cut is not among the

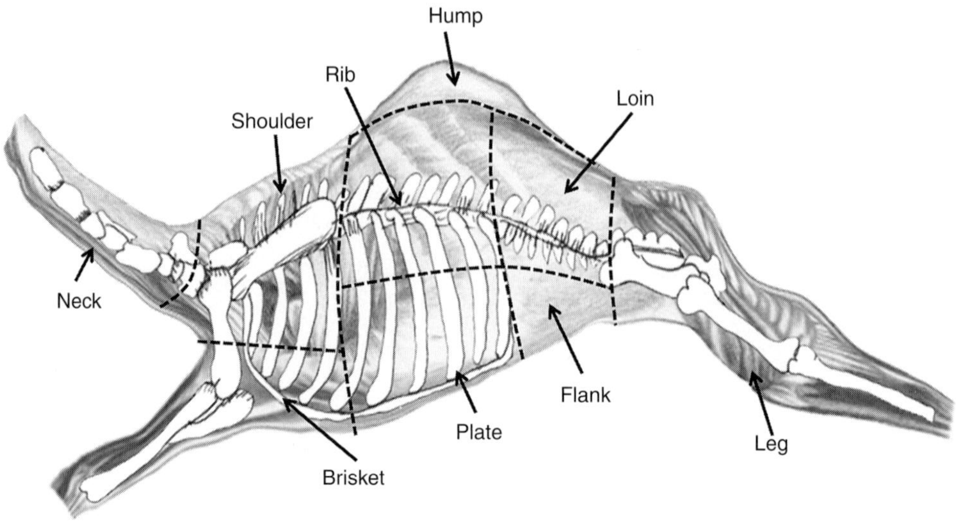

Fig. 9.2. Wholesale carcass cuts in the camel.

high-priced cuts owing to its high bone content. This is very pronounced in the camel with its extra-long neck.

Generally, the most marked feature is that the camel has higher muscle proportions in the proximal forelimb and muscles connecting the thorax to the forelimb, as well as in the intrinsic neck muscles. This results in the forequarter of the camel containing more muscles than those of other livestock (Table 9.1). The camel forequarter has been reported to be much heavier than the hindquarter, although some workers attributed this to the presence of the hump (Kadim *et al.*, 2008). The work summarized in Table 9.1 indicates that the muscle content of the forequarter is higher than that of the hindquarter without including the hump, which is mainly made up of fat in camels in contrast to that in zebu cattle (O. Mahgoub, unpublished data).

The expensive muscle group (EMG) in the camel carcass seems to form a slightly higher proportion than that in other livestock (Table 9.1). This can be mainly attributed to the higher proportions of the muscles in the proximal forelimb and muscles connecting the forelimb to the thorax. These muscles are well developed in the camel to be able to carry the heavy neck. Larger proportions of high-priced muscle groups indicate the good economic value of the camel carcass, although it looks disfigured with slender distal limbs and a sloping hindquarter.

Muscle distribution over various parts of the carcass should be taken into consideration in camel meat processing and marketing (Table 9.2), including in devising a system of camel carcass grading, in carcass cutting and in pricing commercial cuts.

One of the few studies on muscle distribution in the camel carcass was that of Elgasim and El-Hag (1992). They reported that the hindlegs, forelegs, ribs plus backbone, and neck contained 28%, 22%, 30% and 8% of the total carcass muscle. This is in line with the data described above (Mahgoub *et al.*, 2012; Table 9.1).

Table 9.2 shows that muscle group 1 is one of the most important muscle groups containing large-sized muscles. The largest muscle in this group is the *Gluteoceps*, which includes a *Gluteus* part plus the *Biceps femoris* and it accounts for about 5% of the total carcass muscle, which makes it one of the largest muscles of the body. The second largest muscle in the group is the group of muscles known as

Table 9.2. Mean and standard deviation of some individual muscles expressed as a percentage of the side carcass muscle.

Muscle	Mean	SD
Muscle Group 1		
Tensor facia lata	1.16	0.090
Gluteoceps	5.14	0.613
Gluteus medius	1.43	0.110
Gluteus profundus	0.59	0.076
Gluteus accessorius	0.17	0.037
Rectus femoris	1.77	0.203
Vastus lateralis	3.62	
Vastus intermedius	1.03	0.124
Vastus medialis	0.85	0.175
Semitendinosus	1.47	0.109
Semimembranosus	3.69	0.194
Adductor femoris	2.20	0.164
Sartoius	0.32	0.100
Pectineu	0.64	0.080
Gracilis	0.92	0.170
Iliacus	0.41	0.140
Obturatorii externi et interni	0.72	0.120
Muscle Group 2		
Gastrocnemius et soleus	1.79	0.300
Popliteus	0.19	0.030
Flex digit longus	0.22	0.031
Extn digit communi	0.39	0.047
Extn digit longus	0.49	0.038
Extn digit lateralis	0.18	0.065
Tibialis cranialis	0.14	0.025
Tibialis caudalis	0.15	0.085
Peroneus longus	0.17	0.055
Muscle Group 3		
Iliocastali thoracis	0.90	1.387
Psoas major	2.10	0.377
Psoas minor	0.16	0.018
Quadratus lumborum	0.33	0.034
Longissimus thoracis et lumborum	5.61	0.539
Spinalis et spinalis	2.05	1.476
Multifidus	1.80	0.305
Muscle Group 4		
Rectus thoracis	0.08	0.021
Rectus abdominis	2.20	0.231
Serratus dorsalis caudalis	0.11	0.026
Retractor costae	0.12	0.061
Transversus abdominis	0.92	0.165
Obliquus internus abdominis	0.98	0.254
Obliquus externus abdominis	1.64	0.409
Muscle Group 5		
Teres major	0.52	0.053
Teres minor	0.26	0.040
Coracobrachialis	0.25	0.026
Subscapularis	1.08	0.172
Supraspinatus	1.98	0.134
Infraspinatus	1.53	0.089
Deltoidius	0.86	0.110
Brachialis	0.71	0.046
Biceps brachii	1.36	0.110
Triceps caput medialis	0.35	0.060
Triceps caput longus	5.44	0.453
Triceps caput lateralis	1.81	0.204
Tensor fascia antebrachium	0.27	0.121
Muscle Group 6		
Flexor group	1.336	0.074
Extensor group	2.11	0.086
Ulnaris lateralis	0.13	0.133
Anconeus	0.25	0.047
Muscle Group 7		
Trapezius thoracis	0.62	0.173
Pectoralis	5.55	0.532
Latissimus dorsi	2.44	0.258
Serratus ventralis thoracis	2.00	0.371
Muscle Group 8		
Trapezius cervicalis	0.23	0.331
Serratus ventralis	0.97	0.702
cervicalis	0.64	0.234
Rhomboideus		
Brachiocephalicus	0.58	0.528
Omotran-sversarius	0.04	0.085
Muscle Group 9		
Serratus dorsalis cranialis	0.59	0.955
Transversus thoracis	0.07	0.082
Intercostalis externi et interni	1.80	0.875
Longissimus capitis	0.20	0.070
Longisimus atlantis	0.30	0.077
Longissimus cervicalis	1.02	0.135
Intertransversalis dorsalis	0.12	0.167
Intertransversalis ventralis	1.02	0.753
Complexus	1.17	0.620
Rectus capitis dorsalis major	0.25	0.158
Obliquus capitis caudalis	0.33	0.180
Multifidus cervicalis	0.96	0.309
Scalenus ventralis	0.53	0.161
Longus coli	0.87	0.496
Transversus thoracis	0.08	0.136
Sternocephalicus	0.70	0.564
Scalenus medius	0.22	0.266
Scalenus dorsalis	0.24	0.201
Rectus cap ventralis major	0.05	0.069

quadriceps. This includes the large *Vastus lateralis* (3.6%), *Rectus femoris, Vastus medialis* and *Vastus intermedius* muscles. Other large muscles in muscle group 1 include the *Semimembranosus, Adductor femoris* and *Semitendinosus* muscles.

The largest muscle in muscle group 2 is the *Gastrocnemius et soleus* muscle, which accounts for 1.8% of total muscle, with the rest of the muscles all comprising less than 0.5% of total carcass muscle. This renders the cut less important from a carcass quality point of view.

The most important muscle in muscle group 3 is the *Longissimus thoracis et lumborum* muscle, which contributes about 5.6% to the whole side muscle, making it the largest in the carcass. The *Spinalis* muscle is also important and the *Psoas major* muscle and *Multifidus* muscle make a significant contribution.

The largest muscles in muscle group 4 are the *Rectus abdominis* muscle and the *Obliquus externus abdominis* muscle.

Group 5 is a very important group because it is classified as an expensive muscle group (Butterfield, 1988). It contains the large *Triceps brachii* muscle group, which accounts for about 7.5% of the carcass muscle content. The largest among these is the *Triceps femoris, caput longus* muscle (5.4%). The *Supraspinatus, Infraspinatus* and *Biceps brachii* muscles also make good contributions of 2%, 1.5% and 1.4% to the carcass muscle.

Muscle group 6 includes the flexor and extensor groups of the forelimb, which account for 1.3% and 2.1%. These seem to be larger than those of the hindlimb muscles, reflecting their role of carrying the heavier forequarter.

Muscle group 7 includes three large muscles, the *Pectoralis* group, the *Latissimus dorsi* and the *Serratus ventralis thoracis* muscles contributing 5.6, 2.4 and 2.0% of the side muscle content, respectively. The *Pectoralis* group include the *Pectoralis profundus* and *Pectoralis superficialis* muscles.

The muscle group 8 contribution is small, with the largest muscle being the *Serratus ventralis cervicalis* muscle, contributing about 1% to the total side carcass muscle.

Muscle group 9 contains intrinsic muscles of the neck and thorax and contributes a significant proportion to the side muscle of about 13.6%. A total of 19 muscles were identified but individual muscle weight was small. The largest muscle forms about 1.8% (*Intercostalis externi et interni*).

9.4 Fat Partitioning in the Camel Carcass

In general camels are reported to have less fat in their carcasses than other livestock (Kadim *et al.*, 2008). Within the carcass, fat partitioning differs between various sites of the carcass. It seems, however, that the distribution of the adipose tissue in the camel carcass is unique. Table 9.3 describes fat

Table 9.3. The distribution of fat in the carcass of camels, cattle, sheep and goats (% in total body fat).

	Camel[a]		Cattle[b]		Sheep[c]		Goats[c]	
Fat depot	Mean	SD	Mean	SD	Mean	SE	Mean	SE
Kidney + pelvic	11.45	1.927	12.97	0.88	10.33	0.35	11.51	0.37
Omental	3.97	1.783	13.72	0.74	13.87	0.38	15.39	0.40
Channel	4.93	3.646	1.28	0.16	1.04	0.07	1.25	0.07
Abdominal wall	16.77	3.411	NA		NA		NA	
Total non-carcass	37.12	7.685	43.13	1.25	39.49	0.61	45.59	0.64
Hump	30.34	7.234	NA		NA		NA	
Subcutaneous	10.92	3.477	18.08	1.07	31.81	0.62	25.14	0.65
Intermuscular	21.62	4.629	34.87	1.06	28.69	0.55	29.27	0.58
Total carcass fat	62.88	7.685	53.69	1.25	60.51	0.61	54.41	0.64

[a]Mahgoub *et al.* (2012); [b]Mahgoub *et al.* (1995); [c]Mahgoub and Lodge (1998).

partitioning in the camel body (Mahgoub *et al.*, 2012; O. Mahgoub, unpublished data) in comparison with that of other farm animals. The proportions of total carcass fat in the camel are much higher than in the non-carcass. The largest proportion of fat in the carcass is in the hump (30%) followed by that in the intermuscular space. In the non-carcass, the abdominal flap of fat contributes a significant portion (16.7%).

The proportion of the kidney plus pelvic fat is also significant (11.5%). Similarly, reports indicated that Sudanese camel's bodies contained more kidney fat than mesenteric and omental fat (Eltahir *et al.*, 2012), with animals depositing more fat when supplemented with a molasses-based diet. The authors reported values of 1.24%, 0.56% and 0.14% of empty body weight for kidney, mesenteric and omental fat, respectively.

The hump (Fig. 9.3) contributes about 30% of the total carcass fat in the camel (Mahgoub *et al.*, 2012). This is the largest proportion of fat deposited in one site. It should be noted, however, that the size of the hump is not fixed because it changes with body condition in the camel as a result of changes in feed supply according to season and grazing range conditions. Some reports indicated that the hump may contribute up to 9% of the total carcass weight (Kadim *et al.*, 2008). This would have a serious implication on marketing camel meat especially if a standard method of carcass cutting is adopted for camels. The camel hump would fall in the loin and rack (rib) cuts. The way hump fat is removed (prior to or after cutting) would influence the cut tissue composition and the customer's impression of the camel meat.

Fig. 9.3. Large amounts of carcass fat can be deposited in the camel hump.

One interesting characteristic of camel fat partitioning is the significant fat depot found on the abdominal floor. A thick sheet of fat covers the abdominal muscles (*Rectus abdominis* and *Transversus abdominis*), extending backward to meet the kidney fat. It accounts for 16.8% of total body fat. This seems to be unique to the camel (Figs 9.4 and 9.5) and it might be an adaptation feature. This fat depot at the floor of the abdomen would be close to the ground when the camel crouches in its normal sitting pose. The fat layer would provide good insulation against the heat radiated from the hot sand in the desert.

It should be noted that the camel carcasses can become extremely fat under intensive management. When the hump fat extends over the cutis and a fat abdominal flap, a camel carcass may be classified as over-fat upon carcass grading.

9.5 Bone Weight Distribution in the Camel Carcass

The proportion of bone in the carcass is a very important trait in assessment of carcass quality. Although bone is not an edible tissue, it affects the proportions of other edible tissues in the carcass such as muscle and fat. Bone proportions differ between various sites in the carcass. For instance the proximal hindlegs and forelegs have less bone in relation to muscle than do the distal limbs.

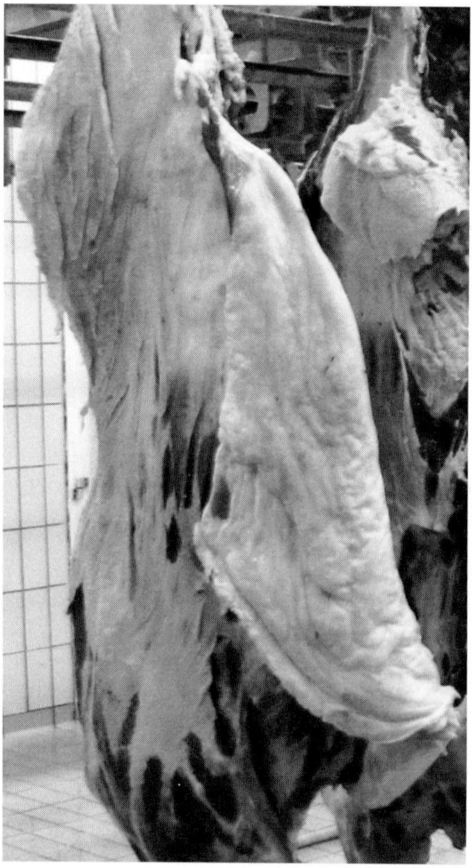

Fig. 9.4. Abdominal fat flap covering the lower abdominal muscles in the camel.

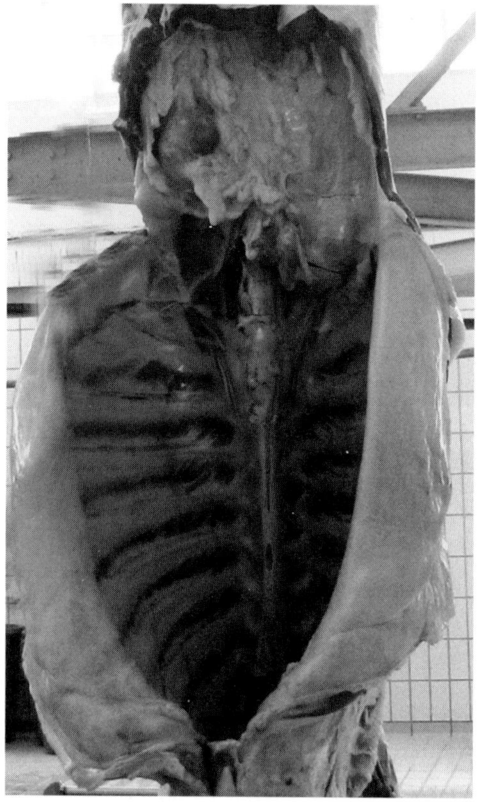

Fig. 9.5. Abdominal and kidney fat in a camel carcass. Note also fat deposition in the channel.

Approximately one-quarter of the camel carcass weight is bone (Kadim *et al*., 2008; Mahgoub *et al*., 2012). The proportion of bone in the carcass is important because it affects other components such as muscle and fat. Bone distribution in the camel carcass is also important because it affects carcass conformation and quality.

The axial skeleton contained 45% of the camel carcass, whereas the forelimb and hindlimb contained similar proportions (27 and 28% of the side total bone (Table 9.4)). The forequarter (the cervical and thoracic vertebrae plus ribs, sternum and forelimb) constitutes about 62%, whereas the hindquarter (lumber and sacral vertebrae plus pelvis and hindlimb) constitutes 38% of the carcass bone. Most of the extra weight of the forequarter comes from the long heavy neck. The forequarter is larger in the camel carcass than the hindquarter (Kadim *et al*., 2008). This large proportion of bone affects the distribution of other tissues in the carcass. The vertebral column makes up about 22% of the carcass bone, with the highest contribution from the neck.

The limb long bones (humerus, radius and ulna, femur and tibia) each provide a similar contribution to the total bone weight of 8–10%.

The distribution of bone in the camel carcass is comparable to that in cattle with a few differences (Table 9.5). The proportion of the vertebral column is greater in cattle carcasses than in camels. The proportions of the forelimb in the bone carcass are, however, greater in camels than in cattle (Table 9.5). This is congruent with previously mentioned higher proportions of

Table 9.4. Weight of bone and percentage of bone in the total bone of Omani camels (Mahgoub *et al.*, 2012).

Parameter	Mean	SD	Maximum	Minimum
Slaughter weight (kg)	257	32.28	322	218
Carcass weight (kg)	121	21.26	169	95
Total vertebral column	22	1.64	24.5	19.7
Ribs	9	0.82	10	7
Sternum	7	1.14	9	5
Axial skeleton	45	2.61	48	39
Scapula	5	0.47	6	5
Humerus	10	0.52	11	10
Radius and ulna	10	0.50	10	9
Carpus	2	0.58	2	1
Forelimb	27	1.02	29	26
Pelvis	6	0.38	6	5
Femur	10	0.74	12	10
Tibia	8	0.78	9	6
Patella	1	0.08	1	1
Tarsus	3	0.56	4	2
Hindlimb	28	1.88	32	26
Forequarter	62.3	2.11	63.2	64.7
Hindquarter	37.7	2.11	36.8	41.1

Table 9.5. A comparison of camel and beef cattle carcass bone contents.

	Omani camel		Omani Dhofari cattle	
Parameter	Mean	SD	Mean	SE
Slaughter weight (kg)	257	32.28	210	1.77
Carcass weight (kg)	121	21.26	115.9	1.64
Bone (% of total carcass bone)				
Total vertebral column	22.5	1.63	26.8	0.08
Ribs	9	0.82	14	0.04
Sternum	7	1.14	7.9	0.18
Axial skeleton	45	2.61	51.1[a]	0.61
Scapula	5	0.47	4.5	0.14
Humerus	10	0.52	8.8	0.19
Radius and ulna	10	0.50	6.7	0.16
Carpus	2	0.58	1.2	0.09
Forelimb	27	1.02	21.6	0.40
Pelvis	6	0.38	2.4	0.30
Femur	10	0.74	11.5	0.20
Tibia	8	0.78	7.7	0.15
Patella	1	0.08	0.76	0.04
Tarsus	3	0.56	3.51	0.16
Hindlimb	28	1.88	23.5	0.36

[a]Including the pelvis.

muscle in the forequarter; bone and muscle growth are linked because they stimulate one another (Mahgoub, 1988). The proportion that the front long bones of the limbs contributes to the bone carcass is higher in camels than in cattle, as is the proportion of the hindlimb. This difference is attributed to the heavier pelvis in camels.

Table 9.6. A comparison of the linear dimensions of bones in Omani camels and cattle (Mahgoub *et al.* 2012).

Parameter	Omani camel		Omani Dhofari cattle	
	Mean	SD	Mean	SE
Slaughter weight (kg)	257	32.28	210	1.77
Carcass weight (kg)	121	21.26	115.9	1.64
Humerus length (cm)	34.0	0.96	24.2	0.34
Radius and ulna length (cm)	41.7	0.96	30.0	0.37
Femur length (cm)	44.4	2.37	29.3	0.33
Tibia length (cm)	51.9	1.05	28.4	0.35
Humerus diameter (cm)	39.0	3.14	35.8	1.2
Radius and ulna diameter (mm)	35.8	4.93	40.5	0.6
Femur diameter (mm)	33.6	1.99	33.0	0.6
Tibia diameter (mm)	41.3	4.70	33.5	0.8
Humerus circumference (mm)	13.6	0.82	9.9	0.24
Radius and ulna circumference (mm)	11.4	0.52	10.3	0.16
Femur circumference (mm)	11.0	0.61	9.5	0.20
Tibia circumference (mm)	12.2	0.84	9.5	0.19

9.6 Bone Linear Measurements

The proportions of muscle in the camel carcass are comparable with those of cattle but the proportions of bone are higher in the camel carcass (Kadim *et al.*, 2008), resulting in a lower muscle-to-bone ratio in camels than in cattle (Babiker, 1984). This difference is attributed to the longer bones of the camel. Camels have longer limb bones, which include the humerus, radio-ulna, femur and tibia (Mahgoub *et al.*, 2012), than those of Omani Dhofari cattle of 210 kg body weight (Mahgoub *et al.*, 1995). The diameter of the long bones is, however, comparable in camels and cattle (Table 9.6). The tibia is the longest and thickest bone in the carcass followed by the femur (Table 9.6).

9.7 Conclusion

This chapter discusses the distribution and partitioning of tissues in the camel carcass. The major tissues in the carcass are muscle, fat and bone. Very little work had been published on carcass tissue distribution in the camel compared with that in beef, sheep and goats. This is a very important aspect because it affects carcass and meat quality. The shape of the camel body and carcass suggest a unique distribution of carcass tissue. The camel is characterized by a long neck and well-developed forequarter. The limbs have to be equally developed to support the extra weight of the neck and hump. Some data were obtained from Omani camels raised under intensive management and slaughtered at 218–322 kg body weight. Camel carcass muscle distribution was found to be comparable to those in other farm animal species, with a pattern of more muscle in the forequarter. This results in high proportions of muscle in the expensive muscle group (proximal hind and lower legs plus muscles around the vertebral column). The camel carcass contains 24.1% bone. The forequarter bone content (62%) is much higher than that of the hindquarter (38%). The axial skeleton, forelimb and hindlimb constituted 45%, 27% and 28% of the total carcass bone. Bone distribution in the camel carcass might affect carcass conformation and quality.

References

Babiker, M.M. (1984) Abundance and economic potential of camels in the Sudan. *Journal of Arid Environment* 7, 377.

Butterfield, R.M. (1988) New concepts of sheep growth. The Department of Veterinary Anatomy, University of Sydney, Australia.

Butterfield, R.M. and May, N.D.S. (1966) *Muscles of the Ox*. University of Queensland Press, Australia. pp 164.

Charles, D.D. and Johnson, E.R. (1976) Muscle weight distribution in four breeds of cattle with reference to individual muscles, anatomical groups and wholesale cuts. *Journal of Agricultural Science, Cambridge*, 86, 435–442.

Congiu, S. (1953) Dressing percentage, weight distribution and correlation between different parts of the body in Somali dromedary. *Zootecnica Veterinaria* 8, 188–191.

Elgasim and El-Hag, G.A. (1992) Characteristics of the Arabian camel. *Camel Newsletter* 9, 20–24.

Eltahir, I.E., Mohammed, A.M. and El Khidir, O.A. (2012) Feedlot performance and carcass characteristics on one-humped camel fed molasses and sorghum grain diets. Directorate of Animal Research, Ministry of Animal Wealth, Sudan.

Kadim, I.T., Mahgoub, O. and Purchas, R.W. (2008) A review of the growth and carcass and meat quality characteristics of the one-humped camel (*Camelus dromedarius*). *Meat Science* 80, 555–569.

Kulaeva, V. (1964) The production of the Bactrian camel. *Animal Breeding Abstracts*, 32.

Kurtu, M.Y. (2004) An assessment of the productivity for meat and carcass yield of camel (*Camelus dromedarius*) and the consumption of camel meat in the Eastern region of Ethiopia. *Tropical Animal Health and Production* 36, 65–76.

Mahgoub, O. (1988) Studies in normal and manipulated growth of sheep with special reference to skeletal growth. PhD thesis, Lincoln College, University of Canterbury, New Zealand.

Mahgoub, O. and Lodge, G.A. (1998) A comparative study on growth, body composition and carcass tissue distribution in Omani sheep and goats. *The Journal of Agricultural Science, Cambridge* 131, 329–340.

Mahgoub, O., Olvey, F.H. and Jeffrey, D.C. (1995) Growth and composition of the Omani Dhofari cattle. 2. Distribution of carcass tissues. *Asian-Australasian Journal of Animal Sciences* 8, 617.

Mahgoub, O., Kadim, I.T., Al-Marzooqi, W., Al-Lawatia S.A. and Al-Abri A.S. (2012) Performance of Omani camels raised under stall-feeding management. *East African Agricultural and Forestry Journal* (in press).

10 Structure and Quality of Camel Meat

Isam T. Kadim and Osman Mahgoub
Department of Animal and Veterinary Sciences, College of Agricultural and Marine Sciences, Sultan Qaboos University, Muscat, Sultanate of Oman

10.1 Introduction

Camel body mass contains around 70% skeletal muscle, which is responsible for posture and movement. Meat is an excellent source of many nutrients, especially amino acids, B vitamins, iron and zinc. The structure and metabolism of camel muscle is the fundamental basis for understanding the transformation of muscle to meat (Geay *et al.*, 2001). Muscle has many biological functions such as contraction, deposition of proteins and protection (Hocquette *et al.*, 1988). Muscle structure is an important factor to consider when discussing meat quality, which has been recognized as one of the most important social and economic challenges for meat producers. Muscle physiology also plays an important role in the post-mortem conversion of muscle to meat (Hwang *et al.*, 2004).

Slaughtering of animals is accompanied by a discontinued supply of oxygen to skeletal muscle cells, which stops normal cell processes. Many of the biochemical reactions present in the muscle living cells retain some degree of activity in the non-living cells. Rigor mortis is a biochemical reaction that is responsible for profound quality changes during post-mortem and storage. The rate and extent of muscle post-mortem metabolism is dependent on the availability of glycogen in the skeletal muscles at slaughter (Janz *et al.*, 2001), the temperature of storage (Newbold, 1996) and procedures to accelerate metabolic reactions (Kadim *et al.*, 2009a,b). Initially, muscles become stiff and hard for the first 24–48 h post-mortem but gain some softness after hanging and conditioning (ageing).

Although, camel carcasses may provide ample quantities of quality meat, the meat is commonly perceived as tough, coarse, watery and sweetish in taste compared with meats from other animals (Kadim *et al.*, 2008). This may be partly attributed to the fact that camel meat is usually a by-product of traditional production systems. Meat is mainly obtained from old animals that have become less effective in their primary roles of providing transportation, milk or as breeding females. However, the quality of camel meat produced by younger animals was comparable to beef (Kadim *et al.*, 2010). With increasing age, there is an increase in meat toughness with meat becoming less palatable and of inferior quality (Kadim *et al.*, 2006). In general, camel owners are reluctant to sell their young stock. Although, the potential of camel as a meat source has received recognition, data on the composition and quality characteristics are not widely published (Kamoun, 1995a,b ; Kadim *et al.*, 2006, 2008, 2009a,b).

This chapter summarizes the recent advances in our understanding of muscle structure, physiology, metabolism and their effects on camel meat quality characteristics. The first part of this chapter deals with the structure, and muscle physiology. The second part addresses the development of rigor mortis and the relationships between post-rigor changes and meat quality traits. The chapter also highlights the general meat quality characteristics of the dromedary camel.

10.2 Muscle Structure, Physiology and Biochemistry

According to Roberts *et al.* (2000), the skeletal muscle is made up of numerous longitudinal muscle fibres bound together in groups by connective tissues through which nerves and blood vessels run (Fig. 10.1). Camel skeletal muscle is composed of subunits (fascicles), which are bundles of adhesive contractile individual muscle fibre that makes up muscle tissue. The muscle fibre diameter varies from 10 to 100 μm but is dependent on the health status of the animal, as well as the breed, sex, age and plane of nutrition (Lawrie, 2006). The skeletal muscle fibre is made up of thousands of smaller myofibrils packed together lengthwise that are composed of thousands of protein filaments (Colville *et al.*, 2002). Myofibrils are made up of two protein filaments: thick and thin filaments that are arranged in units called sarcomeres (Fig. 10.1). Each sarcomere consists of one set of thick (myosin) filaments and two set of thin (actin) filaments (Helliwell, 1999). Myosin is the most abundant muscle protein and it has the ability to convert chemical energy into mechanical energy through structural change. Troponin and tropomyosin are the major proteins of the thin filament in myofibrils. They are calcium-activated complexes and are crucial for muscle structure. They are essential components of the regulatory machinery of muscle contraction (Clark *et al.*, 2002). Skeletal muscle is therefore responsible for producing movement of the animal body by elucidating contractility. Skeletal muscles fibres have many nuclei located beneath the sarcolemma (cell membrane), which surrounds the sarcoplasm, or cytoplasm of the muscle fibre (Colville *et al.*, 2002). The fibre has a characteristic transmembrane potential owing to the distribution of positive and negative charges across the cell membrane (Martini, 2001). The skeletal muscle fibre contracts by a signal conducted through the transverse tubules. Transverse tubules are narrow tubes that are continuous with the sarcolemma and extend into the sarcoplasm at right angles to the cell surface. The transverse tubules have the same general properties as the sarcolemma and electrical impulses conducted by the sarcolemma travel along the transverse tubules. These impulses are the trigger for muscle fibre contraction (Martini, 2001).

Camel muscle structure is similar to those of other farm animals. The muscle structure is composed of muscle fibres and intramuscular connective tissue (Warris, 2000). The connective tissue encompasses collagen and elastin fibres. Each skeletal muscle fibre is surrounded by a collagen membrane that is synthesized by muscle fibres, endomysial fibroblasts and endothelial cells (Abrahamson, 1986). The forces of the skeletal muscle contraction are transferred to bones through the collagenous structures of tendons and fasciae (Helliwell, 1999). Each muscle is surrounded in a layer of thick epimysium connective tissue. The bundle of muscle fibre is surrounded by a thinner perimysium connective tissue, whereas individual muscle fibres are covered by delicate endomysium connective tissue (Leeson *et al.*, 1985). The three types of connective tissues contain different types of collagen: epimysium is type one collagen, perimysium includes types one and three collagen, whereas endomysium contains types three, four and five collagen. Types four and five collagens are associated with the membranes of the muscle fibres (Duance *et al.*, 1980). The arrangement of myofibres, myofibrils and myofilaments generates the appearance and texture of meat (Swatland, 1984).

Electron microscopy is a powerful tool for providing visual information on meat structure and appearance. Under the electron microscope, each myofibril is divided

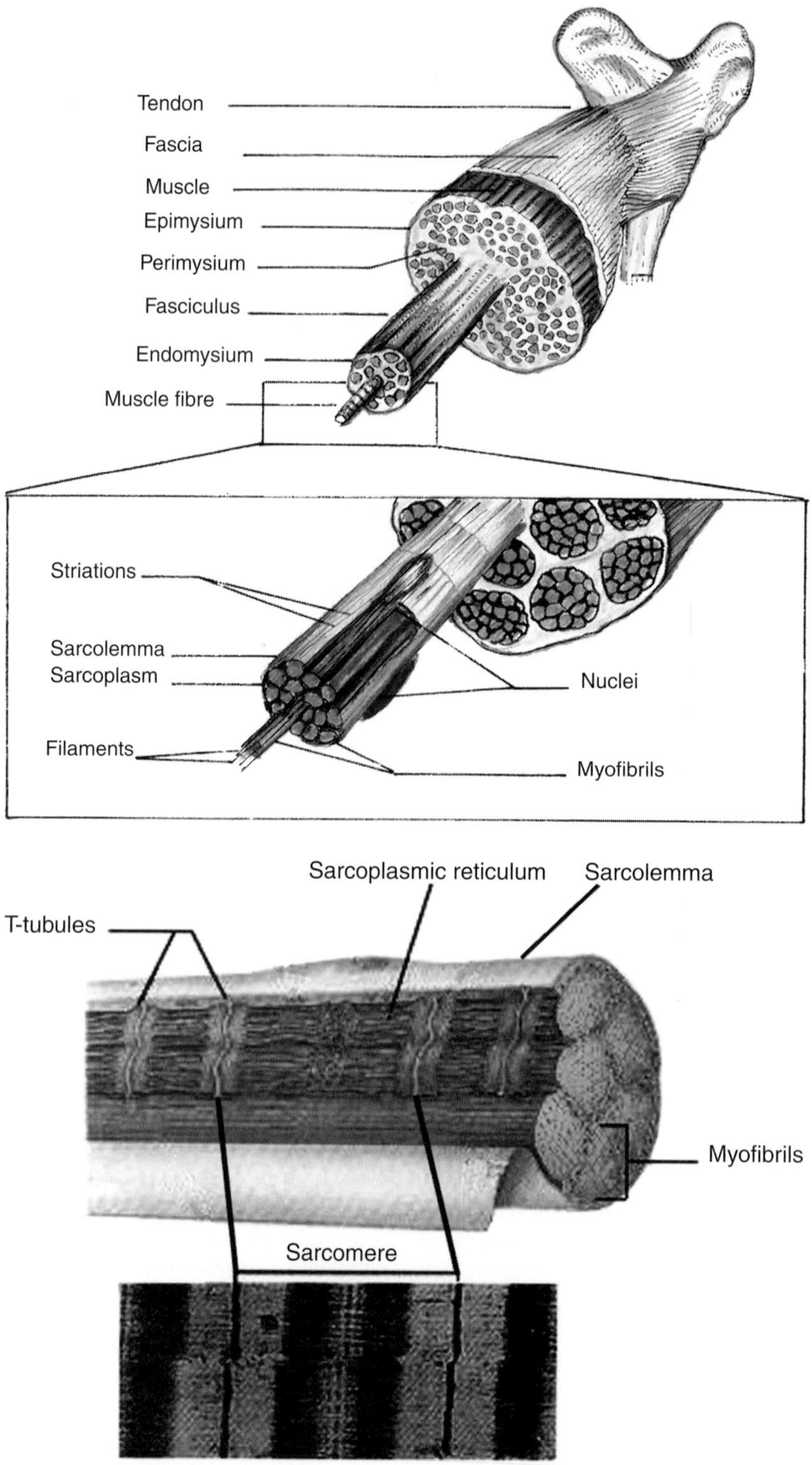

Fig. 10.1. Microstructure of animal muscles.

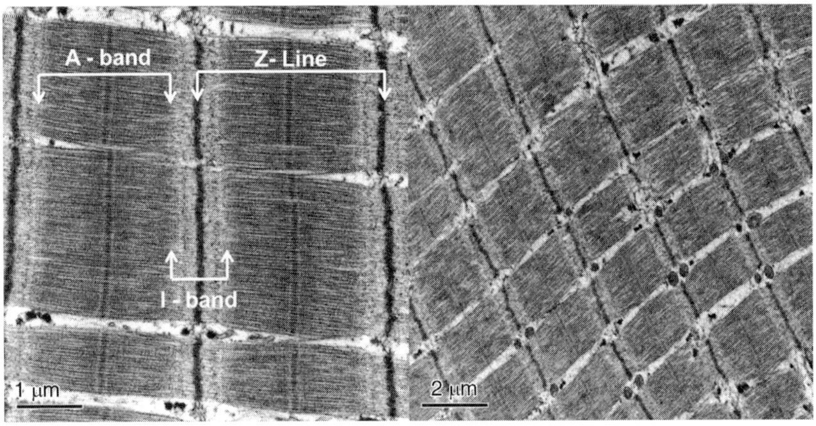

Fig. 10.2. Long-section of a myofibril showing sarcomeres, Z lines, A bands and I bands in *Longissimus thoracis* camel muscle.

up into light bands (I bands) and dark bands (A bands). The dark band contains a light region in the middle (H zone) and in the middle of the light band is a darker area (Z zone). The distance from the Z line to the next Z line is called a sarcomere and is the basic contracting unit of the skeletal muscle (Colville *et al.*, 2002). Using the electronic microscope to examine the ultra-structural features of dromedary camel meat (*Longissimus thoracis*) revealed that camel muscles displayed all the major ultra-structural features, i.e. Z line, A band, H band and M line (Fig. 10.2).

10.3 Muscle Fibre Types

Skeletal muscles contain fibres with various contractile and metabolic characteristics (Saltin *et al.*, 1994). Skeletal muscle fibres can be distinguished by their oxidative capacity using the enzyme succinate dehydrogenase (SDH), adenosine 5-triphosphate substrate (ATPase) and glycolytic capacity by using enzyme NADH-tetrazolium reductase (Brooke and Kaiser, 1970; Rahelic and Puac, 1981; Baker and Santer, 1990). Many animal species have been thoroughly studied, including various breeds of pigs, cattle, sheep and goat. Little data is available on camel energy-related enzyme activities and substrate levels within muscle. Studies with dromedary camels revealed that their muscle is composed of three fundamental fibre types categorized with biochemical and physiological methods, namely slow-twitch oxidative, fast-twitch oxidative glycolytic and fast-twitch glycolytic fibres (Kassem *et al.*, 2004; Kadim *et al.*, 2009a,b; Al-Kharusi, 2011; Fig. 10.3). There are three ways to describe the skeletal muscle fibre types. Brooke and Kaiser (1970) classified them as types I, IIA and IIB, Ashmore *et al.* (1972) used type βR (red), αR (intermediate) and αW (white), and types IR, IIR and IIW were used by Khan (1976) or types I, IIA, IIX, IIB using antibody and mRNA staining.

Muscle fibre type I, or slow-twitch fibres or βR, produce energy for ATP re-synthesis by aerobic energy transfer. This fibre type has a low myosin ATPase activity level and a less developed glycolytic capacity than fast-twitch fibres. The contraction speed of the fast-twitch fibre is approximately three times faster than the slow-twitch fibre (Schiaffino and Reggiani, 1996). Fibre type I contains relatively large and numerous mitochondria, myoglobin, lipid and iron with lower amounts of glycogen and glucose than type IIB fibres (Hintz *et al.*, 1984). Muscle fibres IIB (fast-twitch fibres) use glucose as fuel and have a more developed sarcoplasmic reticulum and T-tubule system. Muscle type IIB fibres rapidly transfer energy for muscle contractions (Stienen

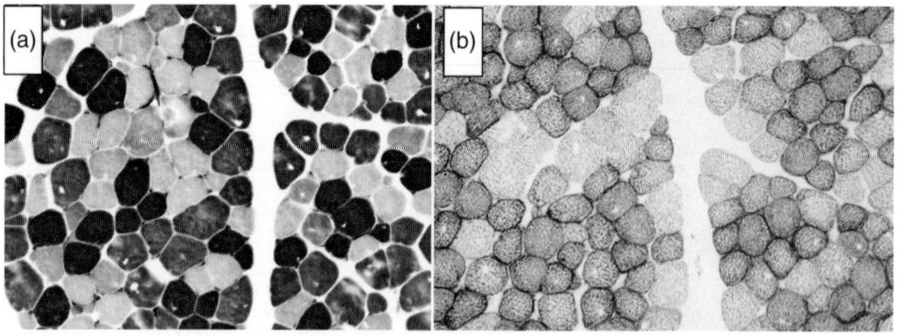

Fig. 10.3. Type I, IIA and IIB fibres. (a) Myosin ATPase of *Longissimus thoracis* muscle of dromedary camel at pH 4.35 and (b) succinate dehydrogenase enzyme activity associated with oxidative phosphorylation. Magnification 100 ×. (Al-Kharusi, 2011.)

et al., 1996). Type IIA (IIX, αR) muscle fibres are similar with a slower speed of contraction than type IIB fibres (Schiaffino and Reggiani, 1996).

Variations in the relative occurrence of type I (slow-twitch), IIA (fast-twitch) and IIB (fast-twitch) fibres between camel muscles were reported by Al-Kharusi (2011), Kadim *et al.* (2009a,b), Kassem *et al.* (2004) and Saltin *et al.* (1994). In the study of Saltin *et al.* (1994), the *Gluteus medius* muscle had a clear predominance of muscle fibre type I (73.6%), whereas the *Semitendinosus* had only 19.4% I fibres (Table 10.1). Camel *Supraspinatus* muscle contained an average of 93.6% type I fibres, whereas the *Biceps brachii* had 35.9% type I fibres. In contrast, Al-Kharusi (2011) found that camel *Supraspinatus* had 34.4% type I (Table 10.2). The enzyme activities as well as substrate levels were high in all muscles of the dromedary camel.

The type IIA muscle fibre constituted the dominant proportion of the fibres in all the muscles studied by Saltin *et al.* (1994), because the type IIB fibres were scarce in most of the muscles. In the study of Al-Kharusi (2011), no dominant fibre types in six camel muscles were reported. Similar results in the camel were reported by Kadim *et al.* (2009a,b) for *Longissimus thoracis* muscles. Muscle fibre type IIB in the *Biceps brachii*, *Supraspinatus* and the *Gluteus medius* muscles were very rare (1%). Saltin *et al.* (1994) reported that the proportion of muscle fibre type IIB in the *Biceps brachii*, *Semitendinosus* and *Vastus lateralis* range from 10 to 21%. Compared with horse and donkey (Snow and Guy, 1981, Snow and Harris, 1985), the variation between muscles in the camel is quite substantial. Differences between the findings of Saltin *et al.* (1994) on the one hand, and Kadim *et al.* (2009a,b) and Al-Kharusi (2011) and Kassem *et al.* (2004) on the other hand might be attributed to the heterogeneity of dromedary camels because there are no pure camel breeds as in other species. The sampling technique is another factor that might explain the differences. For example, there may be some inconsistency in measurements in the study of Saltin *et al.* (1994), when muscle fibre composition is based on a small tissue sample (biopsy). In a homogeneous distribution, muscles of the various fibre types do not cause a problem with sampling (Ariano *et al.*, 1973). In muscles of camels, fibre types are heterogeneously mixed (Henriksen-Larsen *et al.*, 1983, Saltin *et al.*, 1994). The high coefficient of variation values for type I, IIA and IIB fibres relate to a heterogeneous distribution of muscle fibre types in camel (Table 10.3). The distribution of muscle fibre types in the *Longissimus dorsi* muscle of two camel types indicated that the black camel has slightly more type IIB, less type I and less type IIA fibres than the red camel (Kassem *et al.*, 2004).

Table 10.1. Muscle composition in different locomotor muscles in the racing camel (Saltin *et al.*, 1994).

	Muscle[a] ($N = 6$, mean ± standard error)							
Measurement	DB	TB	BB	SI	GM	ST	BF	VL
Proportion (%)								
Type I	49.3 ± 5.0	35.9 ± 3.7	66.6 ± 16.6	93.6 ± 3.2	73.6 ± 2.9	19.4 ± 2.6	50.8 ± 3.6	27.6 ± 2.6
Type IIA	41.2 ± 5.0	44.9 ± 4.4	32.4 ± 17.6	4.9 ± 2.9	25.6 ± 2.6	63.0 ± 6.6	41.4 ± 2.7	50.8 ± 3.1
Type IIB	9.5 ± 6.0	19.2 ± 3.7	1.1 ± 1.1	1.2 ± 0.9	0.8 ± 0.8	17.6 ± 5.1	7.8 ± 4.0	21.5 ± 2.1
Area (μm^2)								
Type I	4806 ± 241	5100 ± 563	4108 ± 44	7283 ± 643	5817 ± 140	4604 ± 391	4287 ± 339	4572 ± 722
Type IIA	7179 ± 933	6170 ± 1112	3888 ± 838	8099 ± 1198	6061 ± 585	9396 ± 1082	4938 ± 388	7232 ± 803
Type IIB	8256 ± 1333	7821 ± 1006		5364 ± 668		10708 ± 1180	6379 ± 504	7848 ± 535

[a]Muscle: DB, *Deltoid brachii*; TB, *Triceps brachii*; BB, *Biceps brachii*; SI, *Supraspinatus*; GM, *Gluteus medius*; ST, *Semitendinosus*; BF, *Biceps femoris*; VL, *Vastus lateralis*.

Table 10.2. Muscle composition in six locomotor muscles in the dromedary camel (Al-Kharusi, 2011).

	Muscle[a] (*N* = 10, mean ± standard error)					
Measurement	SI	TB	LT	ST	SM	BF
Proportion%						
Type I	33.9 ± 2.29	32.3 ± 1.99	32.2 ± 2.01	37.0 ± 3.15	30.1 ± 1.76	33.4 ± 2.33
Type IIA	31.2 ± 1.8	32.4 ± 2.21	32.6 ± 2.54	30.5 ± 2.91	36.4 ± 3.31	35.5 ± 2.65
Type IIB	34.9 ± 2.17	35.3 ± 3.10	35.2 ± 3.31	32.5 ± 2.54	33.5 ± 2.21	31.1 ± 2.22
Area (μm^2)						
Type I	5393 ± 437	6588 ± 399	7487 ± 410	6357 ± 321	4451 ± 334	6878 ± 445
Type IIA	5892 ± 460	7157 ± 471	8165 ± 411	7640 ± 367	5502 ± 291	7189 ± 411
Type IIB	6392 ± 317	7671 ± 482	8649 ± 448	8031 ± 417	6455 ± 357	7756 ± 391

[a]Muscle: SI, *Supraspinatus*; TB, *Triceps brachii*; LT, *Longissimus thoracis*; ST, *Semitendinosus*; SM, *Semimembranosus*; BF, *Biceps femoris*.

Table 10.3. Coefficients of variation for different measurements on the *Gluteus medius* (Saltin *et al.*, 1994).

Variable	Coefficient of variation (%)
Type I fibres (%)	14.8 ± 8.5
Type IIA fibres (%)	52.9 ± 25.9
Type IIB fibres (%)	83.0 ± 79.0
Area – type I fibres	15.9 ± 2.4
Area – type IIA fibres	24.7 ± 8.7

N = 6, mean ± standard error.

The proportions of Type I, IIA and IIB were 33.1, 25.2 and 41.2%, respectively, in the dromedary camel and 33.9, 25.4 and 40.8%, respectively, in beef *Longissimus thoracis* muscles (Kadim *et al.*, 2009a).

In muscles of horses, a close correlation has been found between type I fibres and oxidative enzymes with an inverse relationship between the percentages of type I fibres and the activity of glycolytic enzymes (Essen-Gustavsson, 1986; Hoppeler, 1986). Such relationships may also be found in muscles of camel, but they are less apparent. Eight muscles in camels were examined by Saltin *et al.* (1994) who found no relationship with an r-value above 0.5–0.6. The lack of a relationship between fibre types and enzyme levels of camel muscles may be due to functions of the performance demands on muscles. In horses, there are a number of type II fibres in all locomotor muscles, which may be required to move fast. In the racing camel, Saltin *et al.* (1994) found that the *Biceps femoris, Deltoid brachii* and *Triceps brachii* muscles have similar proportions of type I and II fibres, whereas the *Semitendinosus* and *Vastus lateralis* muscles have a predominance of type II fibres. The high proportion of type I fibres might explain the camel's survival capacity at high levels of exercise. Glycogen depletion in the skeletal muscles provides an insight into muscle-fibre recruitment during exercise (Rose *et al.*, 1994). The glycogen depletion patterns of type I fibres indicates that the *Gluteus medius* muscle is actively involved in locomotion at a moderate exercise intensity (Saltin *et al.*, 1994). Although there was some variation between individual camels, the overall pattern was distinctive. Kiessling and Kiessling (1984) reported that reindeer muscle contained high proportions of type II fibres that atrophy during winter when feed supply is limited. They suggested that type II fibres serve as an energy store during starvation. In this respect, the dromedary camel could also have a large proportion of type II fibres because it can survive during starvation periods. The camel has other means to handle long periods of feed deprivation.

The area of different muscle fibre types varied between muscles within the same breed of camel and between different muscle fibre types. In the dromedary camel skeletal muscles, the type I fibre is the smallest in size with type IIA in the middle and type IIB the largest (Al-Kharusi, 2011; Kadim *et al.*, 2009a,b). Saltin *et al.* (1994) found that the *Supraspinatus* muscle had the smallest type IIA fibre, whereas the type IIA fibre was smaller than the type I in the *Biceps brachii*. The largest sizes of the fibre types were in the *Supraspinatus* and *Semitendinosus* muscles. The type IIB fibre was the largest fibre in the *Semitendinosus* muscle, whereas the same fibre type was the smallest in the *Supraspinatus* muscle (Table 10.1). Although there is a distinct variation in camel type I fibre proportion, there are some differences in fibre size when substrate and enzyme levels are similar (Saltin *et al.*, 1994). This could result in small differences in metabolic profiles of the various muscle fibres found in camel, which could be due to glycogen rather than enzyme activities. On the basis of staining with SDH, the difference in staining intensity is quite marked between type I and type II fibres, with less differences between the subgroups of the type II fibres. This is apparent from the fibres depicted in Fig. 10.3, where type I fibres are intensely stained and type II fibres are pale in comparison.

Camels have larger mean muscle fibre areas than beef, goats, sheep, dogs or horses (Snow and Harris, 1985; Essen-Gustavsson, 1986; Kadim *et al.*, 2009a,c, 2010). The significantly larger size of muscle fibre in camels than in beef is most probably related to animal size (141 kg compared with 304 kg carcass weight for cattle and camel, respectively) (Kadim *et al.*, 2009a). This is mainly due to the considerable size in many muscle fibre types. Female and male camels seem to have quite similar fibre type sizes. Saltin *et al.* (1994) compared muscle fibre sizes between breeding camels and racing camels and found that the racing camel had a larger muscle fibre area. They concluded that the breeding camels may have smaller muscle fibre areas because they had been out of training for some time (>1 year).

Skeletal muscle fibre appearance is affected by neuromuscular junction activity, exercise, stress, hormones and ageing (Pette and Staron, 1990). The quality characteristics of meat are therefore determined by the muscle structure, fibre types, architecture, connective tissue and ageing. The effect of muscle fibre type on meat quality of camels might be due to the muscle fibre size. The larger the size of muscle fibre is then the tougher the meat. Type II fibres (fast-contracting fibres) with a glycolytic metabolism are larger than type I (slow and oxidative red fibres). The type I fibres are rich in lipids and red in colour, therefore contributing to taste and colour quality, which are also related to metabolic differences. Camel meat quality parameters such as pH, water-holding capacity, colour and tenderness are influenced by muscle fibre type. The relationship between meat quality and proportion of various muscle fibre types was reported by Calkin *et al.* (1981). When the proportion of α-white fibre increases in muscles, there will be more connective tissue, less intramuscular fat and less tenderness than muscles with more β-red fibres.

10.4 Rigor Mortis

Camel muscle will be converted to meat by going through rigor mortis after slaughter. This involves physiological, biophysical and biochemical changes that are influenced by muscle temperature and pH. Although the exact point of conversion of muscle to meat is not easy to determine, the metabolic muscle activity of the skeletal muscle would not stop. In this respect, biochemical components necessary for anaerobic metabolism in muscle cells are present and functional, and consequently glycolysis proceeds depending on the glycogen level and temperature (Warris *et al.*, 1987). Concentrations of glycolytic substrates will keep changing until the reaction in the glycolytic process is constrained. Metabolism then ceases and the ability to produce adenosine triphosphate (ATP) is completely lost and rigor mortis is established. At this stage, a greater percentage of myosin heads remain attached to actin. Rigor mortis is a stiffening of muscle occurring during the post-mortem glycolysis when all supplies of energy are exhausted (Honikel *et al.*, 1983). This does not occur across all muscles simultaneously; with a concomitant fall in pH for single fibres, there is a contracture as the final ATP disappears and each fibre has its own time course depending on initial glycogen levels. The myofibril myosin and actin molecules remain locked together when ATP is exhausted and yield the stiff nature of muscle in rigor. The development of rigor mortis has been evaluated by several methods including loss of extensibility, muscle shortening (Honikel *et al.*, 1983), tension development (Nuss and Wolfe, 1981), resistance to strain and by a combination of muscle tension and shortening (Olsson *et al.*, 1994).

Changes in concentrations of hydrogen ion, acid-labile phosphorus, creatine phosphate and extensibility from slaughtering time until the onset of rigor mortis are illustrated in Fig. 10.4. The concentration of ATP does not start decreasing immediately after slaughter but remains at physiological levels for a brief period before declining because ATP can be generated from creatine phosphate during anaerobic glycolysis. In the first phase of post-mortem, the glycogen concentration and pH decreases and the lactate increases; however, ATP remains constant (Lawrie, 2006).

During rigor mortis, the production of H^+ leads to more acidic conditions resulting in a decrease in muscle pH. Fast glycolyzing muscles cause lower muscle pH values compared with slow glycolyzing muscles. Muscle protein denaturation and myofibrillar shrinkage is influenced by muscle temperature and declining pH during the onset of rigor and has an important influence on meat quality characteristics. At pH 6.2, muscle temperature has been used as an important threshold in meat quality because it could be an indirect indication of cold and heat shortening (Pearson and Young, 1989). In general, an increase of 10°C results in a doubling of the reaction rate. The temperature of the carcass at slaughter is 38–40°C and immediately after processing, the carcass is placed into a cooler at 4°C. Temperature had a greater influence on glycolytic reactions and the time course of rigor onset. Carcass fat acts as an insulator and can slow the rate of post-mortem temperature decline. In the camel, however, most of the carcass fat is accumulated in the hump and a thin layer of fat covers the remainder of the carcass. The rate of post-mortem temperature decline is therefore fast, which causes a slow decline of the rate of glycolysis. Various muscles within a

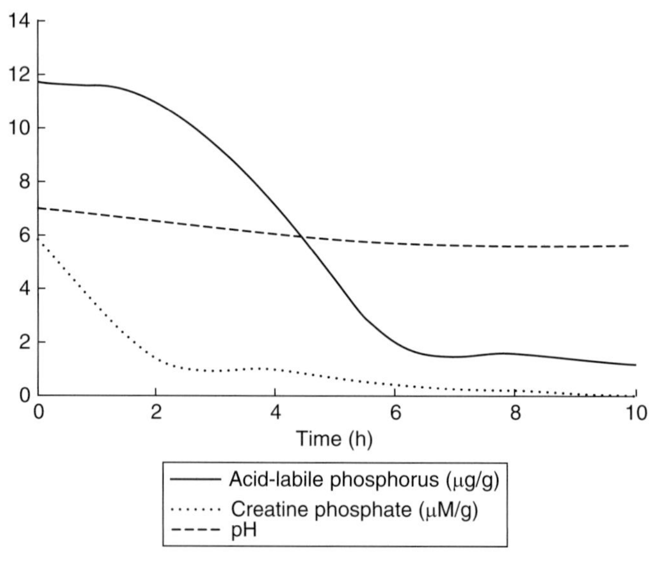

Fig. 10.4. Changes in biochemical metabolites during the onset of rigor mortis (Newbold, 1996).

given camel carcass will display different cooling rates on the basis of their location.

Pre-rigor muscle has a higher water-holding capacity and better fat emulsifying properties than post-rigor meat, which is more suitable for processed meats such as sausages (Hamm, 1981). The water content of meat can be maintained if pre-rigor meat is frozen quickly to temperatures below –20°C, but during thawing, meat will lose its water through thaw shortening. Adding 1.8% salt to pre-rigor meat helps to maintain the pre-rigor attributes for several days when chilled (Boles and Swan, 1996). With pre-rigor salting, the water-holding capacity is maintained because of a strong electrostatic repulsion between protein molecules caused by an initial combined effect of relatively high ATP concentration, high pH and ionic strength (Hamm, 1977). Adding salt pre-rigor inhibits ATP turnover but does not affect the rate of glycogen breakdown (Hamm, 1977).

Cold shortening is a phenomenon where meat is frozen prior to rigor onset resulting in the contractile apparatus of the muscle, the sarcomeres, shortening markedly. The pre-rigor freezing of meat might damage the sarcoplasmic reticulum and destroy its ability to regulate calcium concentrations within the myofibre. Calpains and myosin-ATPase enzymes require Ca^{2+} ions in the cytoplasm for their activities (Celio *et al.*, 1996). Cornforth *et al.* (1980) stated that calcium-reserving organelles lose their function at abnormal cellular temperatures. During the process of thawing, all the components necessary for muscle contraction exist, but control of the reactions is lost.

The degree of cross-bridges between myosin and actin filaments contributes to meat toughening (Tornberg, 1996). Changes in angles of criss-cross connective tissue and fold length may be responsible for the relationship between sarcomere length and meat tenderness (Renerre *et al.*, 1999). However, the toughness of cold-shortened meat is largely affected by an endogenous enzymatic tenderization mechanism rather than shortened sarcomere length (Hwang *et al.*, 2004). Hertzman *et al.* (1993) stated that sarcomere shortening alone does not cause meat toughness because heat-shortened sarcomeres have a limited effect on shear force. This suggests that there is a more direct cold shortening/toughening relationship in camel carcasses with less subcutaneous fat that are exposed to rapid chilling early post-mortem.

10.5 Electrical Stimulation

Electrical stimulation is the post-slaughter application of a specifically designed electric current to a carcass of freshly slaughtered animals to prevent muscle contraction during the onset of rigor mortis and accelerate post-mortem glycolysis. It should be in a series of short pulses, each of which causes the muscles to contract violently, but between pulses, the muscles return to their normal relaxed state (Hwang *et al.*, 2003). Electrical stimulation plays a positive role in preventing cold shortening when the muscle temperature drops below 10°C, while the muscle still contains sufficient energy to cause contraction (Simmons *et al.*, 2008). In general, carcasses must be chilled quickly following slaughter to minimize microbiological growth and to reduce weight loss. If the carcass is cooled too soon, however, before rigor mortis is achieved, muscle groups near the surface can become tough through exposure to the cold temperature (cold shortening). The lower the temperature is at which cold shortening is achieved, then the more severe the effect and consequent toughness. Electrical stimulation accelerates the decline in pH and the onset of rigor mortis (Simmons *et al.*, 2008). It causes muscle contractions that convert muscle glycogen to lactic acid. The acid build-up reduces the pH and the muscles enter rigor mortis. In this state the muscles have no further energy to contract and will not cause cold shortening when chilled to 12°C. If the pH decline is too rapid, muscles enter rigor mortis (pH 6) at above 30°C and cold shortening cannot occur (Simmons *et al.*, 2008). Effective electrical stimulation requires slaughter floor managers to ensure pH decline and chilling occur at appropriate rates. The correct integration of all these processes is essential to produce consistent results.

Two electrical stimulation systems are in use commercially by applying electrodes to different regions of the carcass to deliver an electric current that can take different forms; low (less than 150 V) and high (up to 1130 V), but also a wide range (2.5 to 9000 V) has been used at an experimental level (Petch, 2001). Both result in effective stimulation and both have advantages and disadvantages. The difference between the two electrical stimulation systems is the route of the stimulation. For high voltages, direct muscle stimulation is achieved, whereas for low voltages, the nervous system transfers the low voltage stimulation to muscles (Petch, 2001). Low-voltage systems have lower initial costs, are safer and result in extra blood yield, if applied as soon as possible after slaughter. High-voltage systems require greater initial outlay, primarily for safety reasons, but can be incorporated into a high-speed chain. It can be successfully applied up to 60 min after slaughter, giving greater flexibility with the setting of the electrical stimulation module.

The mechanism of electrical stimulation is to improve camel meat tenderness by sufficient muscular contraction to cause physical disruption of the myofibrillar matrix in muscles (Fig. 10.5; Kadim *et al.*, 2009a). Physical disruptions of the sarcomeres (Takahashi *et al.*, 1987) and accelerated proteolysis (Lee *et al.*, 2000) are the main

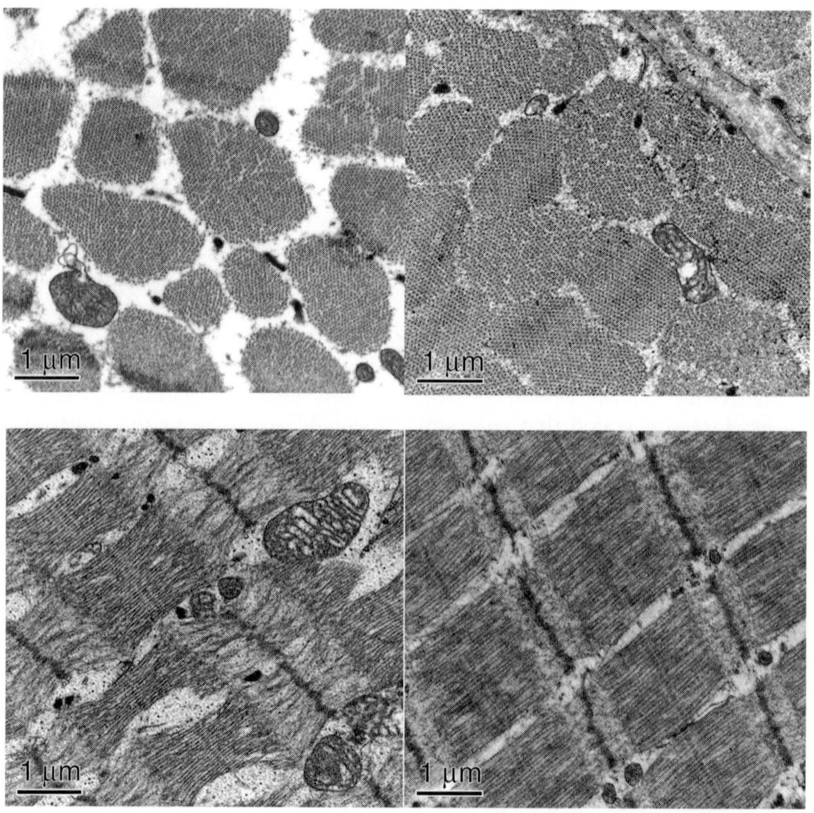

Fig. 10.5. Micrograph from sections of electrically stimulated and non-stimulated sections of camel *Longissimus thoracis* muscle (magnification 2500 ×) showing a region of super-contracture, swollen mitochondria, stretched sarcomere (bottom left) and pronounced transverse element (muscle fibre bundles distorted and disintegrated and spaces between fibres and disintegration at interfibrillar bridges; top left) (Kadim *et al.*, 2009a,b).

mechanisms of the effect of stimulation on meat tenderness.

Takahashi *et al.* (1987) investigated types of stimulation, formation of contracture bands and their relation to tenderness. They concluded that stimulation at 50–60 Hz with 500 volts, 40 min post-mortem resulted in severe structural alteration and improved tenderness, whereas 2 Hz failed to improve meat tenderness with no structural alteration. It seems that ultrastructural alteration takes place in stimulated muscles, which results in an improvement in meat tenderness either through its effects on physical alteration and/or acceleration of energy turnover during and after the treatment (Hwang *et al.*, 2003). Contracture bands are not a direct consequence of an electrical current passing through the muscle, but rather due to the supercontracture caused through localized excessive calcium ions released from the sarcoplasmic reticulum. It could be this extra calcium that also causes the tenderization to proceed. Kadim *et al.* (2009a,b; 2010) first reported that electrical stimulation resulted in ultrastructural changes in camel *Longissimus thoracis* muscles. Histological images showed the appearance of contracture bands containing predominantly stretched, ill-defined and disrupted sarcomeres similar to those illustrated (Fig. 10.5). This led to the hypothesis that physical disruption per se lowers the resistance to the mechanical shearing force. Similarly, other studies have advocated the link between physical disruption and improved tenderness for high voltage (300–500 V; Will *et al.*, 1980; Takahashi *et al.*, 1987) and for intermediate voltage (145–250 V; Sorinmade *et al.*, 1982; Ho *et al.*, 1996) systems. As in the other animal species, an improvement in dromedary camel meat quality does not result from low stimulation unless it markedly accelerates post-mortem glycolysis (Kadim *et al.*, 2009a). The positive effect of low voltage on the rate of *Longissimus thoracis* muscle pH decline in camels is caused by accelerating muscle glycogen degradation and thus increasing the concentration of lactic acid. Lactic acid is accumulated; therefore, the pH of stimulated muscle can reach a pH of 6.0 in several hours instead of the 12–16 h that may be required for non-stimulated muscles. Kadim *et al.* (2009a) showed that electrical stimulation consistently produced a more rapid glycolysis early post-mortem in muscle samples from four age groups (Fig. 10.6). The greatest pH fall, 40 min post-mortem, resulted from electrical stimulation applied 20 min after slaughter (Kadim *et al.*, 2009a,b), which was explained by the fact that 20 min after death, the camel's muscle is more responsive to stimulation. At 40 min post-mortem, the average pH decline values in stimulated camel muscle were 0.12 below the non-stimulated group (Kadim *et al.*, 2009a). After a relatively fast fall within the first 4 h, the mean pH values of camel muscles underwent a slow decline until the ultimate pH at 24 h post-mortem (Kadim *et al.*, 2009a,b).

Marketing of camel meat benefits from using low voltage electrical stimulation because of the improving meat-quality characteristics. Meat from electrically stimulated dromedary camel carcasses has a brighter colour and more desirable overall appearance than that from non-stimulated carcasses (Kadim *et al.*, 2009a,b). The desirable appearance might be due to early post-mortem conditions of stimulated muscles that favour protein denaturation (Warris and Brown, 1987). A combination of high muscle temperatures and low muscle pH are associated with increased protein denaturation. Ashmore *et al.* (1973) reported that low muscle glycogen stores at slaughter do not allow the development of a desirable pH (approximately 5.5) of the lean tissue after slaughter. Moreover, muscles from stimulated camel carcasses had a significantly lower shear force value compared with non-stimulated carcasses (Kadim *et al.*, 2009a,b).

The reason for improvement tenderness might be the physical alteration and/or acceleration of energy turnover during and after electrical stimulation (Kadim *et al.*, 2009a,b). Hopkins *et al.* (2002) stated that physical disruption lowers the resistance to mechanical shearing force and therefore increases tenderness. Kadim *et al.* (2009a,b) found that stimulated camel carcasses had significantly longer sarcomere length and

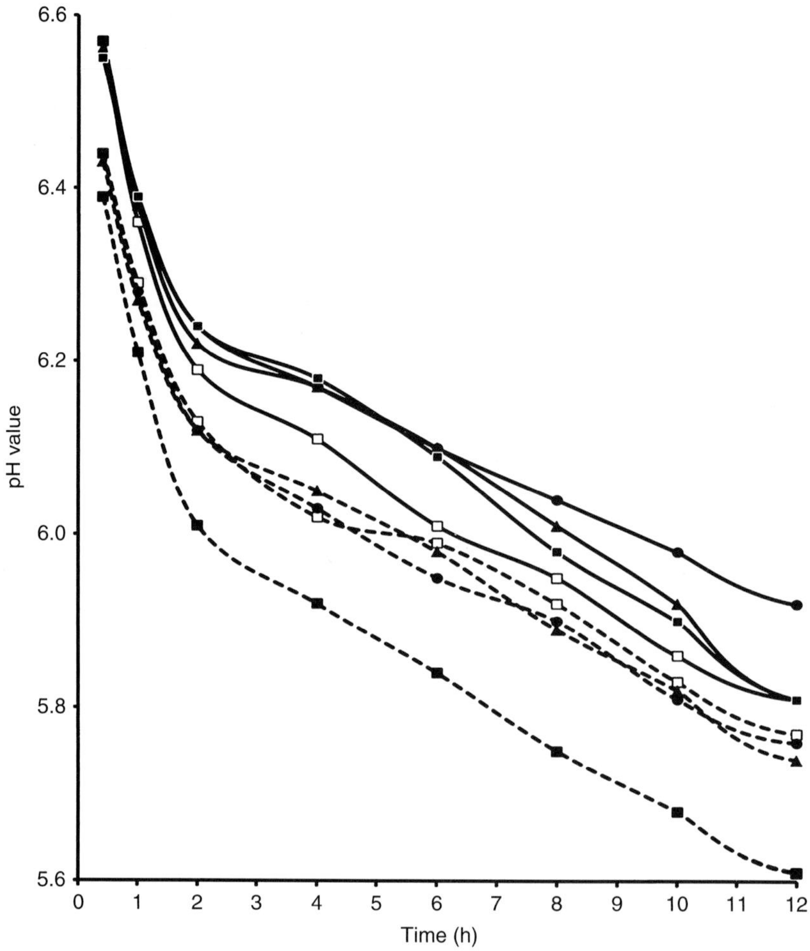

Fig. 10.6. Mean changes in pH within the *Longissimus thoracis* muscle for carcasses from four age groups (group 1: 1–3 years; group 2: 4–6 years; group 3: 7–9 years; and group 4: 10–12 years) of dromedary camels with electrical stimulation or control (group 1 stimulated: --□--, or control: –□–; group 2 stimulated: --•--, or control: –•–; group 3 stimulated: --▲--, or control: –▲–; group 4 stimulated: --■--, or control –■– (Kadim *et al.* 2009a).

less water-holding capacity than non-stimulated muscles. The difference between the stimulated and non-stimulated camel muscles is probably a result of shrinkage of the myofibrils caused by pH fall post-mortem and denaturation of protein (low pH and high temperature; Offer and Knight, 1988). Water-holding capacity is also affected by the integrity of the muscle cell membranes and the rate of fluid migration within the meat (Den Hertog-Meischke *et al.*, 1997).

Kadim *et al.* (2009a) reported that age had no impact on the rate of pH decline of carcasses (Fig. 10.6). However, carcasses from the 10–12 year age group had a faster pH decline than other stimulated groups within the following 48 min. After a relatively fast fall within the first 4 h, the mean pH values of all the carcasses underwent a slow decline until they reached the ultimate pH at 48 h post-mortem. Although pH declining followed a similar pattern from 1–4 h post-mortem, the average difference in pH between the stimulated and non-stimulated camel carcasses was from 0.12 to 0.39 units (Kadim *et al.*, 2009a). The difference in pH between electrically stimulated and non-stimulated carcasses decreased with time as the difference was 0.39 units at 4h and 0.19 units at 10h post-mortem. The overall rate of pH decline variation between the two treatments was higher in the 10–12 year age group and lower in the 1–3 year age group, whereas the values of the variation for the other age groups were in between these values. Variation in ultimate pH among the four age group muscles might be attributed to differences in proportions of muscle fibre types and, consequently, differences in patterns of energy metabolism during both ante-mortem and post-mortem (Swatland, 1982). Petersen and Blackmore (1982) found that the increased muscle activity in animals that have been electrically stimulated is probably responsible for more rapid post-mortem pH decline. Similarly, Vergara and Gallego (2000) also observed that in stimulated animals, pH decreases more rapidly and ageing starts earlier.

10.6 Meat Quality Characteristics

Many subjective and objective procedures for meat-quality evaluation have been developed to achieve the comprehensive assessment of quality attributes (Mullen *et al.*, 2002). The subjective evaluation depends on appearance (colour, shape and integrity), texture (tenderness, firmness, mouthful, bite and chewing ability) and flavour (taste and odour). These assessments require trained panels of judges to minimize subjectivity (Singh *et al.*, 1997). The objective evaluation depends on instruments to determine certain parameters related to quality characteristics. Meat quality parameters including ultimate pH, Warner–Bratzler shear force, sarcomere length, myofibrillar fragmentation index (MFI), water-holding capacity, cooking loss percentage and colour. These attributes in various muscles in the dromedary camel are summarized in Table 10.4. The quality of camel meat has received little attention; it was described as tough, coarse and varying in colour from raspberry red to brown red with white intramuscular fat (Kadim *et al.*, 2008). In general, camel meat is often labelled as inferior in urban societies, and its consumption considered fit only for low socioeconomic members of society. This might be attributed to the great reluctance of camel owners to sell their young stock and they are usually slaughtered at the end of their productive life. Most camel meat trade consists of meat from old camels of low quality, which has a direct bearing on the extent of demand for meat outside the camel-herding societies. Numerous studies have reported, however, that meat-quality characteristics from the young camel are comparable to those of beef (Leupold, 1968; Fischer, 1975; Knoess, 1977; Mukasa-Mugerwa, 1981; Kadim *et al.*, 2006, 2009b; Shariatmadari and Kadivar, 2006). Nevertheless, camel meat had a significantly lower level of sarcoplasmic proteins as a proportion of total proteins than beef in the study of Babiker and Tibin (1986).

Camels at 2–4 years old and beef at 2–3 years old had similar meat-quality characteristics of the *Longissimus thoracis* muscle (Kadim and Mahgoub, 2008). The camel *Longissimus thoracis*, *Semitendinosus* and *Triceps brachii* muscles have been reported to lose more water during cooking than beef (48% versus 37%), whereas no tenderness differences were observed between the two species (Kamoun, 1995a,b). In contrast, Babiker and Tibin (1986) reported that camel meat has less cooking losses and a higher water-holding capacity than beef meat. Table 10.5 depicts the effect of camel age on meat-quality parameters and shows that meat becomes less tender and of inferior quality with increasing animal age (Kadim *et al.*, 2006). Kamoun (1995a,b) noted, however, that age is not a predominant factor in meat quality, in the case of dromedaries fed the same diet and slaughtered between 1 and 4 years of age. Effect of age on meat quality is discussed in order to optimize the best age for slaughtering camels for high-quality meat. Kadim *et al.* (2006) suggested that the young males should be slaughtered between 1 and 3 years of age. This is in agreement with the conclusion of Dina and Klintegerg (1977). At this age the animals

Table 10.4. Meat-quality characteristics of various muscles of the dromedary camel.

Parameter	Al-Kharusi (2011)						Kadim *et al.* (2006)	Kadim *et al.* (2009a)	Kadim *et al.* (2009b)	Youssif & Babiker (1989)			Suliman *et al.* (2011)	
	Muscle						Muscle	Muscle	Muscle	Muscle			Muscle	
	IS	TB	LT	ST	SM	BF	LT	LT	LT	TB	LT	ST	LT	BF
Age (year)			2				3–5	1–3			5		1	
pH	5.72	5.67	5.61	5.77	5.78	5.74	5.84	5.85	5.79				5.90	5.76
WB-SF	6.3	6.7	6.5	9.0	12.9	10.3	8.11	8.10	6.83	5.8	4.8	5.7	11.3	22.2
SL (μm)	1.49	1.50	1.46	1.27	1.58	1.48	1.24	1.71	1.71				1.67	1.68
MFI (%)	73.2	72.8	70.0	75.2	79.3	65.3	73.3	73.5	75.8				67.6	62.0
WHC (cm^2/g)	34.8	42.1	41.8	36.8	42.4	40.2	27.4	37.2	35.8					
CL (%)	31.6	29.2	33.5	28.5	30.6	29.5	34.0	24.9	24.3	37	38	33	22.7	28.5
Colour														
*L**	31.7	29.2	33.5	28.5	30.6	29.6	34.0	39.8	39.2				31.9	29.7
*a**	12.7	12.6	14.0	10.5	13.6	13.3	13.8	15.7	17.4	15.9	17.2	13.8	13.1	13.4
*b**	2.57	3.74	4.07	2.18	2.90	3.77	3.78	5.51	6.16				4.51	4.03

Quality parameters: pH, ultimate pH; WB-SF, Warner–Bratzler shear-force value (kg); SL, sarcomere length (μm); MFI, myofibrillar fragmentation index (%); WHC, water-holding capacity (expressed juice); CL, cooking loss (%); colour, lightness (*L**); redness (*a**) and yellowness (*b**). Muscles: IS, *Infraspinatus*; TB, *Triceps brachii*; LT, *Longissimus thoraces*; ST, *Semitendinosus*; SM, *Semimembranosus*; BF, *Biceps femoris*.

Table 10.5. Effect of age on some meat-quality characteristics of the dromedary camel *Longissimus thoracis* muscle.

	Kadim *et al.* (2006)			Kadim *et al.* (2009b)	
	Age group (year)			Age group (year)	
Parameter	1–3	3–5	5–8	1–2	8–10
Ultimate pH	5.91	5.84	5.71	5.68	5.65
W-B shear force value (Newton)	68.4	79.5	131.9	6.74	8.90
Sarcomere length (μm)	1.85	1.24	1.06	1.66	1.60
Myofibrillar fragmentation index (%)	80.99	73.3	60.4	72.2	67.3
Expressed juice (cm²/g)	29.6	27.36	21.26	38.1	37.4
Cooking loss (%)	26.06	23.72	22.42	23.4	22.0
Colour parameters					
L^* (lightness)	37.74	34.03	31.69	39.1	38.1
a^* (redness)	13.37	13.82	16.18	16.5	15.6
b^* (yellowness)	6.09	6.78	7.26	5.58	6.29

were not yet fully grown; they averaged about 60–70% of full live weight and therefore their meat is tender.

Meat-quality parameters of four Indian camel breeds were compared by Suliman *et al.* (2011) using the *Longissimus thoracis* and *Biceps femoris* muscles (Table 10.6) and the results indicate little variation between the four breeds. The range of shear force values in *Longissimus thoracis* muscles was from 6.45 kg in Magahem to 14.32 kg in Shoal, whereas in *Biceps fermoris* muscles the values were between 19.44 kg for Wodoh and 23.3 kg for Shoal. On the other hand, various breeds exhibited a similar myofibrillar fragmentation index, ultimate pH and sarcomere length for both the *Longissimus thoracis* and *Biceps femoris* (Table 10.6). Muscles of the loin region were more tender than those from the leg.

The eating quality of six muscles of the dromedary camel was studied by Kamoun (1995b) who concluded that the *Vastus lateralis* muscles had the highest weight and volume losses (51.1 and 47.8%, respectively), whereas the *Psoas major* muscles had the lowest (44.6 and 41.1%, respectively; Table 10.7). The *Triceps brachii* and *Vastus lateralis* muscles contained more soluble collagen than the *Semitendinosus*, *Psoas major*, *Longissimus thoracis* and *Semimembranosus* muscles, possibly indicating a less stable thermal bond between collagen molecules and weaker connective tissue structures of those muscles (Kamoun, 1995b). Although all six muscles studied by Kamoun (1995b) were ranked acceptable for tenderness, the *Longissimus thoracis* muscle was more tender and had less detectable connective tissue than other muscles. The *Longissimus thoracis* muscle had the highest juiciness score and the *Semitendinosus* and *Vastus lateralis* muscles were less juicy than the *Psoas major*, *Semimembranosus* and *Triceps brachii* muscles.

10.7 Ultimate Muscle pH

The ultimate pH of muscles is a consequence of lactic acid accumulation via glycogen glycolysis (metabolic substrate) that affects meat-quality characteristics (Watanabe *et al.*, 1996; Simek *et al.*, 2003). According to Laack *et al.* (2001), 40–50% of variation in ultimate pH is determined by glycogen concentration. Lowering the pH of 1 kg of muscle from 7.2 to 5.5 requires 0.81g/100g of glycogen (Warris, 1990). The ultimate pH of camel muscles is the result of a combination of many factors including

Table 10.6. Effect of camel breed on some meat-quality characteristics of the dromedary camel *Longissimus thoracis* and *Biceps femoris* muscles (Suliman *et al.*, 2011).

	Breed							
	Magahem		Wodoh		Shoal		Sofor	
Parameter	LT	BF	LT	BF	LT	BF	LT	BF
Ultimate pH	5.76	5.90	5.87	5.90	5.91	5.82	6.07	6.03
W-B shear force value (kg)	6.45	23.32	13.73	19.44	14.32	23.25	10.40	22.77
Sarcomere length (μm)	1.69	1.68	1.68	1.66	1.64	1.69	1.65	1.67
Myofibrillar fragmentation index (%)	71.6	62.4	66.2	60.9	65.4	63.0	67.0	61.2
Cooking loss (%)	23.7	28.3	21.9	26.0	22.9	31.2	22.4	28.7
Colour parameters								
L^* (lightness)	31.6	30.0	33.4	28.1	31.6	31.2	31.2	29.7
a^* (redness)	11.8	13.3	13.0	13.6	12.8	13.4	15.0	13.1
b^* (yellowness)	4.03	4.07	4.74	3.91	4.43	4.26	4.85	3.91

LT, *Longissimus thoracis*; BF, *Biceps femoris.*

Table 10.7. Eating-quality attributes of the six major muscles (Kamoun 1995b).

	Muscle					
Parameter	PM	LT	SM	ST	VL	TB
Myoglobin (mg/g)	3.9	4.1	5.8	3.4	4.1	5.1
Collagen (mg/g)	3.3	4.1	5.0	7.5	6.6	5.6
Sensory tenderness	7.2	6.6	3.7	3.6	1.9	3.9
Collagen soluble (%)	29	29	30	34	42	41
Sensory juiciness	6.2	6.8	5.2	3.8	4.1	5.8
Cooking weight loss (%)	45	45	49	48	51	51
Cooking volume loss (%)	41	42	46	44	48	45

Muscles: PM, *Psoas major*, LT, *Longissimus thoracis*; SM, *Semimembranosus*; ST, *Semitendinosus*; VL, *Vastus lateralis*; TB, *Triceps brachii.*

pre-slaughter handling, post-mortem treatment, glycogen storage and muscle physiology (Thompson, 2002). Low muscle glycogen stores at slaughter prevent the development of a desirable pH post-mortem (Ashmore *et al.*, 1973). A high ultimate pH in camel muscles is a consequence of low muscle glycogen as a result of pre-slaughter stress, including poor nutrition, rough handling and long transportation. The ultimate pH has an effect on several properties such as colour, tenderness, water-holding capacity, cooking time, flavour and drip loss, all of which influence consumer acceptance of meat palatability. Glycogen degradation speed differs between 'red' and 'white' muscles. Red muscles have many red fibres, which contract slowly, have an oxidative metabolism and a low concentration of glycogen, which is actively degraded to glucose. White muscles contract rapidly and have a high concentration of glycogen, normally with a glucolytic metabolism and an active degradation to lactic acid (Lawrie, 2006). There is, however, a variation in the pH between the muscles in different parts of the carcass; also the position of the muscle in the body affects its final pH (Beriain *et al.*, 2000; Al-Kharusi, 2011).

The ultimate pH of dromedary camel meat ranges between 5.5 and 6.6 (Babiker and Yousif, 1990; Kadim *et al.*, 2006, 2009a,b, 2010; Kadim and Mahgoub, 2007). Generally, young camels tend to produce meat with a higher pH than older camels owing to lower levels of glycogen. In this respect, Kadim *et al.* (2006) found that the meat of camels younger than 3 years old had a pH value (5.91) that was higher than that of camels older than 6 years (5.71). The ultimate pH of the *Longissimus thoracis* muscles varied between 5.53 and 5.75 and between 5.68 and 5.80 for electrically stimulated and non-stimulated camel carcasses, respectively (Kadim *et al.*, 2009a). The mean ultimate pH of 5.64 for the electrically stimulated carcass samples was significantly lower than the 5.74 for non-stimulated samples. Ageing of camels had no effect on pH values of camel *Longissimus thoracis* muscles, which indicated that the lowest pH values were attained at 48 h post-mortem (5.69 versus 5.69 for 2 and 7 days ageing, respectively; Kadim *et al.*, 2009a). There were no variations in the ultimate pH in the *Longissimus thoracis* and *Biceps femoris* muscles in terms of the breed of camel (Suliman *et al.*, 2011). In the non-stimulated group, the pH values remained constant in the final period, but the values in the electrically stimulated group increased after 48 h. One factor that obscures the direct effect of stimulation is the accelerated development of rigor mortis so that ageing commences at higher temperatures and is therefore more rapid (Davey *et al.*, 1976).

10.8 Tenderness (Shear Force Value)

Tenderness of meat is the most important organoleptic characteristic and is the predominant quality determinant of red meat at the expense of flavour and colour (Koohmaraie, 1988). Muscle characteristics, glycogen content, collagen content, solubility, and the activities of proteases and their inhibitors are the most important physiological parameters that determine meat tenderness (Hocquette *et al.*, 2005). The amount of alkali-insoluble protein, the shear force value and the diameter of the fibres are inversely proportional to the tenderness of the meat. The sensory assessment of tenderness is supported by Warner–Bratzler shear force data (Wheeler and Koohmaraie, 1994). One-quarter to one-third of the variability in tenderness is related to the variability of various muscle characteristics (Renand *et al.*, 2001). The *Longissimus thoracis* muscle had more soluble collagen than the *Semitendinosus* and *Triceps brachii* muscles (Kamoun *et al.*, 1995b). The *Triceps brachii* muscle had the highest shear force values, maximum connective tissue strength and lowest collagen solubility compared with the *Longissimus thoracis*, *Semitendinosus*, *Semimembranosus*, *Psoas major* and *Vastus lateralis* muscles in camel, indicating that it is the toughest muscle in this group (Babiker and Youssif, 1990). The *Psoas major* and *Longissimus thoracis* muscles were the most tender and had less detectable connective tissue than other muscles. Tenderness increased with larger sarcomere lengths (Davis *et al.*, 1979). The tenderization process starts after slaughter and it varies among individual carcasses (Veiseth *et al.*, 2001). It is dependent on the post-mortem activity of the calpain proteolytic enzymes that include calpastatin (Parr *et al.*, 1999). The most marked difference in meat-quality characteristics between camel meat and other livestock is largely believed to be tenderness (Mukasa-Mugerwa, 1981). Camels are usually slaughtered at the end of their productive life (>10 years), which is the reason that camel meat is classified as of low quality compared with other meat animals. In Kenya, the average age for camels slaughtered was 14.5 years (Mukassa-Mugerwa, 1981). Average shear force value of camel meat at 5–8 years was 48 and 40% higher than those of 1–3 and 3–5 year olds, respectively (Table 10.5; Kadim *et al.*, 2006). A number of studies have also shown that shear values of meat increase with increasing camel age (Dawood, 1995). Differences owing to age may be related to changes in muscle structure and composition as the animal matures, particularly in the connective tissue (Asghar and Pearson, 1980). This suggests that the increase in toughness of older camels is due to change in the nature and quantity of connective tissue in their meat. There is also another factor of differences between individual muscles. Muscle fibre strength, as measured by shear force, was lower in *Longissimus dorsi* than in the *Semitendinosus* and *Triceps brachii* muscles from well-finished dromedary camels (Babiker and Yousif, 1990), which reflects the importance of location and function of individual muscles. Significant differences ($p < 0.05$) were found between the different ages (8, 16 and 26 months of age) and cuts (chuck, rib-eye and leg) for shear force values of male Nahdi camels (Dawood, 1995).

10.9 Ageing and Meat Tenderization

Historically, meat has been aged to improve its quality characteristics because meat is often unacceptably tough immediately following rigor onset. Ageing is the process that causes an improvement in tenderness, flavour, colour and texture over time and involves specific degradation of structural proteins (Hwang *et al.*, 2003; Jaturasitha *et al.*, 2004). The ageing process is normally carried out in the form of either wet ageing or dry ageing. The wet-ageing process consists of vacuuming and storing the meat at temperatures between -1 and 4°C to prevent microbial growth to achieve a required level of tenderization. The dry-ageing process involves refrigeration of meat without vacuum. The time required for ageing varies with the type, size, species and age of the animal. Moderate temperature storage can accelerate the ageing process by keeping carcasses at temperatures of 15°C or greater (Petrovic *et al.*, 1993). According to Ingene and Pearson (1979), ageing between 6 and 43°C had significant effects on the shear force values of meat. This type of conditioning may be applied in the pre- or post-rigor state and is very effective in improving meat tenderness. The ageing processes originate within the myofibres and are responsible for degradation of cellular constituents. This resembles the method adopted by Kadim *et al.* (2009a), where camel *Longissimus thoracis* muscles were stored at a temperature of 2–3°C for 7 days (Table 10.8). The results showed that ageing at 2–3°C for 7 days improved camel meat quality characteristics. This implies that ageing is one of the post-mortem treatments that increases camel meat tenderness that might be adopted in the camel meat industry. According to George-Evins *et al.* (2004) and Lagerstedt *et al.* (2008), increasing ageing time from 4 to 7 days may cause more cooking losses in beef meat. However, Kadim *et al.* (2009a) found no differences in cooking loss with ageing of camel *Longissimus thoracis* muscles from 2 to 7 days. The average time for ageing meat cuts varies from 2 to 61 days with an average range of 17–19 days (Brooks *et al.*, 2000). They stated that beef steaks that are intended to be quality guaranteed needed more than 32 days of ageing.

Table 10.8. Effects of age and ageing on meat-quality attributes of the *Longissimus thoracis* of the dromedary camel (Kadim *et al.*, 2009a).

	Age (year)							
	1–3		4–6		7–9		10–12	
Parameter	Ageing 7-day	Ageing 2-day	Ageing 7-day	Ageing 2-day	Ageing 7-day	Ageing 2-day	Ageing 7-day	Ageing 2-day
Ultimate pH	5.86	5.85	5.79	5.78	5.71	5.71	5.60	5.61
Expressed juice (cm^2/g)	38.6	37.2	37.2	36.6	30.8	30.3	21.3	21.1
Cooking loss (%)	25.7	25.0	23.9	22.7	21.3	19.8	18.9	17.8
W-B shear force (kg)	7.28	8.10	8.41	8.97	9.14	9.76	11.29	12.79
Sarcomere length (μm)	1.73	14.71	1.65	1.67	1.48	1.47	1.39	1.37
Myofibrillar fragmentation index (%)	77.9	73.5	71.6	69.8	66.9	64.5	62.7	60.2
Lightness (L^*)	40.5	39.80	38.71	36.86	35.31	33.72	30.15	28.47
Redness (a^*)	15.6	15.7	16.9	16.1	18.2	19.0	19.9	19.5
Yellowness (b^*)	5.40	5.51	6.04	6.03	7.03	7.05	7.93	7.98

However, the level of improvement in tenderness within a certain ageing time varies among different meat cuts, ages of the animal and species owing to differences in the level of endogenous enzymes, contraction status and connective tissue content (George-Evins *et al.*, 2004). In general, ageing can improve quality characteristics of meats that have relatively small amounts of connective tissue and that have not cold-shortened (Wheeler *et al.*, 1999). The ageing process has various problems associated with the extra cost of packaging, labour, chilling and the risk of meat spoilage (Dransfield, 1994).

The ageing process causes several biochemical changes that influence meat colour. The oxygen consumption rate decreases with an increase in ageing time and the autoxidation rate is limited during vacuum packing, which collectively regulates the amount of oxygen uptake. The level of oxidation during ageing time and activity of metmyoglobin reduction will therefore be affected, leading to different colour profiles (Echevarne *et al.*, 1990).

Analysis of muscle proteins along with meat-quality traits during chiller ageing is crucial in understanding the biological basis of changes in meat quality. The proteolytic enzymes in meat that have been most studied are the cathepsins and calpains. As ageing time increased, tenderness improved (Cifuni *et al.*, 2004). The tenderization process involves complex changes in muscle metabolism in the post-slaughter period and is dependent on animal breed, metabolic status and environmental factors such as the rearing system and pre-slaughter stress. During ageing, the structure of myofibrillar, and other associated proteins, undergoes some modifications, and collagen is weakened to a lesser extent (Christensen, 2004). The degradation of nine actin and/or actin-relevant peptides out of 20 identified actins is related to meat-quality traits during ageing. The proteolytic enzymes in meat play a significant role in improving meat tenderness during ageing. Enzymes require specific conditions such as temperature and pH for optimal activity, and these can be determined and maximized in meat to improve meat tenderness. It is possible to breed animals for high proteolytic enzyme activities. Genetic engineering might also be used to achieve more tender meat. Cathepsins are effective proteolytic agents that have been located within lysosomes and operate best at pH <5.2 values. Myofibrillar proteins are degraded when incubated with various cathepsins *in vitro*. It is believed that catheptic enzymes are able to act on the pH to produce tender meat by degrading myofibrillar proteins.

10.10 Myofibrillar Fragmentation Index

The myofibrillar fragmentation index is a useful indicator of the extent of myofibrillar protein degradation of meat post-slaughter in various animal species (Olson *et al.*, 1976; Kadim *et al.*, 2006, 2009a,b,c, 2010; Lametsch *et al.*, 2007). The differences in rates of fragmentation of myofibrillar proteins may account for differences in the rate of post-mortem tenderization of meat (Thomson *et al.*, 1996; Nagaraj *et al.*, 2005). The structural changes occurring in muscle tissue after slaughter are generally believed to be caused by alterations in and interactions of myofibrillar proteins in the tissue (Nagaraj *et al.*, 2006). Claeys *et al.* (1994) reported that, at a higher pH, proteins preferentially solublized were titin, filamin, neubulin and myosin heavy chain. Except for myosin, all are preferentially degraded by calpains (Goll *et al.*, 1983), which has an optimum effect on pH values near neutrality. Similarly, Silva *et al.* (1999) verified that the myofibrillar fragmentation index in meat was significantly higher at ultimate pH 6.5 than at 5.7. There is a correlation between myofibrillar fragmentation index and tenderness of meat (Veiseth *et al.*, 2001). The myofibrillar fragmentation index of camels more than 6 years old was lower than that for camels of 1–3 years of age (Kadim *et al.*, 2008, 2009a). Moreover, the myofibrillar fragmentation index was significantly higher in electrically stimulated than in non-stimulated muscles in dromedary camels, which was attributed to a variation in muscle pH or to enhanced protein degradation (Kadim *et al.*, 2009a). Ho *et al.* (1996) stated that electrically stimulated muscles exhibited faster protein degradation than non-stimulated muscles. A strong relationship between physical disruptions of the myofibrils and tenderness of camel's meat has been established (Kadim *et al.*, 2009a,b).

10.11 Water-holding Capacity (Expressed Juice)

Water retention in meat is primarily caused by the immobilization of tissue water within the myofibrillar system. Applying pressure can cause a shift of water from the intracellular to the extracellular space and then onto the meat surface as a result of structural alterations at the level of the sarcomeres or of the myofilaments structure. Water-holding capacity is the ability of meat to retain water when treatment or external force is applied to it. It affects the retention of minerals, vitamins and volume of water (Beriain *et al.*, 2000). Therefore, water-holding capacity is an important meat-quality characteristic because of its influence on nutritional value, appearance and palatability. Water-holding capacity is affected by muscle pH because of the electrostatic effects of meat proteins (Hamm, 1975). In pale, soft and exudative pork meat, proteins are denatured in muscles and this decreases electrostatic repulsion between myofilaments owing to a low pH and high temperature. This shows that the colour and water-holding capacity of meat are related to protein denaturation and shrinkage of myofibrils (Bendall and Wismer-Pendersen, 1962). The dromedary camel meat contains higher expressed juice than other Camelidae such as llamas and alpacas, possibly because of the lower fat content (Cristofaneli *et al.*, 2004). The amount of loss was probably due to the ultimate pH of the muscle, composition of muscle and denaturation of proteins by the ionic strength of the extracellular fluid and oxidation of lipids, which decreases the solubility of proteins (Dyer and Dingle, 1967). Kadim *et al.* (2006) reported that meat from camels slaughtered at 1–3 years had higher water-holding capacity values than those slaughtered at 5–8 years of age, probably because of variations in fat content and the binding ability of meat. The water-holding capacity decreases as fat levels increase because of an increase in the ratio of moisture to protein (Miller *et al.*, 1968). Dawood (1995) reported that young camel meat (8 months of age) had significantly higher water-holding capacities than meat from 26-month-old camels. The volume of dromedary camel meat was reduced by 44.3% and weight by 48.2% after boiling in water for 40 min (Kamoun, 1995b). Babiker and Yousif (1990) found that the *Semitendinosus* muscle had significantly ($p < 0.05$) less cooking loss than *Longissimus dorsi* or *Triceps brachii*, which coincided with its high water-holding capacity. Similarly, Al-Kharusi (2011) found that *Longissimus thoracis* muscles had more expressed juice than *Infraspinatus* muscles, whereas the variation with *Semitendinosus*, *Semimembranosus*, *Biceps femoris* and *Triceps brachii* were not significant. The thawing loss of camel meat samples stored for 10 weeks at -20°C ranged from 8.2 to 12.3% of the original weight of the meat (Dawood, 1995). The volume lost through cooking and cooking time were the same for *Longissimus thoracis* and *Biceps femoris* for 1-year-old camels (Suliman *et al.*, 2011). Muscle of a high pH has a greater water-holding capacity than low pH muscles, which increase compactness and light absorption (Abril *et al.*, 2001).

Electrical stimulation negatively affects the myofibrillar water-holding capacity of the camel *Longissimus thoracis* muscles (Kadim *et al.*, 2009a,b). Filter paper wetness was significantly higher for stimulated muscle samples than for the control. Water-holding capacity is not only affected by the ability of the myofibrils to hold onto water, but also by other factors such as integrity of the muscle cell membranes or the rate of fluid migration within the meat (Den Hertog-Meischke *et al.*, 1997). The decrease in myofibrillar water-holding capacity of electrically stimulated muscles may be partly due to the presence of denatured sarcoplasmic proteins in the myofibrillar fraction. Den

Hertog-Meischke *et al.* (1997) suggested that the decrease in water-holding capacity of stimulated muscles may be a result of increased denaturation of sarcoplasmic proteins. The lower myofibrillar water-holding capacity of stimulated muscles was not due to differences in pH of the myofibrillar protein suspensions. It is possible that the freezing process causes such a decrease in water-holding capacity that the effect of electrical stimulation is no longer noticeable.

10.12 Colour (*L**, *a**, *b**) Values

Meat colour is one of the most important sensory characteristics by which consumers make judgements on meat quality. Meat colour measurements involve two basic methods: human visual appraisal and instrumental analysis. Both methods inherently involve an assessment of the concentration and the chemical form of myoglobin, the morphology of the muscle structure, and the ability of the muscle to absorb or scatter light. The degree of meat pigmentation is directly related to the chemical structure of myoglobin content. In general, myoglobin concentration within a given muscle will differ according to the species or age and is dependent on the proportions of muscle fibre type (Lawrie, 2006). Muscle comprised predominantly of red-fibre types contains more myoglobin than muscles with white-fibre-type content.

The haem group contains a centrally located iron atom that has six coordination sites available for chemical bonds. Four of these sites bond the iron atom within the haem structure, while the fifth bond links the iron atom to the amino-acid chain. The sixth site bonds the iron atom to a haem chemical group that determines meat colour. Proportions of deoxymyoglobin, oxymyoglobin and metmyoglobin in the meat depend on oxygen availability and determine the colour of fresh meat (Lindahl *et al.*, 2001). The oxygen availability depends on the oxygen partial pressure, penetration and consumption rate of the muscle (Ledward, 1992). The penetration depth of light decreases as an effect of increased light scattering caused by an increased amount of myofibrillar water, pH and the extent of protein denaturation (Feldhusen, 1994). During post-mortem glycolysis, the muscle proteins denature, leading to an increase in light scattering and less light penetration (Joo *et al.*, 1999), and changes in the selective light absorption through chromospheres such as myoglobin and haemoglobin (Feldhusen, 1994).

The colour of meat is determined by measurements that include lightness (L^*), redness (a^*) and yellowness (b^*). The colour is influenced by muscle pH, age, intramuscular fat, muscle texture and species (Gardner *et al.*, 1999). A negative linear relationship was reported between colour values and pH in *Longissimus thoracis* muscles (Menzies and Hopkins, 1996). Meat samples darkened at a decreasing rate in terms of L^*, a^* and b^* values as ultimate pH increased (Jacob, 2003). In general, camel meat is described as raspberry red to dark brown in colour. Babiker and Yousif (1990) reported that dromedary camel *Longissimus dorsi* muscles had higher lightness (L^*), redness (a^*) and yellowness (b^*) values than *Semitendinosus* and *Triceps brachii* muscles. Suliman *et al.* (2011) found that the colour of the *Biceps femoris* muscle was not affected by the breed of camels, whereas the redness (a^*) values of *Longissimus thoracis* muscles appeared different. A high redness (a^*) colour component in the camel *Longissimus thoracis* muscle was associated with a lower lightness (L^*), which might be due to an increase in myoglobin content. The camel *Longissimus thoracis* muscle was lighter with less redness than *Biceps femoris* muscle (Suliman *et al.*, 2011).

The age of the camel has a significant effect on meat colour. Kadim *et al.* (2006) showed that meat from 6–8 and 10–12 year old camels was darker (lower L^*), redder (higher a^*) and yellower (high b^*) than that from 1–3 year old camels because of higher concentrations of myoglobin. Post-mortem colour changes in fresh meat depend on the biochemical characteristics of the tissue and metabolic type (Monin and Ouali, 1991). Post-mortem protein degradation is directly related to the ultimate pH, which increases light scattering properties of meat and thereby increases L^*, a^* and b^* values (Offer, 1991). Meat samples with a low ultimate pH might have more protein degradation, resulting in higher colour values than the high ultimate pH meat samples. Abril *et al.* (2001) reported that the reflectance spectrum value for meat samples was higher for an ultimate pH above 6. Post-mortem glycolysis decreases muscle pH, making muscle surfaces brighter and superficially wet (Swatland, 1989). If the ultimate meat pH is high, the physical state of the proteins will be above their iso-electric point, proteins associate with more water in the muscle and therefore fibres will be tightly packed (Abril *et al.*, 2001).

Colour deterioration and lipid oxidation might be linked, although the precise mechanisms are still unclear. Some control over increased susceptibility to oxidation can be attained by feeding higher levels of vitamin E, as an antioxidant active in meat (Asghar *et al.*, 1991). A delay of myoglobin oxidation is accomplished in a variety of ways, including storage and display of meat under refrigerated conditions, hygienic preparation of meat cuts and selective use of lighting. In addition, the application of antioxidants, such as ascorbic acid or vitamin E, might extend colour shelf life.

10.13 Conclusion

The dromedary camel seems to be the most advantageous animal for the protein supply of populations in arid and semi-arid regions, presenting a viable alternative to cattle. In appearance and colour, texture and palatability, camel meat is very similar to beef. Total collagen content and soluble collagen are important factors relating to cooked-meat tenderness, although a trend was observed for muscles with a higher percentage of fat to be more tender and juicy. Camel meat could therefore be successfully marketed alongside that of cattle, sheep and goat. Pre-mortem and post-mortem factors should be carefully considered in improving meat-quality characteristics.To encounter better post-harvest conditions, technology has been considered to improve camel meat quality through electrical stimulation, ageing and chilling temperatures. Electrical stimulation is effective where cold shortening is an actual risk owing to low chilling temperatures applied in the early post-mortem period.

References

Abrahamson, D.R. (1986) Recent studies on the structure and pathology of basement membranes. *Journal of Pathology* 149, 257–278.

Abril, M., Campo, M.M., Onenc, A., Sanudo, C., Alberti, P. and Negueruela, A.I. (2001) Beef colour evolution as a function of ultimate pH. *Meat Science* 58, 69–78.

Al-Kharusi, A.H.H. (2011) Composition, quality and histochemistry of the Arabian camel (Camelius dromedarius) muscles. MS Thesis, Department of Animal and Veterinary Sciences, Sultan Qaboos University, Oman.

Ariano, M.A., Armstrong, R.B. and Edgerton, V.R. (1973) Hindlimb muscle fiber populations of five mammals. *Journal of Histochemistry and Cytochemistry* 21, 51–55.

Asghar, A. and Pearson, A.M. (1980) Influence of ante- and postmortem treatments upon muscle composition and meat quality. *Advances in Food Research* 26, 53–213.

Asghar, A., Gray, J.I., Booren, A.M., Gomaa, E.A., Abouzied, M.M. and Miller, E.R. (1991) Effects of supr-nutritional dietary vitamin E levels on sub cellular deposition of α-tocopherol in the muscle and on pork quality. *Journal of Science and Food Agriculture* 57, 31–41.

Ashmore, C.R., Tompkins, G. and Doerr, L. (1972) Postnatal development of muscle fibre types in domestic animals. *Journal of Animal Science* 4, 37–41.

Ashmore, C.R., Carroll, F., Doerr, J., Tompkins, G., Stokes, H. and Parker, W. (1973) Experimental prevention of dark-cutting meat. *Journal of Animal Science* 35, 33–36.

Babiker, S.A. and Tibin, I.M. (1986) Comparative study of camel meat and beef. Camel Research Unit, University of Khartoum, Sudan, pp. 73–77.

Babiker, S.A. and Yousif, K.H. (1990) Chemical composition and quality of camel meat. *Meat Science* 27, 283–287.

Baker, D.M. and Santer, R.M. (1990) Development of a quantitative histochemical method for determination of succinate dehydrogenase activity in autonomic neurons and its application to the study of aging in the autonomic nervous system. *Journal Histochemistry Cyotochemistry* 38, 525–531.

Bendall, J.R. and Wismer-Pendersen, J. (1962) Some properties of the fibrillar proteins of normal and watery pork muscle. *Journal of Food Science* 27, 144–159.

Beriain, M.J., Bass, P., Purroy, A. and Treacher, T. (2000) Effect of animal and nutritional factors and nutrition on lamb meat quality. CIHEAM, 75–86.

Boles, J.A. and Swan, J.E. (1996) Effect of post-slaughter processing and freezing on the functionality of hot-boned meat from young bull. *Meat Science* 44, 11–18.

Brooke, M.H. and Kaiser, K.K. (1970) Three myosin ATPase systems, the nature of their pH lability and sulfhydral dependence. *Journal of Histochemistry and Cytochemistry* 18, 670–672.

Brooks, J.C., Belew, J.B., Griffin, D.B., Gwartney, B.L., Hale, D.S., Henning, W.R., Johnson, D.D., Morgan, J.B., Parrish, F.C. Jr., Reagan, J.O. and Savell, J.W. (2000) National tenderness survey -1998. *Journal of Animal Science* 78, 1852–1860.

Calkin, C.R., Dutson, T.R., Smith, G.C., Carpenter, Z.I. and Davis, G.W. (1981) Relationship of fiber type composition to marbling and tenderness of bovine musle. *Journal of Food Science* 46, 708–710.

Celio, M.R., Pauls, T.L. and Schwaller, B. (1996) Guidebook to the calcium-binding proteins. Oxford University Press Inc., New York.

Christensen, M., Larsen, L.M, Ertbjerg, P. and Purslow, P.P. (2004) Effect of proteolytic enzyme activity and heating on the mechanical properties of bovine single muscle fibres. *Meat Science* 66, 361–369.

Cifuni, G.F., Napolitano, F., Riviezzi, A.M., Braghieri, A., and Girolami, A. (2004) Fatty acid profile, cholestrol content and tenderness of meart from Podolian young bulls. *Meat Science* 67, 289–297.

Claeys, E., Uytterhaegen, L., Demeyer, D. and De Smet, S. (1994) Beef myofibrillar protein salt solubility in relation to tenderness and proteolysis. *Proceedings of 40th International Congress of Meat Science and Technology (S-IVB. 09).* The Hague, The Netherlands.

Clark, K.A., McElhinny, A.S., Beckerle, M.C. and Gregorio, C.C. (2002) Striated muscle cytoarchitecture: an intricate web of form and function. *Annual Review of Cell Development Biology* 18, 637–706.

Colville, T. and Bassert, J.M. (2002) *Clinical Anatomy and Physiology for Veterinary Technicians.* Mosby, Inc., St Louis, Missouri.

Cornforth, M.R., Pearson, A.M. and Merkel, R.A. (1980) Relationship of mitochondria and sarcoplasmic reticulum to cold shortening. *Meat Science* 4, 103–121.

Cristofaneli, S., Antonini, M., Torres, D., Polidori, P. and Renieri, C. (2004) Meat and carcass quality from Peruvian llama (*Lama Glama*) and alpaca (*Lama Pacos*). *Meat Science* 66, 589–593.

Davey, C.L., Gilbert, K.V. and Carse, W.A. (1976) Carcass electrical stimulation to prevent cold-shortening toughness in beef. *New Zealand Journal of Agriculture Research* 19, 13–18.

Davis, C.L., Smith, G.C., Carpenter, Z.L., Dutson, T.R. and Cross, H.R. (1979) Tenderness variations among beef steaks from carcass of the same USDA quality grade. *Journal of Animal Science* 49, 103–114.

Dawood, A. (1995) Physical and Sensory characteristics of Najdi camel meat. *Meat Science* 39, 59–69.

Den Hertog-Meischke, M.J.A., Smulders, F.J.M. van, Logtestijn, J.G. and van Knapen, F. (1997) The effect of electrical stimulation on the water-holding capacity and protein denaturation of two bovine muscles. *Journal of Animal Science* 75, 118–124.

Dina, D. and Klintegerg, R. (1977) Proposal for a rural development training project and study concerned with camel utilization in arid lands in Ethiopia. Addis Ababa (Mimeographed). Quoted by Shalah, 1979 and Wilson, 1984.

Dransfield, E. (1994) Optimization of tenderization, ageing and tenderness. *Meat Science* 36, 105–121.

Duance, V.C., Stephens, H.R., Dunn, M. Bailey, A.J. and Dubowitz, V. (1980) A role for collagen in the pathogenesis of muscular dystrophy. *Nature* 284, 470–472.

Dyer, W.J. and Dingle, J.R. (1967) Fish proteins with special reference to freezing. In: Borgstrom, G. (ed.) *Fish as Food* Vol. 1. Academic Press, New York.

Echevarne, C., Renerre, M. and Labas, R. (1990) Metmyoglobin reductase activity in bovine muscles. *Meat Science* 27, 161–172.

Essen-Gustavsson, B. (1986) Activity and inactivity-related muscle adaptation in the animal kingdom. In: Saltin, B. (ed.) *Biochemistry of Exercise*

VI. Human Kinetics Publishers, Champaign, Illinois, pp. 435–444.
Feldhusen, F. (1994) Einflüsse auf die postmortale Farbveränderung der Oberfläche von Schweinemuskulatur. *Fleischwirtschaft* 74, 989.
Fischer, H. (1975) Schaf, Zeigen-Wasserbuffel und Dromedarhaltung. In Kon Kurreuz fat Runderhaltung. Beitrage III veterinaragung, Ouagadougoa, pp. 1–16.
Gardner, G.E., Kennedy, L., Milton, J.T.B. and Pethick, D.W. (1999) Glycogen metabolism and ultimate pH of muscle merino, first-cross, and second-cross wether lambs as affected by stress before slaughter. *Australian Journal of Agricultural Research* 50, 175–181.
Geay, Y., Bauchart, D., Hocquette, J.F. and Culioli, J. (2001) Effect of nutritional factors on biochemical, structural and metabolic characteristics of muscles in ruminants, consequences on dietetic value and sensorial qualities of meat. *Reproduction of Nutritional Development* 41, 1–26.
George-Evins, C.D., Unruh, J.A., Waylan, A.T. and Marsden, J.L. (2004) Influence of quality classification, aging period, blade tenderization, and endpoint cooking temperature on cooking characteristics and tenderness of beef *Gluteus medius* steak. *Journal of Animal Science* 82, 1863–1867.
Goll, E.D., Otsuka, Y. and Muguruma, M. (1983) Role of muscle proteinases in maintenance of muscle integrity and mass. *Journal of Food Biochemistry* 7, 137–177.
Hamm, R. (1975) Water-holding capacity of meat. In: Cole, D.A.J. and Lawrie, R.A. (eds) *Meat*. Butterworth, London, pp. 321–328.
Hamm, R. (1977) Postmortem breakdown of ATP and glycogen in ground muscle: A review. *Meat Science* 1,15–18
Hamm, R. (1981) Postmortem changes in muscle affecting the quality of comminuted meat products. In: *Developments in Meat Science –2*. Elsevier Applied Science, Barking, UK, p 93.
Helliwell, T.R. (1999) Muscle: Part 1- Normal structure and function. *Current Orthopaedics* 13, 33–41.
Heneiksen-Larsen, K.B., Lexell, J. and Sjostrom, M. (1983) Distribution of different fibre types in human skeletal muscles. I. Method for the preparation and analysis of cross-sections of whole *Tibialis anterior*. *Histochemistry Journal* 15,167–178.
Hertzman, C., Olsson, U. and Tornberg, E. (1993) The influence of high temperature, type of muscle and electrical stimulation on the course of rigor, ageing and tenderness of beef muscle. *Meat Science* 35, 119–141.
Hintz, C.S., Coyle, E.F., Kaiser, K.K., Chi, M.M. and Lowry, O.H. (1984) Comparison of muscle fiber typing by quantitative enzyme assays and by myosin ATPase staining. *Journal of Histochemistry and Cyotochemistry* 32, 655–660.
Ho, C.Y., Stromer, M.H. and Robson, R.M. (1996) Effect of electrical stimulation on postmortem titin, nebulin, desmin, and troponin T degradation ultrastructural changes in bovine *Longissimus* muscle. *Journal of Animal Science* 74, 1563–1575.
Hocquette, J.F., Ortigues-Marty, I., Pethick, D.W., Herpin, P. and Fernandez, X. (1988) Nutritional and hormonal regulation of energy metabolism in skeletal muscles of meat-producing animals. *Livestock Production Science* 56, 115–143.
Hocquette, J.F., Malek, I.C., Jailler, R. and Picard, B. (2005) Recent advances in research on bovine muscle and meat sensory quality. II. Influence of rearing factors on muscle characteristics. *Cahiers Agricultures* 14, 365–372.
Hopkins, D.L. and Thompson, J.M. (2002) The relationship between postmortem calcium concentration or pH and indicators of proteolysis in ovine muscle. *Meat Science* 61, 411–414.
Honikel, K.O., Roncales, P. and Hamm, R. (1983) The influence of temperature on shortening and rigor onset in beef muscle. *Meat Science* 8, 221–241.
Hoppeler, H. (1986) Muscle aerobic potential in the animal kingdom. In: Saltin, B. (ed.) *Biochemistry of Exercise VI*. Human Kinetics Publisher, Champaign, Illinois, pp. 417–434.
Hwang, I. H., Devine, C.E. and Hopkins, D.L. (2003) The biochemical and physical effects of electrical stimulation on beef and sheep meat tenderness. *Meat Science* 65, 677–691.
Hwang, I.H., Park, B.Y., Kim, J.H., Cho, S.H. and Lee, J.M. (2004) Assessment of postmortem proteolysis by gel-based proteome analysis and its relationship to meat quality traits in pig *Longissimus*. *Meat Science* 69, 79–91.
Ingene, J.O. and Pearson, A.M. (1979) Role of phospholipids and triglycerides in warmed-over flavor development in meat model systems. *Journal of Food Science* 44, 1285–1290.
Jacob, R.H. (2003) Optimising the concentration of glycogen in lamb meat. PhD thesis, Murdoch University, Western Australia.
Janz, J.A.M, Aalhus, J.L. and Price, M.A. (2001) Blast chilling and low voltage electrical stimulation influences on bison (*Bison bison bison*) meat quality. *Meat Science* 57, 403–411.
Jaturasitha, S., Thirawong, P., Leangwunta, V. and Kreuzer, M. (2004) Reducing toughness of

beef from Bos indicus draught steers by injection of calcium chloride: Effect of concentration and time postmortem. *Meat Science* 68, 61–69.

Joo, S.T., Kauffman, R.G., Kim, B.C. and Park, G.B. (1999) The relationship of sarcoplasmic and myofibrillar protein solubility to color and water-holding capacity in porcine *Longissimus muscle. Meat Science* 52, 291–297.

Kadim, I.T. and Mahgoub, O. (2008) Effect of age on quality and composition of one-humped camel *Longissimus thoracis* muscle. *International Journal of Post-Harvest Technology and Innovation* 1, 327–336.

Kadim, I.T., Mahgoub, O., Al-Marzooqi, W., Al-Zadgali, S., Annamali, K. and Mansour, M.H. (2006) Effects of age on composition and quality of muscle *Longissimus thoracis* of the Omani Arabian camel (*Camelus dromedarius*). *Meat Science* 73, 619–625.

Kadim, I.T., Mahgoub, O. and Purchas, R.W. (2008) A review of the growth, and of the carcass and meat quality characteristics of the one-humped camel (*Camelus dromedarius*). *Meat Science* 80, 555–569.

Kadim, I.T., Mahgoub, O., Al-Marzooqi, W., Khalaf, S.K., Mansour, M.H., Al-Sinani, S.S.H. and Al-Amri, I.S. (2009a) Effects of electrical stimulation on histochemical muscle fiber staining, quality and composition of camel and cattle *Longissimus thoracis* muscles. *Journal of Food Science* 74, S44–S52.

Kadim, I.T., Al-Hosni, Y., Mahgoub, O., Al-Marzooqi, W., Khalaf, S.K., Al-Sinawi, S.S.H., Al-Lawati, A.M. and Al-Amri, I.S. (2009b) Effect of low voltage electrical stimulation on post mortem biochemical and quality characteristics of *Longissimus thoracis* muscle from one-humped camel (*Camelus dromedarius*). *Meat Science* 82, 77–85.

Kadim, I.T., Mahgoub, O., Al-Marzooqi, W., Khalaf, S.K., Mansour, M.H., Al-Sinani, S.S.H., and Al-Amri, I.S. (2009c) Effects of low voltage electrical stimulation and splitting the carcass on histochemical and meat quality characteristics of *Longissimus thoracis* muscles from the one-humped camel (*Camelus dromedarius*). *Journal of Camelid Science* 2, 30-40.

Kadim, I.T., Mahgoub, O., Al-Marzooqi, W., Khalaf, S.K., Mansour, M.H., Al-Sinani, S.S.H. and Al-Amri, I.S. (2010) Effects of transportation during hot season, breed and electrical stimulation on histochemical and meat quality characteristics of goat *Longissimus* muscle. *Animal Science Journal* 81, 352–361.

Kamoun, M. (1995a) Dromedary meat: production, qualitative aspects and capability for transformation. *Option Mediterraneennes. Serie B, Etudes et Recherches* 13, 105–130.

Kamoun, M. (1995b) Evolution de la qualite de la carcasse et de la viande des dromadaires En function de la conduit zootechnique des animaux. Rapport final relative a la bourse de recherché FIS no. 1372-2.

Kassem, A., El Sayed, M.T. and Ahmed, A. (2004) Microstructural characteristics of Arabian camel meat. *Journal of Camel Science* 1, 86–95.

Khan, M.A. (1976) Histochemical characteristics of vertebrates striated muscle: A review. *Progress in Histochemistry and Cytochemistry* 8, 1–48.

Kiessling, K.H. and Kiessling, A. (1984) Fibre composition and enzyme activities in five different muscles from the Svalbard reinder. *Comparative Biochemistry and Physiology* 77A, 75–78.

Knoess, K.H. (1977) The camel as a meat and milk camel. *World Animal Review* 22, 39–44.

Koohmaraie, M. (1988) The role of endogenous proteases in meat tenderness. *Reciprocal Meat Conference Proceedings* 41, 89–100.

Laack, R.V., Kauffman, R. and Greaser, M. (2001) Determinants of ultimate pH of meat. *Proceedings of the 47th International congress of Meat Science and Technology* 1, 22–26.

Lagerstedt, A., Enfalt, L., Johansson, L. and Lundstrom, K. (2008) Effect of freezing on sensory quality, shear force and water loss in beef M. *Longissimus dorsi. Meat Science* 80, 457–461.

Lametsch, R., Knudsen, J., Ertbjerg, P., Ekberg, N. and Therkildsen, M. (2007) Novel method for determination of myofibril fragmentation post mortem. *Meat Science* 75, 719–724.

Lawrie, R.A. (2006) *Lawrie's Meat Science*, 7th edn. Woodhead Publishing Limited, Cambridge, UK.

Ledward, D.A. (1992) Colour of raw and cooked meat. In: Johnston, D.E., Knight, M.K. and Ledward, D.A. (eds) *The Chemistry of Muscle-based Foods.* The Royal Society of Chemistry, Cambridge, UK, pp. 128–144.

Lee, S., Polidori, P., Kauffman, R.G. and Kim, B.C. (2000) Low voltage electrical stimulation effects on proteolysis and lamb tenderness. *Journal of Food Science* 65, 786–790.

Leeson, C.R., Lesson, T.S. and Paparo, A.A. (1985) *Textbook of Histology.* 5th edn, W.B. Saunders Company, Philadelphia, Pennsylvania.

Leupold, J. (1968) Le chameau important animal domestique des pays subtropicaux. *Les Cahiers Bleus Vétérinaires* 15, 1–6.

Lindahl, G., Lundström, K. and Tornberg, E. (2001) Contribution of pigment content, myoglobin forms and internal reflectance to the colour of pork loin and ham from pure breed pigs. *Meat Science* 59, 141–151.

Martini, F.H. (2001) *Fundamental of Anatomy and Physiology.* 5th edn, Prentice-Hall, Upper Saddle River, NJ.

Menzies, D.J. and Hopkins, D.L. (1996) Relationship between color and ph in lamb loins. *Proceedings of the Australian Society of Animal Production* 21, 353.

Miller, W.O., Staffle, R.L. and Zirkle, S.B. (1968) Factors, which influence the water-holding capacity of various types of meat. *Food Technology* 22, 11390.

Monin, G. and Ouali, A. (1991) Muscle differentiation and meat quality. In: Lawrie, R. (ed) *Development in Meat Science-5,* Elsevier Applied Science, Barking, UK.

Mukasa-Mugerwa, E. (1981) The camel (*Camelus dromedarius*): A biographical review. International Livestock Centre for Africa, Addis Ababa, Ethiopia, pp.147.

Mullen, A.M. (2002) New technique for analysing raw meat. In: Kerry, J., Kerry, J. and Ledward, D. (eds) *Meat Processing: Improving Quality.* Woodhead Publishing Ltd, Cambridge, UK.

Nagaraj, N.S., Anilakumar, K.R. and Santhanam, K. (2005) Postmortem changes in myofibrillar proteins of goat skeletal muscles. *Journal of Food Biochemistry* 29, 152–170.

Nagaraj, N.S., Anilakumar, K.R. and Santhanam, K. (2006) Biochemical & physicochemical changes in goat meat during postmortem aging. *Journal of Muscle Foods* 17,198–213.

Newbold, R.P. (1996) Changes associated with rigor mortis. In: Briskey, E.J., Cassens, R.G. and Trautmann, J.C. (eds) *The Physiology and Biochemistry of Muscle as Food.* The University of Wisconsin Press, Madison, WI.

Nuss, J.I. and Wolfe, H. (1981) Effect of postmortem storage temperatures on isometric tension, pH, ATP, glycogen and glucose-6-phosphate for selected bovine muscles. *Meat Science* 5, 201–213.

Offer, G. (1991) Modeling of the formation of pale, soft and exudative meat: effects of chilling regime and rate and extent of glycolysis. *Meat Science* 30, 157–184.

Offer, G. and Knight, P. (1988) The structural basis of water-holding in meat. In: Lawrie, R.A. (ed.) *Development in Meat Science- 4.* Elsevier Applied Science, London, UK, pp. 63.

Olson, D.G., Parrish, F.C.J.R. and Stromer, M.H. (1976) Myofibril fragmentation and shear resistance of three bovine muscles during postmortem storage. *Journal of Food Science* 41, 1036–1041.

Olsson, U., Hertzman, C. and Tornberg, E. (1994) The influence of low temperature, type of muscle and electrical stimulation on the course of rigor mortis, ageing and tenderness of beef muscle. *Meat Science* 37, 115.

Parr, T., Sensky, P.L., Scothern, G.P., Bardsley, R.G., Buttery, P.J., Wood, J.D. and Warkup. C. (1999) Relationship between skeletal muscle-specific calpain and tenderness of conditioned porcine *Longissimus* muscle. *Journal of Animal Science* 77, 661–668.

Pearson, A.M. and Young, R.B. (1989) *Muscle and Meat Biochemistry.* Academic Press, San Diego, California.

Petch, P.E. (2001) Electrical inputs and meat processing. In: Hui, Y.H., Nip, W.K., Rogers, R.W. and Young, O.A. (eds) *Meat Science and Application.* Marcel Dekker, Inc., New York, pp. 319–349.

Petersen, G.V. and Blackmore, D.K. (1982) The effect of different slaughter methods on the post mortem glycolysis of muscle in lambs. *New Zealand Veterinary Journal* 30, 195–198.

Petrovic, L., Grujic, R. and Petrovic, M. (1993) Definition of the optimum freezing rate –2. Investigation of the physico-chemical properties of beef *m. Longissimus dorsi* frozen at different freezing rates. *Meat Science* 33, 319–331.

Pette, C. and Staron, S. (1990) Cellular and molecular diversities of mammalian skeletal muscle fibers. *Review Physiology, Biochemistry and Pharmacology* 116, 1–76.

Rahelic, S. and Puac, S. (1981) Fibre types in *Longissimus dorsi* from wild and highly selected pig breeds. *Meat Science* 5, 451–455.

Renand, G., Picard, B., Touraille, C., Berge, P. and Lepetit, J. (2001) Relationships between muscle characteristics and meat quality traits of young Charolais bulls. *Meat Science* 59, 49–60.

Renerre, M., Poncet, K., Mercier, Y., Gatellier, P. and Metro, B. (1999) Influence of dietary fat and vitamin E on antioxidant status of muscles of turkey. *Journal of Agriculture and Food Chemistry* 47, 237–244.

Roberts, M., Reiss, M. and Monger, G. (2000) *Advanced Biology.* First edn, Nelson House, UK.

Rose, R.J., Evans, D.L., Henckel, P., Knight, P.K., Cluer, D. and Saltin, B. (1994) Metabolic responses to prolonged exercise in the racing camel. *Acta Physiologica Scandinavica, 150 Supplementum* 617, 49–60.

Saltin, B., Rose, R.J. and Henckel, P. (1994) Skeletal muscle morphology and metabolic potential. *Acta Physiologica Scandinavica, 150 Supplementum,* 617, 24–32.

Schiaffino, S. and Reggiani, C. (1996) Molecular diversity of myofibrillar proteins: gene regulation and functional significance. *Physiological Reviews* 76, 371–423.

Shariatmadari, R. and Kadivar, M. (2006) Postmortem aging and freezing of camel meat (a comparative study). In: Proceedings of the 52nd International Congress of Meat Science and Technology, Dublin, Ireland, pp. 673–674.

Silva, J.A., Patarata, L. and Martins, C. (1999) Influence of ultimate pH on bovine meat tenderness during ageing. *Meat Science* 52, 453–459.

Simek, J., Vorlova, L., Malota, L., Steinhauserova, I. and Steinhauser, L. (2003) Post-mortal changes of pH value and lactic acid content in the muscles of pigs and bulls. *Journal of Animal Science* 7, 295–299.

Simmons, N.J., Daley, C.C., Cummings, T.L., Morgan, S.K., Johnson, N.V. and Lombard, A. (2008) Reassessing the principles of electrical stimulation: review. *Meat Science* 80, 110–122.

Singh, K., Dobbie, P. M., Simmons, N. J., Duganzich, D. and Bass, J.J. (1997) The effects of the calpain proteolytic system on meat tenderisation rates in different ovine skeletal muscles. In: *Proceedings of the 43rd International Congress of Meat Science and Technology*, New Zealand, pp. 612–613.

Snow, D.H. and Guy, P.S. (1981) Fibre type and enzyme activities of the *Gluteus medius* in different breeds of horse. In: Poortmans J. and Niset, G. (eds) *Biochemistry of Exercise IV-B.* University of Park Press, Baltimore, Maryland, pp. 275–282.

Snow, D.H. and Harris, R.C. (1985) Thorough breds and greyhounds: biochemical adaptations in creatures of nature and of man. In: R. Gilles (ed.) *Circulation, Respiration and Metabolism.* Springer-Verlag, Berlin Heidelberg, Germany, pp. 227–238.

Sorinmade, S.O., Cross, H.R., Ono, K. and Wergin, W.P. (1982) Mechanisms of ultrastructural changes in electrically stimulated beef *Longissimus. Meat Science* 6, 71–77.

Stienen, G.J., Kiers, J.L., Bottinelli, R. and Reggiani, C. (1996) Myofibrillar ATPase activity in skinned human skeletal muscle fibres: fibre type and temperature dependence. *Journal of Physiology* 493, 299–307.

Suliman, G., Sami, A., Alowaimer, A. and Koohmaraie, M. (2011) Effect of breed on the quality attributes of camel meat. *Indian Journal of Animal Sciences* 81, 407–411.

Swatland, H.J. (1982) The challenges of improving meat quality. *Canadian Journal of Animal Science* 62, 15–24.

Swatland, H.J. (1984) The structure and properties of meat. In: *Structure and Development of Meat Animals.* Prentice-Hall, Inc., Englewood Cliffs, New Jersey, pp. 153–200.

Swatland, H.J. (1989) A review of meat spectrophotometry (300 to 800 nm). *Canadian Institute of Food Science and Technology* 22, 390–402.

Takahashi, G., Wang, S.M., Lochner, J.V. and Marsh, B.B. (1987) Effects of 2-Hz and 60-Hz stimulation on the microstructure of beef. *Meat Science* 19, 65–76.

Thomson, B.C., Dobbie, P.M., Singh, K. and Speck, P.A. (1996) Postmortem kinetics of meat tenderness and the components of the calpain system in bull skeletal muscle. *Meat Science* 44, 151–157.

Thompson, J. (2002) Managing meat tenderness. *Meat Science* 62, 295–308.

Tornberg, E. (1996) Biophysical aspects of meat tenderness. *Meat Science* 43, S175.

Veiseth, E., Shackelford, D.D., Wheeler, T.L. and Koohmaraie, M. (2001) Technical Note: Composition of myofibril fragmentation index from fish and frozen pork and lamb *Longissimus. Journal of Animal Science* 79, 904–906.

Vergara, H. and Gallego, L. (2000) Effect of electrical stunning on meat quality of lamb. *Meat Science* 56, 345–349.

Warris, P. (1990) The handling of cattle pre-slaughter and its effects on carcass and meat quality. *Applied Animal Behaviour Science* 28, 171–186.

Warris, P. (2000) *Meat Science, An Introductory Text.* CAB International, Wallingford, UK.

Warris, P.D. and Brown, S.N. (1987) The relationship between initial pH, reflectance and exudation in pig muscle. *Meat Science* 20, 65–74.

Warris, P., Browns, S., Bevis, E., Kestin, S. and Young, C. (1987) Influence of food withdrawal at various times preslaughter on carcass yield and meat quality in sheep. *Journal of Science Food Agriculture* 39, 325–334.

Watanabe, A., Daly, C.C. and Devine, C.E. (1996) The effects of the ultimate pH of meat on tenderness changes during aging. *Meat Science* 42, 67–78.

Wheeler, T.L. and Koohmaraie, M. (1994) Measurement of Warner-Bratzler shear force. In: Science and Technology Receprocation Fair, *47th Annual Reciprocal Meat Conference Proceedings*, National Livestock and Meat Board, Chicago, Illinois, pp 49–50.

Wheeler, T.L., Shackelford, S.D. and Koohmaraie, M. (1999) Tenderness classification of beef;

IV Effect of USDA quality grade on the palatability of "tender" beef *Longissimus* when cooked well done. *Journal of Animal Science* 77, 882–888.
Will, P.A., Ownby, C.L. and Henrickson, R.L. (1980) Ultrastructural postmortem changes in electrically stimulated bovine muscle. *Journal of Food Science* 45, 21–34.
Wilson, R.T. (1984) *The Camel.* Longman Group Ltd, Essex, UK.
Yousif, O.K. and Babiker, S.A. (1989) The desert camel as meat animals. *Meat Science* 26, 245–254.

11 Interventions to Improve the Tenderness of Fresh Meat: a Future Prospect for Camel Meat Research

Alaa El-Din Bekhit,[1] David L. Hopkins,[2] Mustafa M. Farouk[3] and Alan Carne[4]

[1]Department of Food Science, Division of Sciences, University of Otago, Dunedin, New Zealand; [2]NSW Department of Primary Industries, Centre for Red Meat and Sheep Development, Cowra, Australia; [3]AgResearch Limited, Ruakura Research Centre, Hamilton, New Zealand; [4]Department of Biochemistry, Otago School of Medical Sciences, University of Otago, Dunedin, New Zealand

11.1 Introduction

Tenderness is the most important eating quality trait and consumers, especially in developed countries, might reject meat with inferior tenderness (Jeremiah, 1981; Miller *et al.*, 2001). There is no information from developing countries regarding the tenderness range required for meat acceptability, as well as for the toughness threshold for the rejection of meat. In these countries, it is expected that significant differences would exist in meat tenderness levels owing to differences in cooking style and the use of meat at different rigor stages (Geesink *et al.*, 2011). This is particularly relevant to camel meat because the animal is predominantly reared in arid and semi-arid regions where the cooking of pre-rigor meat is a common practice, limited availability and choice of meat means that expectations about the eating experience are not high and the consumption of dried meat products with a strong texture is common and acceptable.

Various physiological (e.g. genetic background, stress, muscle type and location), biological (e.g. breed, diet and handling) and processing (e.g. post-mortem temperature regime, electrical inputs and storage conditions) factors dictate post-mortem meat tenderization. Less than 10% of a carcass is classified as prime grilling cuts (Polkinghorne *et al.*, 2008). To improve the tenderness of meat and produce a consistent tender meat, several interventions have been investigated in common meats (beef, lamb and pork), which could potentially be useful in improving the tenderness of camel meat. The chemical interventions are discussed in this chapter.

11.2 Meat Toughness

Several physiological, biophysical and biochemical changes take place post-mortem during the conversion of muscles to meat. Meat tenderness is determined by the physical properties of the myofibrillar proteins and the meat connective tissue. Myofibrillar proteins are the major proteins, making up to 80% of muscle fibres, and they are the building block of striated muscle. Muscle striation is due to the cylindrical-shaped myofibrils that are composed of continuous repeats of sarcomeres, the contractile unit of the muscle. The biophysical (muscle contraction and shortening) and biochemical (pH decline, temperature change, endogenous proteases activity and Ca^{2+} leakage) changes that occur in the sarcomere post-mortem determine, to a large extent, the tenderness of meat. The contribution of connective tissue to meat toughness depends on the structure and/or the amount of different collagens and elastin in the meat (Lepetit, 2008). Processing and post-mortem handling can be optimized to achieve better tenderizing conditions for myofibrillar protein, but meat toughness caused by connective tissue is not affected significantly by post-mortem processing and handling practices.

11.3 Biological Variation in Meat Tenderness

The physiological function of muscles in live animals determines the fibre-type composition (fast versus slow twitch, aerobic versus anaerobic) and the connective tissue content and solubility. Also, after slaughter, the rate of chilling of a muscle or part of a muscle will depend on its location in the carcass. These differences will lead to a wide variation in the chemical and biochemical composition, and the eating and keeping qualities of the meat. As a result, some muscles are generally tender (e.g. *Infraspinatus* and *Psoas major*), or tough (e.g. *Semimembranosus*). Other muscles, such as the *Triceps brachii*, vary in tenderness depending on the rearing method and processing conditions used (Belew *et al.*, 2003; Torrescano *et al.*, 2003). Variation in the tenderness of a muscle amongst and within animals has been documented (Shackelford *et al.*, 1995; Devine *et al.*, 2006).

11.4 Interventions to Manipulate the Tenderness of Fresh Meat

Post-mortem interventions used in fresh meat tenderization can be classified into three main categories (physical, chemical and enzymatic) on the basis of their mode of action. An alternative classification can be used based on the mode of action/mechanism of the process as follows:

1. Pre-rigor methods to reduce/prevent muscle contractions (rapid freeze, tender stretch, tender cut, wrapping) and/or cold shortening (electrical stimulation, high-temperature conditioning).
2. Rigor and post-mortem interventions to increase the proteolytic activity via the release and/or the activation of endogenous enzymes or the addition of exogenous enzymes (ultrasound, Ca^{2+}, electrical stimulation, high-temperature conditioning, plant and microbial proteases).
3. Methods to increase the solubilization of proteins and modify the net charge on the proteins surface (salts, acids).
4. Methods that cause weakening and disintegration of the protein network (blade and needle tenderization, hydrodyne and static pressure).

Comprehensive reviews on the enzymatic interventions (Bekhit *et al.*, 2012a) and physical interventions (Bekhit *et al.*, 2012b) are available and only chemical interventions will be covered here. The level of tenderization achieved by these different methods varies tremendously from not having any effect at all to over-tenderization of the meat. The intervention methods have their advantages and disadvantages in terms of their effect on flavour, texture and overall product acceptability. Furthermore, the time to achieve a required level of tenderness through these interventions varies from hours to weeks, thereby offering a variety of options to suit the requirements of a range of meat processors with varying chilling and holding space capacities, and target markets (local versus international).

11.5 Enhancement Systems for Meat Tenderization

11.5.1 Chemical tenderization

Infusion/injection of meat with ionic compounds in solution, termed 'meat enhancement', can manipulate several biochemical processes depending on the post-mortem time of injection. For example, pre-rigor infusion can have a dramatic effect on the rate of glycolysis, the rate and state of contraction, the oxidative processes and the rate of the proteolysis, whereas post-rigor injection will affect mainly proteolysis and oxidative processes. The role of metal ions in muscle function has been extensively studied. The importance of particularly Ca^{2+} ions on tenderization were initially suggested when $CaCl_2$, chelating agents such as EDTA or EGTA were injected into muscle post-mortem. Weiner and Pearson (1969) observed longer sarcomeres in pig muscle injected pre-rigor with 0.1 M EDTA, which consequently resulted in significantly lower shear force values than control muscle at 24-h post-mortem. In contrast, the injection of 0.1 M $CaCl_2$ resulted in significant shortening in rabbit muscle compared with muscles injected with 0.1 M $MgCl_2$. These effects could be attributed to the role of Ca^{2+} ions in muscle contraction. Given the shortening effect of pre-rigor $CaCl_2$ injection, careful design is needed to study the impact of Ca^{2+} ions on tenderization, independently from the effects on contraction (Hopkins and Thompson, 2002).

Several chemical compounds that are permitted as additives and have a generally regarded as safe (GRAS) status have been used to improve the tenderness and the juiciness of different meats. The use of calcium chloride has been by far the most studied because of its effective stimulation of the calpains. Calcium salts in general, and calcium chloride in particular, have been used to improve the tenderness of beef (Koohmaraie *et al.*, 1990; Morgan *et al.*, 1991; Wheeler *et al.*, 1992, 1993, 1996, 1997; Geesink *et al.*, 1994; Boleman *et al.*, 1995; Polidori *et al.*, 2001; Pazos *et al.*, 2002; Jaturasitha *et al.*, 2004), lamb (Koohmaraie *et al.*, 1988, 1989, 1990; Koohmaraie and Shackelford, 1991; Mendiratta *et al.*, 1999; Polidori *et al.*, 2000; Ilian *et al.*, 2004), camel meat (Al-Sheddy and Al-Owaimer, 2000), and pork (Rees *et al.*, 2002). Sodium chloride has been used to improve meat tenderness and juiciness owing to its ability to solubilize myofibrillar proteins and increase the water-holding capacity in beef (Geesink *et al.*, 1994) and lamb (Koohmaraie *et al.*, 1989). Other studies used mixtures containing various compounds (e.g. maltose, dextrose, polyphosphate, glycerine and flavouring mixtures) to improve the palatability of beef (Farouk *et al.*, 1992a,b; Paterson *et al.*, 1988; Lee *et al.*, 2000; Yancey *et al.*, 2002; Dikeman *et al.*, 2003; McGee *et al.*, 2003), lamb (Farouk and Price, 1994; McKenna *et al.*, 2003; Murphy and Zerby, 2004), pork (Wu *et al.*, 1990; Sheard and Tali, 2004; Sheard *et al.*, 2005; Stephens *et al.*, 2006) and other meats (Dhanda *et al.*, 2002, 2003). Also several weak organic acids (lactic, citric and acetic) have been used to improve meat tenderness (Wendham and Locker, 1976; Aktaş and Kaya, 2001; Berge *et al.*, 2001; Burke and Monahan, 2003; Őnenç *et al.*, 2004; Ke *et al.*, 2009). Other unorthodox effective marinating products (soy sauce, miso paste and fermented apple solution) resulted in 20–30% tenderization (Ahmed *et al.*, 2006). The tenderizing effect was substantially increased (30–50%) when the meat was mechanically tenderized before the treatment. Higher tenderization was achieved with fermented apple juice and vacuum packaging for 3 h (70% tenderization).

Apart from the wide range of compounds used, several techniques have been employed such as injection at different post-mortem times, marination with or without mechanical tenderization and pre-rigor infusion. It should be remembered that the compounds mostly do not cause the tenderizing effect directly but they are 'activators' or 'modifiers' of enzymes and proteins, with their effects being dependent on other factors such as pH, temperature and the presence of cofactors. Therefore, the impact of the introduced compounds on meat quality will be greatly dependent upon the post-mortem time of the treatment (reflecting the pH and the temperature of the meat), the concentration of introduced compounds (level of activation or modification) and the method of introduction (the distribution of the compounds in the meat). Each combination of the above factors can lead to unique outcomes for different species and within these combinations a set of factors can be optimized for the best outcome. The literature describes several techniques (marination, tumbling, injection and infusion) for the introduction of the compounds in the meat, but in practice only injection and tumbling have been adopted (possibly because of the familiarity with these methods, which are commonly used for processed meat products and small goods, and the capability of handling a large volume output). The purpose of the injection step is to accelerate the penetration rate and generate a uniform distribution of the infused compounds. The injection sometimes causes localized effect and the addition of tumbling ensures a more uniform distribution of the compounds within the meat cut.

The compounds used in meat enhancement do not have the same mechanism of action or the same impact on meat quality. It would therefore be appropriate to discuss the effects separately for each compound. Several benefits can be gained from meat enhancement with the most obvious being the ability to modify the textural attributes (tenderness and juiciness). It has also been suggested that meat enhancement might have a protective effect by maintaining the desirable palatability even when meat is overcooked (Vote *et al.*, 2000).

11.6 Calcium Salts

11.6.1 Description

Since Koohmaraie *et al.* (1988, 1989) demonstrated the tenderizing effects of calcium chloride through pre-rigor infusion, numerous studies have examined the mechanism leading to meat tenderization. Several calcium salts have been investigated (chloride, lactate, ascorbate and dicalcium hydrogen phosphate) that exhibited equal tenderizing effects (Lawrence *et al.*, 2003) but they have varying effects on other meat-quality attributes (Appendix Tables A11.1 and A11.2). The use of calcium chloride at a concentration of 0.3 M and 10% of the meat/animal weight was adopted by Koohmaraie *et al.* (1988) and that level of use was examined in many subsequent studies (Polidori *et al.*, 2001, 2002; Rees *et al.*, 2002; Dikeman *et al.*, 2003; Ilian *et al.*, 2004; Kong *et al.*, 2006). The key point is to attain a homogenous distribution of the compounds at the cellular level, which can be successfully obtained by infusion or multi-needle injection and tumbling. Calcium chloride is very effective in improving the tenderness of meat with very tender meat obtained as early as 12 h post-mortem by pre-rigor infusion of lamb (Ilian *et al.*, 2004) and it can be employed to mitigate inherent tenderness problems such as in Callipyge lambs (Koohmaraie *et al.*, 1998).

11.6.2 Mechanism of action

The increase in meat tenderness during ageing is associated with the degradation of structural proteins and this is generally attributed to the actions of the calpains (Huff-Lonergan *et al.*, 1996; Koohmaraie and Geesink, 2006; Geesink *et al.*, 2006). This enzyme system requires Ca^{2+} for activation and providing this ion early post-mortem accelerates post-mortem proteolysis (Koohmaraie and Shackelford 1991; Wheeler *et al.*, 1993) and increases the degradation of the Z disc (which forms the border between sarcomeres), leading to the tenderization of meat (Koohmaraie, 1994). Calcium chloride was initially used in pre-rigor infusion studies, and results in vigorous muscle contractions that occur during the infusion process. This has led to the formulation of three hypotheses to explain the effects of calcium chloride in the muscle: (i) increased muscle proteolysis owing to the activation of calpains (Koohmaraie and Shackelford, 1991; Koohmaraie and Geesink, 2006; Geesink *et al.*, 2006); (ii) disruption of the myofibril network by the extreme contractions caused by the excess Ca^{2+} ions similar to electrical stimulation (Morgan *et al.*, 1991); and (iii) structural weakening caused by increased ionic strength (Wu and Smith, 1987; Nishimura *et al.*, 1995; Takahashi, 1996). Calpain inhibitors affected the extent of tenderization and proteolysis of meat (Uytterhaegen *et al.*, 1994) but did not completely eliminate tenderization during ageing (Hopkins and Thompson, 2001). When myofibrils that have been treated to remove endogenous proteolytic enzymes (i.e. calpains) are incubated with 0.3 M $CaCl_2$ some key myofibrillar proteins show no degradation with time, indicating that protein solubility alone cannot explain tenderization. This indicates that calpains play a major role in post-mortem tenderization.

Post-rigor meat does not exhibit any contractions when injected with Ca^{2+} and the meat tenderization process still occurs, although at a lower level (Appendix Table A11.1). Ca^{2+} injection of meat causes about 31–35% shortening in the sarcomere length of unstretched lamb (Bekhit *et al.*, 2005). The negative relationship between muscle shortening and meat tenderness as discussed earlier suggests toughening should take place instead of tenderization, but this is not the case. Sarcomere super-contractions in meat can cause tenderization within the range 40–60% shortening (Marsh *et al.*, 1974); therefore shortening is not involved in the tenderization by Ca^{2+}. Collectively, the above information confirms that the activation of calpains is the key for the tenderness found using Ca^{2+}. Several calpains have been shown to be activated as a result of Ca^{2+} injection/infusion and thus increased myofibrillar protein degradation (Whipple and Koohmaraie, 1993; Ilian *et al.*, 2004). Collagen solubility is unaffected by $CaCl_2$ (Mendiratta *et al.*, 1999; Ostoja and Cierach, 2003; Jaturasitha *et al.*, 2004).

Camel meat was shown to benefit from calcium chloride infusion with a reduction in the shear force at 6-day post-mortem of 23% and 32% at 200 mM and 300 mM $CaCl_2$ infusion concentrations, respectively, compared with control untreated camel meat (Al-Sheddy and Al-Owaimer, 2000).

11.7 Other Compounds Used for Meat Enhancement

11.7.1 Salts

Several formulations containing various ratios of salt, polyphosphate salts, sugars and carbohydrates have been used at experimental and commercial levels to enhance meat eating qualities. Several sodium salts, sodium chloride (numerous publications), sodium pyruvate (Al-Sahal *et al.*, 2007), sodium lactate (Papadopoulos *et al.*, 1991b; Vote *et al.*, 2000; McGee *et al.*, 2003; Knock *et al.*, 2006), sodium acetate (Al-Sheddy *et al.*, 1999; Sheard *et al.*, 2005; Knock *et al.*, 2006; Stephens *et al.*, 2006), sodium ascorbate (Sheard *et al.*, 2005), sodium citrate (Sheard *et al.*, 2005, Stephens *et al.*, 2006), sodium bicarbonate (Sheard and Tali, 2004; Rosenvold *et al.*, 2006) have been investigated for meat enhancement. Potassium lactate (Knock *et al.*, 2006), potassium chloride (Wu *et al.*, 1990) and ammonium hydroxide (Hamling and Calkins, 2008) seem to offer the advantage of improving the tenderness without adding sodium or affecting the sensory attributes.

Sodium chloride has been used at various concentrations (0.5–6% solution) with the highest level reported by Aktaş and Kaya (2001). Lactate salts (sodium and potassium) have been used in beef tenderizing solutions at a level of 2.5–3%. The use of polyphosphates is regulated and the maximum level is 0.5%. Trends toward less sodium intake and the saltiness desired (depending on personal and cultural factors) will limit the level of inclusion and the use of different salts (e.g. potassium). Several functions have been assigned to salts in enhanced meats, such as flavour development, improving the water-holding capacity, increasing ionic strength, and solubilization of myofibrillar proteins. All these factors will promote protein modification including effects on the endogenous enzymes.

11.7.2 Sodium chloride and phosphates

Generally, the addition of salt increases the negative charge on proteins above their isoelectric point, which will increase the electrostatic repulsion forces between myofibrillar proteins. This in turn allows more side groups to be available for interactions with water, removing some of the structural constraints to retain water (lattice swelling). This will lead to a better ability to bind water and increase the water-holding capacity.

Salt injection/marination (0.6 M) has a tenderizing effect which matches that obtained with $CaCl_2$ after ageing. Koohmaraie *et al.* (1989) found that NaCl produced about 20% of the tenderization achieved with $CaCl_2$, and a higher tenderization rate early post-mortem has been reported (Geesink *et al.*, 1994; Thomson and Dobbie, 1999). There is general agreement regarding the tenderizing effects of phosphates and sodium chloride used with other ingredients in several meat cuts (Vote *et al.*, 2000; Robbins *et al.*, 2003; Wicklund *et al.*, 2005; Baublits *et al.*, 2006a,b). Although the phosphate type does not play an important role, the level of pumping seems to play an important role in the sensory assessment (Baublits *et al.*, 2006a). Sensory panellists normally found injected meat to have increased myofibrillar tenderness and less connective tissue present.

This effect can be a result of the ability of salt to solubilize myofibril proteins (Wu and Smith, 1987). However, calpains may also be activated leading to higher proteolysis (Lee *et al.*, 2000). The maximum water-holding capacity can be obtained with 0.8–1.0 M (4.6–5.8%) salt (Offer and Trinick, 1983), although best functionality is achieved at 0.4–0.6 M (Trout and Schmidt, 1983). The activity of the salt can be improved with the addition of polyphosphate salts (Murphy and Zerby, 2004). Offer and Trinick (1983) and Detienne and Wicker (1999) reported a synergistic effect between salts and phosphates and found that the salt requirements to improve the water-holding capacity can be halved by the addition of 10 mM tetrasodium pyrophosphate. Sodium chloride and polyphosphates accelerate the degradation rates of titin and troponin-T as well as the appearance of 95 kDa and 30 kDa degradation products (Lee *et al.*, 2000), which leads to higher tenderization rates. Lee *et al.* (2000) attributed these effects to an increased pH owing to the high buffering capacity of polyphosphates.

Sodium chloride, unlike calcium chloride, does not cause a bitter taste in meat, but there are some conflicting results on the impact of enhancements with sodium chloride and polyphosphates on meat flavour. An increase in beef flavour was reported (Vote *et al.*, 2000; Knock *et al.*, 2006) as well as a decrease or no change in beef flavour (Robbins *et al.*, 2003; Molina *et al.*, 2005; Stetzer *et al.*, 2008). Off-flavours have also been reported in meat injected with high salt concentrations (Murphy and Zerby, 2004). Salt and polyphosphates decreased redness in meat and the effect paralleled an increase in concentration (Aktaş and Kaya, 2001). The impact of sodium chloride and phosphate enhancement is different among different muscles (Molina *et al.*, 2005; Stetzer *et al.*, 2008) and this might explain, in part, some of the reported contradictory outcomes.

11.7.3 Sodium and potassium lactate

Sodium and potassium lactate are both listed as GRAS compounds and their usage in meat and poultry is regulated. The maximum usage level of sodium/potassium lactate is 2.9% (4.8% from commercial products that contain 60% solutions). That limit can be increased to a maximum of 4% if sodium and potassium lactate are used for their antimicrobial effects. The bacteriostatic effect of lactate salts has been reported and represents an advantage for cooked meat products (Papadopoulos *et al.*, 1991a; Miller and Acuff, 1994) even under abused storage conditions (Maca *et al.*, 1999).

Sodium and potassium lactate can enhance meat flavour and reduce off-flavours in cooked beef compared with untreated meat (Papadoupolos *et al.*, 1991a). They can also increase the cooking yield of meat (12% increases resulting from the use of 3% sodium lactate) (Papadoupolos *et al.*, 1991b). As mentioned earlier, lactate salts have a beneficial effect on fresh meat colour (Kim *et al.*, 2009) that deserves further investigation.

11.7.4 Sodium carbonates

Sodium carbonate can increase the tenderness, juiciness and overall palatability of meat (Sheard and Tali, 2004; Rosenvold *et al.*, 2006). The mode of action and the efficacy of carbonates are similar to the actions of salt and phosphates, which are brought about by the ability to solubilize myofibrillar proteins and enhance their electrostatic repulsion primarily through the elevation of the pH. As a result, carbonate treatments allow the lattice expansion of myofibrils and the ability to retain water. The use of sodium carbonate with sodium chloride reduces cooking losses (Sheard and Tali, 2004; Rosenvold *et al.*, 2006) and improves the yield. It has been suggested that carbonates promote protein swelling thereby enhancing the juiciness of cooked products and thus increasing consumer acceptance. Bicarbonate reduced the beef flavour in fried beef and produced a more intense beef flavour in boiled beef, leading to a better sensory perception. The effects of the administration of sodium carbonate in feed/pre-slaughter have been investigated (Ahn *et al.*, 1992; Boles *et al.*, 1994; Sen *et al.*, 2006; Bodas *et al.*, 2007) with impacts that seem to be species dependent. For example, the use of sodium carbonate caused significant changes in pork (Ahn *et al.*, 1992; Boles *et al.*, 1994) but not in lamb (Sen *et al.*, 2006; Bodas *et al.*, 2007).

11.7.5 Acids

The use of natural citrus juices solely or with other additives (herbs, sugar and other culinary condiments) to marinate meat is traditionally practiced in many parts of the world, especially Mediterranean countries. This is done to improve the eating quality of the meat when ageing was not normally practiced because of the tradition of home killing, the lack of refrigerating facilities and the high ambient temperature. The use of acids/acidic solutions, including citrus fruit juices, through injection or marination (Wenham and Locker, 1976; Aktaş and Kaya, 2001; Berge *et al.*, 2001; Burke and Monahan, 2003; Őnenç *et al.*, 2004; Ke *et al.*, 2009) has been shown to improve the tenderness of meat. Acids upon injection in meat will reduce the meat pH and this in turn will create swelling of the meat fibres and activate cathepsins, leading to increased protein degradation (Berge *et al.*, 2001), then tenderization and better moisture retention. Wendham and Locker (1976) observed that lactic acid is more effective with meat cuts that contain high levels of connective tissues. Pre-rigor injection with lactic acid activates lysosomal enzymes in beef and increases the solubilization of collagen following heat treatment at 60°C (Ertbjerg *et al.*, 1999; Berge *et al.*, 2001). Both citric and lactic acids (in the range of 0.5–1.5%) were more efficient in reducing the temperatures for the onset and peak denaturation of intramuscular connective tissues compared with $CaCl_2$ and NaCl (in the range of 50–150 mM $CaCl_2$ and 2–6% salt; Aktaş and Kaya, 2001). This information confirms the potential use of lactic acid to improve the tenderness of inherently tough meat cuts. Because the effect of acid injection is largely dependent on the buffering capacity and overall pH reduction in meat, the type of acid used is very important to achieve the desired quality attributes. Lactic acid lowers the pH compared with citric and acetic acids, and elicits more dramatic changes in the meat. Because the actions of the acids are brought about by lowering the meat pH, the use of salts that increase the pH (e.g. sodium tri-polyphosphate) tends to cancel the tenderizing effect of the acid (Ke *et al.*, 2009). It is worth mentioning that not all the improvements in the meat-keeping qualities are attributed to the lower pH generated by citric acid. For example, the lipid inhibition effect of citric acid remained, even if the pH of the meat was readjusted to its normal pH (Ke *et al.*, 2009). Acids generally affect colour negatively and cause lower redness (a^*) values (Aktaş and Kaya, 2001; Őnenç *et al.*, 2004).

Organic acid salts (sodium acetate, potassium sorbate, sodium lactate and trisodium citrate) have been investigated to extend the shelf life of camel meat and only sodium acetate significantly increased the microbial shelf life by 6 days (Al-Shaddy *et al.*, 1999). The effect of these compounds on camel meat tenderness is yet to be determined.

11.8 Conclusion

Many methods are used to tenderize meat, but there is no generic solution to solve tenderness problems. The reason for this is that meat toughness/tenderness is determined by the amount and maturity of connective tissues, the level of muscle contraction and the level of tenderization during ageing (determined by the level of proteases and inhibitors). Depending on the muscle type, species, animal age, processing conditions and cooking method, the relative contribution of these factors to toughness varies. The optimal tenderization strategy might therefore differ depending on the meat cut and the recommended cooking method. This is particularly important for camel meat where the physiological factors affecting the muscles and the carcass size are unique. Also, cooking conditions in camel-consuming regions are different from other western societies, which should be taken into account in evaluating the final tenderization level achieved by the different strategies.

The application of electrical stimulation would improve the tenderness of short-aged camel meat (Kadim *et al.*, 2009) and is recommended in combination with suitable ageing periods; there is scope to study whether pre-rigor stretching techniques would also confer benefits to camel meat. Where the meat is to be consumed fresh, the use of exogenous proteases offers the possibility to selectively degrade connective tissues or proteins involved in muscle contraction. Therefore, the right mix of proteases can, in principle, solve tenderness problems of different cuts of meat. This approach cannot, however, be applied without stringent control of the process. Possible adverse effects include over-tenderization and the development of off-flavours. One of the obstacles in taking full advantage of the available commercial proteases is the lack of information on their specific activity against the different muscle proteins. Several methods or a combination of the methods described above can be used to improve the tenderness of camel meat, especially from old animals.

References

Ahmed, A.M., Matsumoto, N., Kawahara, S., Ohta, K., Kuroda, R., Okayama, T., Nakade, K., Numata, M., Nakamura, T. and Muguruma, M. (2006) Effects of various treatments on the texture softening of post-breeding mature cows meat. *In Proceedings of the 52nd International Congress of Meat Science and Technology*, Dublin, pp. 431–432.

Ahn, D.U., Patience, J.F., Fortin, A. and McCurdy, A. (1992) The influence of pre-slaughter oral loading of acid or base on postmortem changes in *Longissimus dorsi* muscle of pork. *Meat Science* 32, 65–79.

Aktaş, N. and Kaya, M. (2001) The influence of marinating with weak organic acids and salts on the intramuscular connective tissue and sensory properties of beef. *European Food Research and Technology* 213, 88–94.

Al-Sahal, A.A., Dikeman, M.E. and Kastner, C.I. (2007) Quality assessment of bovine muscle injected with sodium pyruvate. In: *Proceedings of the 53rd International Congress of Meat Science and Technology*, Beijing, pp. 399–400.

Al-Sheddy, I.A. and Al-Owaimer, A.N. (2000) Postmortem injection of calcium chloride improves camel meat tenderness. Research Bulletin No. 98, Research Centre of College of Agriculture, King Saud University, Saudi Arabia, pp. 5–16.

Al-Sheddy, I., Al-Dagal, M. and Bazaraa, W.A. (1999) Microbial and sensory quality of fresh camel meat treated with organic acid salts and/or bifidobacteria. *Journal of Food Science* 64, 336–339.

Baublits, R.T., Pohlman, F.W., Brown, A.H. and Johnson, Z.B. (2006a) Enhancement with varying phosphate types, concentrations, and pump rates, without sodium chloride on beef *Biceps femoris* quality and sensory characteristics. *Meat Science* 72, 404–414.

Baublits, R.T., Pohlman, F.W., Brown, A.H., Yancey, E.J., Johnson, Z.B. and Dias-Morse, P. (2006b) Solution enhancement and post-enhancement storage effects on the quality, sensory, and retail display characteristics of beef *Triceps brachii* muscles. *Journal of Food Science* 71, S91–S96.

Behrends, J.M., Goodson, K.J., Koohmaraie, M., Shackelford, S.D., Wheeler, T.L., Morgan, W.W., Reagan, J.O., Gwartney, B.L., Wise, J.W. and Savell, J.W. (2005) Beef customer satisfaction:

USDA quality grade and marination effects on consumer evaluations of top round steaks. *Journal of Animal Science* 83, 662–670.

Bekhit, A.E.D., Ilian, M.A., Morton, J.D., Vanhanan, L., Sedcole, J.R. and Bickerstaffe, R. (2005) Effect of calcium chloride, zinc chloride, and water infusion on metmyoglobin reducing activity and fresh lamb color. *Journal of Animal Science* 83, 2189–2204.

Bekhit, A.E.D., Hopkins, D.L., Geesink, G., Bekhit, A.A. and Franks, P. (2012a) Exogenous proteases for meat tenderization. *Critical Reviews in Food Science and Nutrition* (in press).

Bekhit, A.E.D., Carne, A., Ha, M. and Franks, P. (2012b) Physical interventions to manipulate texture and tenderness of fresh meat: A review. *International Journal of Food Properties* (in press).

Belew, J.B., Brooks, J.C., McKenna, D.R., Savell, J.W. (2003) Warner-Bratzler shear evaluations of 40 bovine muscles. *Meat Science* 64, 507–512.

Berge, P., Ertbjerg, P., Larsen, L.M., Astruc, T., Vignon, X., Møller, A.J. (2001) Tenderization of beef by lactic acid injected at different times post mortem. *Meat Science* 57, 347–357.

Bodas, R., Rodríguez, A.B., López, S., Fernández, B., Mantecón, A.R. and Giráldez, F.J. (2007) Effects of the inclusion of sodium bicarbonate and sugar beet pulp in the concentrate for fattening lambs on acid–base status and meat characteristics. *Meat Science 77*, 696–702.

Boleman, S.J., Boleman, S.L., Bidner, T.D., McMillin, K.W. and Monlezun, C.J. (1995) Effects of postmortem time of calcium chloride injection on beef tenderness and drip, cooking and total loss. *Meat Science* 39, 35–41.

Boles, J.A., Patience, J.F., Schaefer, A.L. and Aalhus, J.L. (1994) Effect of oral loading of acid or base on the incidence of pale soft exudative pork (PSE) in stress-susceptible pigs. *Meat Science* 37, 181–194.

Burke, R.M. and Monahan, F.J. (2003) The tenderisation of shin beef using a citrus juice marinade. *Meat Science* 63, 161–168.

Carr, M.A., Crockett, K.L., Ramsey, C.B. and Miller, M.F. (2004) Consumer acceptance of calcium chloride-marinated top loin steaks. *Journal of Animal Science* 82, 1471–1474.

Detienne, N.A. and Wicker, L. (1999) Sodium chloride and tripolyphosphate effects on physical and quality characteristics of injected pork loins. *Journal of Food Science* 64, 1042–1047.

Devine, C.E., Wells, R.W. and North, M. (2006) Variation of meat shear force measurements-A characteristic of meat. In: *Proceedings of the 52nd International Congress of Meat Science and Technology*, Dublin, pp. 575–576.

Dhanda, J.S., Pegg, R.B., Janz, J.A.M., Aalhus, J.L. and Shand, P.J. (2002) Palatability of bison *Semimembranosus* and effects of marination. *Meat Science* 62, 19–26.

Dhanda, J.S., Pegg, R.B. and Shan, P.J. (2003) Tenderness and chemical composition of Elk (*Cervus elaphus*) meat: Effects of muscle type, marinade composition, and cooking method. *Journal of Food Science* 68, 1882–1888.

Dikeman, M.E., Hunt, M.C., Schoenbeck, J., Addis, P.B., Katsanidis, E., Pullen, M. and Yancey, E.J. (2003) Effects of postexsanguination vascular infusion of cattle with a solution of saccharides, sodium chloride, and phosphates or with calcium chloride on meat quality and sensory traits of steaks and ground beef. *Journal of Animal Science* 81, 156–166.

Ertbjerg, P., Larsen, L.M. and Møller, A.J. (1999) Effect of prerigor lactic acid treatment on lysosomal enzyme release in bovine muscle. *Journal of the Science of Food and Agriculture* 79, 95–100.

Farouk, M.M. and Price, J.F. (1994) The effect of postexsanguination infusion on the composition, exudation, color and postmortem metabolic changes in lamb. *Meat Science* 38, 477–496.

Farouk, M.M., Price, J.F. and Salih, A.M. (1992a) Post-exsanguination infusion of ovine carcasses: Effect on tenderness indicators and muscle microstructure. *Journal of Food Science* 57, 1311–1315.

Farouk, M.M., Price, J.F., Salih, A.M. and Burnett, R.J. (1992b) The effect of postexsanguination infusion of beef on composition, tenderness and functional properties. *Journal of Animal Science* 70, 2773–2778.

Geesink, G., Sujang, S. and Koohmaraie, M. (2011) Tenderness of pre- and post rigor lamb *Longissimus* muscle. *Meat Science* 88, 723–726.

Geesink, G.H., Smulders, F.J.M. and van Laack, R.L.J.M. (1994) The effects of calcium-, sodium- and zinc-chlorides treatment on the quality of beef. *Sciences des Aliments* 14, 485–502.

Geesink, G.H., Kuchay, S., Chishti, A.H. and Koohmaraie, M. (2006) μ-Calpain is essential for postmortem proteolysis of muscle proteins. *Journal of Animal Science* 84, 2834–2840.

Got, F., Rousset-Akrim, S., Bayle, M.C. and Culioli, J. (1996) Effects of injection of calcium as a lactate salt on tenderness and flavour of beef. In: *Proceedings of the 42nd International Congress of Meat Science and Technology*, Lillehammer, Norway, pp. 394–395.

Hamling, A.E. and Calkins, C.R. (2008) Enhancement of beef chuck and loin muscles with ammonium hydroxide and salt. *Journal of Animal Science* 86, 967–971.

Hopkins, D.L. and Thompson, J.M. (2001) Inhibition of protease activity 2. Degradation of myofibrillar proteins, myofibril examination and determination of free calcium levels. *Meat Science* 59, 199–209.

Hopkins, D.L. and Thompson, J.M. (2002) Factors contributing to proteolysis and disruption of myofibrillar proteins and the impact on tenderisation in beef and sheep meat. *Australian Journal of Agricultural Research* 53, 149–166.

Huff-Lonergan, E., Mitsuhashi, T., Beekman, D.D., Parrish, F.C. Jr, Olson, D.G. and Robson, R.M. (1996) Proteolysis of specific muscle structural proteins by m-calpain at low pH and temperature is similar to degradation in postmortem bovine muscle. *Journal of Animal Science* 74, 993–1008.

Ilian, M.A., Bekhit, A.E.D., Stevenson, B. Morton, J.D., Isherwood, P. and Bickerstaffe, R. (2004) Up- and down-regulation of *Longissimus* tenderness parallels changes in the myofibril-bound calpain 3 proteins. *Meat Science* 67, 433–445.

Jaturasitha, S., Thirawong, P., Lengwunta, V. and Kreuzer, M. (2004) Reducing toughness of beef from *Bos indicus* draught steers by injection of calcium chloride: effects of concentration and time postmortem. *Meat Science* 68, 61–69.

Jeremiah, L.E. (1981) Factors affecting consumer selection and acceptability of beef in central Alberta. *Journal of Consumer Studies and Home Economics* 5, 257–268.

Kadim, I.T., Mahgoub, O., Al-Marzooqi, W., Khalaf, S.K., Mansour, M.H., Al-Sinani, S.S.H. and Al-Amri, I.S. (2009) Effects of electrical stimulation on histochemical muscle fiber staining, quality, and composition of camel and cattle *Longissimus thoracis* muscles. *Journal of Food Science*, 74, S44–S52.

Ke, S., Huang, Y., Decker, E.A. and Hultin, H.O. (2009) Impact of citric acid on the tenderness, microstructure and oxidative stability of beef muscle. *Meat Science* 82, 113–118.

Kim, Y.H., Keeton, J.T., Smith, S.B., Maxim, J.E., Yang, H.S. and Savell, J.W. (2009) Evaluation of antioxidant capacity and colour stability of calcium lactate enhancement on fresh beef under highly oxidising conditions. *Food Chemistry* 115, 272–278.

Knock, R.C., Seyfert, M., Hunt, M.C., Dikeman, M.E., Mancini, R.A., Unruh, J.A., Higgins, J.J. and Monderen, R.A. (2006) Effects of potassium lactate, sodium chloride, and sodium acetate on surface shininess/gloss and sensory properties of injection-enhanced beef strip-loin steaks. *Meat Science* 74, 319–326.

Kong, B., Diao, X. and Xiong, Y.L. (2006) Postmortem calcium chloride injection alters ultrastructure and improves tenderness of mature Chinese yellow cattle *Longissimus* muscle. *Journal of Food Science* 71, C124–C129.

Koohmaraie, M. (1994) Muscle proteinases and meat aging. *Meat Science* 36, 93–104.

Koohmaraie, M. and Geesink, G.H. (2006) Contribution of postmortem muscle biochemistry to the delivery of consistent meat quality with particular focus on the calpain system. *Meat Science* 74, 34–43.

Koohmaraie, M. and Shackelford, S.D. (1991) Effect of calcium chloride infusion on the tenderness of lambs fed β-adrenergic agonist. *Journal of Animal Science* 69, 2463–2471.

Koohmaraie, M., Babiker, A.S., Schroeder, A.L., Merkel, R.A. and Dutson, T.R. (1988) Acceleration of postmortem tenderization in ovine carcasses through activation of Ca^{2+}-dependent proteases. *Journal of Food Science* 53, 1638–1641.

Koohmaraie, M., Crouse, J.D. and Mersmann, H.J. (1989) Acceleration of postmortem tenderization in ovine carcasses through infusion of calcium chloride: effect of concentration and ionic strength. *Journal of Animal Science* 67, 934–942.

Koohmaraie, M., Whipple, G. and Crouse, J.D. (1990) Acceleration of postmortem tenderization in lamb and Brahman-Cross beef carcasses through infusion of calcium chloride. *Journal of Animal Science* 68, 1278-1283.

Koohmaraie, M., Shackelford, S.D. and Wheeler, T.L. (1998) Effect of prerigor freezing and postrigor calcium chloride injection on the tenderness of Callipyge *Longissimus. Journal of Animal Science* 76, 1427–1432.

Lawrence, T.E., Dikeman, M.E., Hunt, M.C., Kastner, C.L. and Johnson, D.E. (2003) Effects of calcium salts on beef *Longissimus* quality. *Meat Science* 64, 299–308.

Lee, S., Stevenson-Barry, J., Kauffman, R.G. and Kim, B.C. (2000) Effect of ion fluid injection on beef tenderness in association with calpain activity. *Meat Science* 56, 301–310.

Lepetit, J. (2008) Collagen contribution to meat toughness: theoretical aspects. *Meat Science* 80, 960–967.

Maca, J.V., Miller, R.K., Bigner, M.E., Lucia, L.M. and Acuff, G.R. (1999) Sodium lactate and storage temperature effects on shelf life of vacuum packaged beef top rounds. *Meat Science* 53, 23–29.

Marsh, B.B., Leet, N.G. and Dickson, M.R. (1974) The ultrastructure and tenderness of highly cold-shortened muscle. *Journal of Food Technology* 9, 141–147.

McGee, M.R., Henry, K.L., Brooks, J.C., Ray, F.K. and Morgan, J.B. (2003) Injection of sodium chloride, sodium tripolyphosphate, and sodium lactate improves Warner–Bratzler shear and sensory characteristics of pre-cooked inside round roasts. *Meat Science* 64, 273–277.

McKenna, D.R., Strachan, D.S., Miller, R.K., Acuff, G.R. and Savell, J.W. (2003) Cranberry juice marinade improves sensory and microbiological properties of vacuum-packed lamb chops. *Journal of Muscle Foods* 14, 207–220.

Mendiratta, S.K., Lakshmanan, V. and Kesava, R.V. (1999) Effect of postmortem calcium chloride injection on tenderization of sheep meat. *Journal of Food Science and Technology* 36, 149–151.

Miller, M.F., Carr, M.A., Ramsey, C.B., Crockett, K.L. and Hoover, L.C. (2001) Consumer thresholds for establishing the value of beef tenderness. *Journal of Animal Science* 79, 3062–3068.

Miller, R.K. and Acuff, G.R. (1994) Sodium lactate affects pathogens in cooked beef. *Journal of Food Science* 59, 15–19.

Molina, M.E., Johnson, D.D., West, R.L. and Gwartney, B.L. (2005) Enhancing palatability traits in beef chuck muscles. *Meat Science* 71, 52–61.

Morgan, J.B., Miller, R.K., Mendez, F.M., Hale, D.S. and Savell, J.W. (1991) Using calcium chloride injection to improve tenderness of beef from mature cows. *Journal of Animal Science* 69, 4469–4476.

Murphy, M.A. and Zerby, H.N. (2004) Prerigor infusion of lamb with sodium chloride, phosphate, and dextrose solutions to improve tenderness. *Meat Science* 66, 343–349.

Nishimura, T., Hattori, A. and Takahashi, K. (1995) Structural weakening of intramuscular connective tissue during conditioning beef. *Meat Science* 39, 127–133.

Offer, G. and Trinick, J. (1983) On the mechanism of water holding in meat: the swelling and shrinking of myofibrils. *Meat Science* 8, 245–281.

Őnenç, A., Serdaro'lu, M. and Abdraimov, K. (2004) Effect of various additives to marinating baths on some properties of cattle meat. *European Food Research and Technology* 218, 114–117.

Ostoja, H. and Cierach, M. (2003) Effect of calcium ions on the solubility of muscular collagen and tenderness of beef meat. *Food* 47, 388–390.

Papadopoulos, L.S., Miller, R.K., Acuff, G.R., Lucia, L.M., Vanderzant, C. and Cross, H.R. (1991a) Consumer and trained sensory comparisons of cooked beef top rounds treated with sodium lactate. *Journal of Food Science* 56, 1141–1146, 1153.

Papadopoulos, L.S., Miller, R.K., Ringer, L.J. and Cross, H.R. (1991b) Sodium lactate effect on sensory characteristics, cooked meat color and chemical composition. *Journal of Food Science* 56, 621–626, 635.

Paterson, B.C., Parrish, F.C. Jr, and Stromer, M.H. (1988) Effects of salt and pyrophosphate on the physical and chemical properties of beef muscle. *Journal of Food Science* 53, 1258–1265.

Pazos, A.A., Salitto, V.A., Lasta, J.A., Bolondi, A., Guidi, S.M., Carduza, F.J. and Gonzalez, C.B. (2002) Tenderness Improvement of bovine *Biceps femoris* muscle by calcium chloride injection. *Journal of Muscle Foods* 13, 171–187.

Polidori, P., Trabalza Marinucci, M., Fantuz, F., Renieri, C. and Polidori, F. (2000) Tenderization of wether lambs meat through pre-rigor infusion of calcium ions. *Meat Science* 55, 197–200.

Polidori, P., Trabalza Marinucci, M., Fantuz, F. and Polidori, F. (2001) Post mortem proteolysis and tenderization of beef muscle through infusion of calcium chloride. *Animal Research* 50, 223–226.

Polkinghorne, R., Philpott, D.J., Gee, A., Doljanin, A. and Innes, J. (2008) Development of a commercial system to apply the Meat Standards Australia grading model to optimise the return on eating quality in a beef supply chain. *Australian Journal of Experimental Agriculture* 48, 1451–1458.

Rees, M.P., Trout, G.R. and Warner, R.D. (2002) Effect of calcium infusion on tenderness and ageing rate of pork m. *longissimus thoracis et lumborum* after accelerated boning. *Meat Science* 61, 169–179.

Robbins, K., Jensen, J., Ryan, K.J., Homco-Ryan, C., McKeith, F.K. and Brewer, M.S. (2003) Consumer attitudes toward beef and acceptability of enhanced beef. *Meat Science* 65, 721–729.

Rosenvold, K., Clausen, I. and Madsen, N.T. (2006) Mechanical tenderization and enhancement improve eating quality of beef from dairy cows. In: *Proceedings of the 52nd International Congress of Meat Science and Technology*, Dublin, Ireland, pp. 611–612.

Rousset-Akrim, S., Got, F., Bayle, M.C. and Culioli, J. (1996) Influence of $CaCl_2$ and $NaCl_2$ injection on texture and flavour of beef. *International Journal of Food Science and Technology* 31, 333–343.

Scanga, J.A., Delmore, R.J. Jr, Ames, R.P., Belk, K.E., Tatum, J.D. and Smith, G.C. (2000) Palatability of beef steaks marinated with solutions of calcium chloride, phosphate, and (or) beef flavoring. *Meat Science* 55, 397–401.

Sen, A.R., Santra, A. and Karim, S.A. (2006) Effect of dietary sodium bicarbonate supplementation on carcass and meat quality of high concentrate fed lambs. *Small Ruminant Research* 65, 122–127.

Shackelford, S.D., Wheeler, T.L. and Koohmaraie, M. (1995) Relationship between shear force and trained sensory panel tenderness ratings of 10 major muscles from *Bos indicus* and *Bos taurus* cattle. *Journal of Animal Science* 73, 3333–3340.

Sheard, P.R. and Tali, A. (2004) Injection of salt, tripolyphosphate and bicarbonate marinade solutions to improve the yield and tenderness of cooked pork loin. *Meat Science* 68, 305–311.

Sheard, P.R., Nute, G.R., Richardson, R.I. and Wood, J.D. (2005) Effects of breed and marination on the sensory attributes of pork from Large White and Hampshire-sired pigs. *Meat Science* 70, 699–707.

Stephens, J.W., Dikeman, M.E., Unruh, J.A., Haub, M.D. and Tokach, M.D. (2006) Effects of pre-rigor injection of sodium citrate or acetate, or post-rigor injection of phosphate plus salt on postmortem glycolysis, pH, and pork quality attributes. *Meat Science* 74, 727–737.

Stetzer, A.J., Cadwallader, K., Singh, T.K., Mckeith, F.K. and Brewer, M.S. (2008) Effect of enhancement and ageing on flavour and volatile compounds in various beef muscles. *Meat Science* 79, 13–19.

Swanek, S.S., Morgan, J.B., Owens, F.N., Gill, D.R., Strasia, C.A., Dolezal, H.G. and Ray, F.K. (1999) Vitamin D_3 supplementation of beef steers increases *Longissimus* tenderness. *Journal of Animal Science* 77, 874–881.

Takahashi, K. (1996) Structural weakening of skeletal muscle tissue during postmortem ageing of meat: the non-enzymatic mechanism of meat tenderization. *Meat Science* 43, 67–80.

Thomson, B.C. and Dobbie, P.M. (1997) The effect of calcium chloride treatment on shear force and weight loss in *Gluteus medius* and *Longissimus* muscles from pasture fed bulls. *New Zealand Journal of Agricultural Research* 40, 507–512.

Torrescano, G., Sánchez-Escalante, A., Giménez, B., Roncalés, P. and Beltrán, J.A. (2003) Shear values of raw samples of 14 bovine muscles and their relation to muscle collagen characteristics. *Meat Science* 64, 85–91.

Trout, G. and Schmidt, G.R. (1986) Effect of phosphates on the functional properties of restructured beef rolls: The role of pH, ionic strength, and phosphate type. *Journal of Food Science* 51, 14116–1423.

Uytterhaegen, L., Caleys, E. and Demeyer, D. (1994) Effects of exogenous protease effectors on beef tenderness development and myofibrillar degradation and solubility. *Journal of Animal Science* 72, 1209–1223.

Vote, D.J., Platter, W.J., Tatum, J.D., Schmidt, G.R., Belk, K.E., Smith, G.C. and Speer, N.C. (2000) Injection of beef strip loins with solutions containing sodium tripolyphosphate, sodium lactate, and sodium chloride to enhance palatability. *Journal of Animal Science* 78, 952–957.

Weiner, P.D. and Pearson, A.M. (1969) Calcium chelators influence some physical and chemical properties of rabbit and pig muscle. *Journal of Food Science* 34, 592–596.

Wendham, L.M. and Locker, R.H. (1976) The effect of marinading on beef. *Journal of the Science of Food and Agriculture* 27, 1079–1084.

Wheeler, T.L., Crouse, J.D. and Koohmaraie, M. (1992) The effect of postmortem time of injection and freezing on the effectiveness of calcium chloride for improving beef tenderness. *Journal of Animal Science* 70, 3451–3457.

Wheeler, T.L., Koohmaraie, M., Lansdell, J.L., Siragusa, G.R. and Miller, M.F. (1993) Effects of postmortem injection time, injection level, and concentration of calcium chloride on beef quality traits. *Journal of Animal Science* 71, 2965–2974.

Wheeler, T.L., Koohmaraie, M. and Shackelford, S.D. (1996) Effect of vitamin C concentration and co-injection with calcium chloride on beef retail display color. *Journal of Animal Science* 74, 1846–1853.

Wheeler, T.L., Koohmaraie, M. and Shackelford, S.D. (1997) Effect of postmortem injection time and postinjection aging time on the calcium-activated tenderization process in beef. *Journal of Animal Science* 75, 2652–2660.

Whipple, G. and Koohmaraie, M. (1993) Calcium chloride marination effects on beef steak tenderness and calpain proteolytic activity. *Meat Science* 33, 265–275.

Wicklund, S.E., Homco-Ryan, C., Ryan, K.J., McKeith, F.K., McFarlane, B.J. and Brewer, M.S. (2005). Aging and enhancement effects on quality characteristics of beef strip steaks. *Journal of Food Science* 70, S242–S248.

Wu, C.K., Ramsey, C.B. and Davis, G.W. (1990) Effects of infused glucose, sodium and potassium chlorides and polyphosphates on palatability of hot-boned pork. *Journal of Animal Science* 68, 3212–3216.

Wu, F.Y. and Smith, S.B. (1987) Ionic strength and myofibrillar protein solubilization. *Journal of Animal Science* 65, 597–608.

Yancey, E.J., Dobbels, T.E., Dikeman, M.E. and Katsanidis. E. (1999) Effects of vascular infusion of cattle after exsanguinations with a solution of saccharides, sodium chloride, and phosphates, with or without vitamin C, on carcass and meat palatability traits. *Journal of Animal Science* 77, 47 (Abstract).

Yancey, E.J., Hunt, M.C., Dikeman, M.E., Addis, P.B. and Katsanidis, E. (2001) Effects of postexsanguination vascular infusion of cattle with a solution of saccharides, sodium chloride, phosphates, and vitamins C, E, or C + E on meat display-color stability. *Journal of Animal Science* 79, 2619–2626.

Yancey, E.J., Dikeman, M.E., Addis, P.B., Katsanidis, E. and Pullen, M. (2002a) Effects of vascular infusion with a solution of saccharides, sodium chloride, and phosphates with or without vitamin C on carcass traits, Warner-Bratzler shear force, flavour-profile, and descriptive-attribute characteristics of steaks and ground beef from Charolais cattle. *Meat Science* 60, 341–347.

Yancey, E.J., Dikeman, M.E., Addis, P.B., Katsanidis, E. and Pullen, M. (2002b) Effects of vascular infusion with a solution of saccharides, sodium chloride, phosphates, and vitamins C, E, or both on carcass traits, Warner-Bratzler shear force, and palatability traits of steaks and ground beef. *Journal of Animal Science* 80, 1904–1910.

Appendix

Table A11.1. Summary of effects of various meat calcium infusion tenderisation treatments during post-mortem period.

Dose of $CaCl_2$ (level of injection of green weight)	Muscle	Ageing time (days)	Post-mortem time of injection (h)	Ageing temp (°C)	Shear force (%)[a]	Flavour	Colour	Sensory tenderness	Other	Reference
0.3M (10%)	SM	0	1	2	–24.8	–	–	–	Higher drip loss in injected and higher total loss with higher injection time	Boleman *et al.* (1995)
			12		–1.8					
			24		–8.5					
		10	1		–33.1					
			12		–20.3					
			24		–13.5					
0.2M (5%)	LTL	7	48	1	–23.6	↑Off-flavour with ↑ injection time and with ↑ageing time	No effect for injection time on a^* value. No effect up to 3 days display then $CaCl_2$ exhibits lower a^*	17.3%[b]		Wheeler *et al.* (1997)
		35			–23.3			12.1%		
		7	336		–19.4			12.7%		
		35			–4.3			13.1%		
		1	48		–12.7					
		2			–17.7					
		6			–13.3					
		8			–11.2					
		12			–17.1					
		14			–21.5					
0.175M (10%)	LM	7	0.5	2	–21.9	Beef flavour intensity $CaCl_2$ ≥ control ≥ water. No effect for injection time	Darker colour with $CaCl_2$ (LD and TB) and better colour with H_2O (LD and SM) in pre-rigor but no effect at 24 h injection time	10.2%	Water injection pre-rigor produced tougher and drier meat. Higher microbial count with injected meat compared with control and with pre-rigor compared with 24 h but no effects for interactions.	Wheeler *et al.* (1993)
	SM				–2.9			–		
	TB				–12.6			–		
Water	LM				2.9			–26.5%		
	SM				–7.9			–		
	TB				–			–		
0.175M (10%)	LM	6	24		–29.2			19.6%		
	SM				–25.6			–		
Water	TB				–15.0			–		
	LM				–22.4			9.8%		
	SM				–13.1			–		
	TB				–			–		

0.2 M (5%)	LM	6	24	2	–21.7	Less beef flavour and	Some effects	9.8	No effects on juiciness	
	SM				–25.6	more off-flavour	depending on the	6.7	or microbial count	
0.2 M (10%)	LM				–23.7	with 10%	muscle	11.8		
	SM				–10.9	compared with 5%		6.4		
0.25M (5%)	LM	6	24		–31.4	and with 0.25M		13.8		
	SM				–17.1	compared with		8.9		
	TB				–15.6	0.2 M.		3.7		
0.25M (10%)	LM				–25.0			15.7		
	SM				–8.5			–4.3		
	TB				–3.4			1.8	–	
0.075M	LD	1	0.5	1–2	–1.7	–	–	–		Koohmaraie
		6			11.7					*et al.*
0.15M		1			–28.1					(1989)
		6			–14.3					
0.3M		1			–60.0					
		6			–45.5					
0.3 M $CaCl_2$		1			–50.3					
(10%)		7			–35.5					
0.6 M NaCl		1			–9.0					
(10%)		7			–39.3					
0.3 M $CaCl_2$	LD	14	0.1	2–4	65.7	No effect on flavour	No effect on colour	–19.7	No effect on juiciness	Dikeman
(10%)	SM				–2.1	intensity		1.8		*et al.*
										(2003)
0.1 M $CaCl_2$	LD	2	1	4	–70.6[c]	Abnormal flavour with		16.9	Higher drip losses with	Rousset-
(10%)		6			–50.0	$CaCl_2$, especially		4.4	the injected samples,	Akrim
		14			–43.9	with pre-rigor		–9.6	especially with	*et al.*
		2	24		–53.6	injection		40.8	pre-rigor $CaCl_2$	(1996)
		6			–44.2			40.0	injection	
		14			–39.3			13.5		
0.15 M (NaCl		2	1		–33.1			2.8		
(10%)		6			–18.1			13.3		
		14			–35.0			8.7		
		2	24		–36.6			5.6		
		6			–23.7			18.9		
		14			–30.2			1		

Continued

Table A11.1. Continued.

Dose of $CaCl_2$ (level of injection of green weight)	Muscle	Ageing time (days)	Post mortem time of injection (h)	Ageing temp (°C)	Shear force (%)[a]	Flavour	Colour	Sensory tenderness	Other	Reference
0.3 M $CaCl_2$ (10%)	Strip loin	1	0.5	2	–52.8	Flavour intensity increased in $CaCl_2$ injected meat	–	90.9	No effect on juiciness or connective tissue	Morgan *et al.* (1991)
		7			–51.4			96.3		
		14			–41.6			95.6		
	Top sirloin	1			–62.6			147.7		
		7			–43.1			58.5		
		14			–40.4			49.9		
	Top round	1			–36.8			94.2		
		7			–20.8			62.0		
		14			–15.4			30.5		
0.3 M $CaCl_2$	LD	2	0.5	2	–71.4	–	–	–	Injection after freezing can improve tenderness but caused more cooking losses (8.5%↑)	Wheeler *et al.* (1992)
		5			–66.3					
		2	24		–24.3					
		5			–25.1					
Injected fresh		6	24		–37.5					
Injected after freezing		6	24		–46.4					
Injected before freezing for 1 day		6	24		–12.2					
Injected fresh		7	0.5		–61.2					
Fresh		6	24		–46.8					
Injected after freezing		6	24		–49.7					

[a]Difference from control (–, more tender; +, tougher); [b]tenderness rating (difference from control: +, more tender); [c]myofibrillar resistance by compression method of Lepetit.

Table A11.2. Summary of selected non-meat compounds used in research to improve the eating quality of meat.

Animal	Treatment / infused compounds	Experimental parameters	Impact on meat: Tenderness	Impact on meat: Colour	Impact on meat: Sensory flavour/ acceptability	Impact on meat: Other	Outcome	Reference
Sheep	0.3M $CaCl_2$	LTL from sheep were fed with 4ppm β-adrenergic agonist (BAA) for 6 weeks before slaughter. Carcasses were in cooler at −1.1°C for 24h, and then stored at 2°C	$CaCl_2$ increased meat tenderness and marbling	$CaCl_2$ increased meat lean colour score	–	–	$CaCl_2$ can overcome the negative effect of BAA	Koohmaraie and Shackelford (1991)
	0.1, 0.2 and 0.3M $CaCl_2$	Sheep hind legs at 3-4h PM were injected 10% of the weight with the tested solutions	↑ Tenderness in treated samples	No effect on colour	No effect on flavour. Higher acceptability for $CaCl_2$ treated samples	No effect on juiciness, WHC, microbiological status, or collagen solubility	Improvement of the tenderization without negative effects	Mendiratta *et al.* (1999)
	0.3M $CaCl_2$ water	After artery infusion in 10% live weight, carcasses (n = 12 for each treatment) for each treatment were placed in a cold room at 2°C for 24 h. LTL were vacuum packed, stored at 1°C for 2 and 6 days	Increased tenderness in $CaCl_2$ infused samples but not water	–	–	–	Improved tenderness at 2 and 6 days post-mortem for $CaCl_2$ infused samples	Polidori *et al.* (2000)
	0.3M $CaCl_2$	PM infusion through artery at 10 % of live weight level. 4 hrs at 15°C and then 4°C for 7 days LTL used for the study	↑ Rate of tenderization	↓ L^* and a^* values that persisted after aging	–	Increased titin and nebulin degradation in Ca^{2+} owing to activation of calpains	Lower sarcomere length	Bekhit *et al.* (2005)

Continued

Table A11.2. Continued.

Animal	Treatment / infused compounds	Experimental parameters	Impact on meat: Tenderness	Colour	Sensory flavour/ acceptability	Other	Outcome	Reference
	Water		Improved tenderness and rate of tenderization	↑ *L**, *a** and *b** values that persisted after ageing.	–	↑ Degradation of titin and neublin	Higher initial sarcomere length	
Cattle	0.15 M $CaCl_2$	LD obtained at 6 days PM. Steaks were either marinated for 48 hrs at 4°C directly or after being frozen at –30°C for 6 weeks	Higher tenderness in $CaCl_2$ marinated meat and in frozen-marinated compared with fresh-marinated	–	–	–	Prior freezing improved the tenderness by decreasing the activity of calpastatin	Whipple and Koohmaraie (1992)
	0.1 M or 0.3 M $C_6H_{10}CaO_6$ (calcium lactate)	LD from 6–8 year old animals were injected either pre-rigor (VP, and maintained in water bath 15°C for 24 h) or post rigor (VP, and kept at 4°C). Measurements at 2, 6 and 14 days PM	Lower initial WBSF values with 0.1 M pre-rigor at 2 days PM and no further reduction with further aging. 0.1M post-rigor had lower initial values and continued to decrease with ageing. Sensory: Pre-rigor injected samples were rated tougher whereas post-rigor injected samples were rated more tender over the aging time period	–	Pre-rigor injection caused the development of off-flavour and bitterness over all the ageing period. Post-rigor injected samples had significantly off-flavour and bitterness only at 6 and 14 days PM and the level of these parameters was higher with high lactate concentration	Severe contraction in pre-rigor	Post-rigor injection with calcium lactate is better than pre-rigor	Got *et al.* (1996)

0.2 or 0.3 M $CaCl_2$ 0.4 or 0.6 M NaCl	LTL from Friesian bulls at 1 h PM were allocated to one of five treatments: control (0 treatment), 10% (w/w) water, 10% 0.4M NaCl, 10% 0.2 M $CaCl_2$, and 20% 0.2 M $CaCl_2$. *Gluteus medius* (n = 14) control (no treatment) or injected with 0.2 M $CaCl_2$ (10% w/w) 2 h (Ca_2) or 24 h (Ca_{24}) post-slaughter. The remaining quarter was injected with 0.4 M sodium chloride (10% w/w) 2 h post-slaughter (NaCl, n = 8) or $CaCl_2$ 48 h post-slaughter (Ca_{48}, n = 6)	Tougher meat with $CaCl_2$ and more tender meat with NaCl	–	–	Higher purge loss in injected samples but overall losses (purge+ cooking) seemed to be the same	The species and the finishing diet may be an important factor for Ca^{2+} injection	Thomson and Dobbie (1997)
1) $CaCl_2$, 2) flavouring/ seasoning mixture, 3) $CaCl_2$ and beef-flavouring mix 4) sodium phosphate and beef- flavouring mix, and, 5) tap water.	24 PM *Gluteus medius*. All solutions were applied to the steaks in vacuum pouches at 25% solution added, VP 7 days at 2°C	Beef flavouring increased objective tenderness (about 19%) $CaCl_2$ at 150 mM = no effect		Beef flavouring increased subjective tenderness (about 20%)	Beef flavouring increased beef and salty flavour, and decreased metallic and bitterness. $CaCl_2$ at 150mM increased metallic and bitterness	Beef flavouring more connective tissue more fibre tenderness about 9% increase in yield	Scanga *et al.* (2000)

Continued

Table A11.2. Continued.

Animal	Treatment / infused compounds	Experimental parameters	Impact on meat: Tenderness	Colour	Sensory flavour/ acceptability	Other	Outcome	Reference
	0.3 M $CaCl_2$	After completion of artery infusion process (within 45 min after slaughter), carcasses were placed in a cold room at 2°C. LTL muscles (n = 24) injected. Shear force at 2 and 8 days PM	Increase tenderness in injected samples	–	–	$CaCl_2$ infusion improves beef tenderness through activation of µ- and m-calpain during post-mortem proteolysis and tenderization, $CaCl_2$ has no effect on sarcomere length	Improved tenderness at 2 and 8 days post-mortem for injected samples	Polidori *et al.* (2001)
	0.25 M $CaCl_2$	BF (n = 15) from 36–42 month old steers obtained at 24 h PM. Muscles injected with 0.25 M $CaCl_2$ to 110% of initial weight. Samples aged for 0–7 days at 1°C	Tenderness ↓WBSF for Ca treated samples and ↑ tenderness scores from panellists	–	71% of the panellists preferred Ca treated samples	–	The level of Ca^{2+} used improved the tenderness of BF. The level of tenderization was greatly improved by combining the injection with ageing	Pazos *et al.* (2002)

MPSC, 0.3M $CaCl_2$ ($CaCl_2$)	LTL, ST After vascular infusion at 10% live weight, carcasses were chilled at 2°C using a 1-min spray-chill cycle every 15 min for 8 h, then followed by 16 h of air chilling. Meat aged for 14 days	$CaCl_2$ infusion decreased LL tenderness due to severe muscle contraction early post-mortem. $CaCl_2$ has no effect on ST tenderness	–	$CaCl_2$ infusion decreased LL flavour intensity. $CaCl_2$ caused ↑chemical/ soapy flavour. MPSC infusion has no effect on flavour quality	–	–	Dikeman *et al.* (2003)
Distilled water or a 0.1, 0.2, or 0.3 M solution of calcium ascorbate, calcium chloride, or calcium lactate	*Longissimus* (unknown PM?) injected (11% by weight), VP, tumbled (15 min) stored for 14 days at -1°C	↑ Tenderness in all treated samples. Tenderization increase with increasing the concentration of calcium salts	Discoloration ↑ ≥0.2 $CaCl_2$ Ca ascorbate ↑discoloration	All three calcium salts equally increased sensory-tenderness scores, however tenderness of 0.3 M > 0.1 and 0.2 M. Juiciness, no impact	The CaLac inhibited microbial growth more than CaAsc or $CaCl_2$ treatments. Lipid oxidation↑ with ↑ CaLac and $CaCl_2$ but not with CaAsc	CaLac seems to be a better Ca source than $CaCl_2$ and CaAsc	Lawrence *et al.* (2003)
3.0 M $CaCl_2$ and $C_6H_{10}CaO_6$ (calcium lactate)	48 h PM ST injected (10% of the weight)/pickled (2:1, P:M) and kept for 2 days at 4°C	↑ Tenderness in treated samples	–	–	–	27–34% increase in tenderness without any changes in muscle collagen	Ostoja and Cierach (2003)
200 mM $CaCl_2$	LD steaks injected (5%) at 72 h post-mortem. Stored at 2°C for 7 d after injection	Sensory more tender compared to control	–	More juicy compared to control ($p > 0.05$) – trained panellists and consumers More beef flavour compared to control ($p > 0.05$) – trained panellists and consumers	Increased mouth-feel compared to control – trained panellists Better overall quality compared to control – consumers	Consumers, on a national basis, could detect improve-ments in tenderness, juiciness and flavour when $CaCl_2$ is injected into *Longissimus lumborum* steaks	Carr *et al.* (2004)

Continued

Table A11.2. Continued.

Animal	Treatment / infused compounds	Experimental parameters	Impact on meat: Tenderness	Colour	Sensory flavour/ acceptability	Other	Outcome	Reference
	$CaCl_2$ solution in concentrations of 0, 0.2, 0.3 and 0.4 M	10% (wt/wt) either 45 min or 24 h post-mortem into *Longissimus dorsi* (LD) muscles After injection, samples were stored for 7 days at 4°C in sealed but not vacuumed plastic bags and afterwards cut into 2.5 cm thick slices	Sensory tenderness scores were higher by 50% with all $CaCl_2$ concentrations, but only with pre-rigor treatment	$CaCl_2$ reduced redness and luminosity	A higher concentration of $CaCl_2$ solution did not further improve juiciness and acceptance. In the LD samples treated with solution containing 0.4 M $CaCl_2$, a bitter taste was remarked in 50% of all samples	Pre-rigor treatment was twice as efficient as post-rigor injection	–	Jaturasitha *et al.* (2004)
	0.2M $CaCl_2$	Top round (n = 200) including SM and *Adductor* muscles were injected at 2 or 3 days of PM. Meat aged for 2–3 weeks	No difference in tenderness	–	↓ Overall liking and flavour in injected samples when cooked medium or less, but increase in flavour in medium well and more	–	0.2 M $CaCl_2$ injection at 2–3 days PM has no effect on meat tenderness after 2–3 weeks of storage	Behrends *et al.* (2005)
	0.3 M $CaCl_2$ Water	LD muscles from 6–7-year old steers (*N* = 4) were obtained at 2 h PM and were split into 4 sections and randomly assigned into one of 4 treatments (control, injected with water, 0.3M $CaCl_2$ or 50mM $ZnCl_2$). The samples were wrapped with PVC film and stored at 4°C for 1, 4, 7 or 10 days)	↑ Tenderization rate in $CaCl_2$ injected samples but not in water	No effect on colour	–	Higher cooking losses in the injected samples	Activation of calpains as indicated by the ultra-structural changes of $CaCl_2$ treated samples	Kong *et al.* (2006)

Pigs	0.3M $CaCl_2$	Hot boned LTL muscles injected at 0.5 h pm or 6 h pm, then incubation at 0 or 14°C. Infusion was 10% body weight	$CaCl_2$ increase meat tenderness	$CaCl_2$ had no effect on colour except increase occasional L^* values	–	$CaCl_2$ had a negative effect on water holding capacity and increased drip loss	Infusion time has no effect on ageing rate	Rees *et al.* (2002)
Sheep	0.23% dextrose, 0.21% glycerine, 0.14% phosphate blend and 0.1% maltose (tenderizing blend, No Ca^{+2}, NCa), tenderizing blend+ 0.015M $CaCl_2$ (with Ca^{2+}, WCa)	LTL (loin), IS (shoulder), leg after artery infusion in 10% live weight, carcasses were kept in a holding cooler at 2–4°C	–	Infusion samples both fresh and frozen had higher lightness and yellowness than control, WCa had less red colour than NCa and control	–	Drip and cooking loss: WCa > NCa > control, Infusion has no effect on drip/cooking loss in refrigerated samples	–	Farouk and Price (1994)
	Marinade contained cranberry juice (36.9%), water (61.52%), salt (1.54%), black pepper oleoresin (0.0068%), onion oleoresin (0.0059), marjoram oleoresin (0.0146) and rosemary oleoresin (0.0008).	Loins were injected with the marinade to a level equivalent to 115% of the green weight. After injection, loins were cut into 2.54-cm thick chops. Chops from treated loins were coated with a spice rub containing black pepper and garlic powder. Chops from the same anatomical location of control and treated sides of each loin were paired and assigned randomly to a retail display day (0, 7 or 14 days)	Consumers indicated that tenderness was higher for treated chops than for controls ($p < 0.05$)	Cooked appearance was higher for treated chops than for controls ($p < 0.05$). Consumers preferred the appearance of raw control chops over the appearance of raw treated chops ($p < 0.05$)	Consumers indicated that odour, juiciness, flavour and overall-like were higher for treated chops than for controls $p < 0.05$)	Chops from loins injected with marinade had lower TBARS values than control chops. Control chops had Incrementally higher TBARS values after 7 and 14 days of retail display than did treated chops. Marination extended the shelf-life of lamb chops by reducing lipid oxidation and microbial growth	Marinating lamb chops improved palatability traits of cooked lamb chops and extended the shelf-life characteristics of raw lamb chops	McKenna *et al.* (2003)

Continued

Table A11.2. Continued.

Animal	Treatment / infused compounds	Experimental parameters	Impact on meat: Tenderness	Colour	Sensory flavour/ acceptability	Other	Outcome	Reference
	NaCl, Tri-polyphosphate (TPP), Dextrose (DX)	The carcasses (n = 6 each treatment) were split along the vertebral axis. Half the carcasses were injected pre-rigor with one of the following treatments while the other half served as a control, 1) deionized H_2O, 2) 2% NaCl, 3) 0.5% TPP, 4) 3% DX, 5) 2% NaCl + 0.5% TPP, 6) 2% NaCl + 3% DX, 7) 0.5% TPP + 3% DX and 8) 2% NaCl + 3% DX + 0.5% TPP. The target weight was 120% the green weight, left at 0–4°C for 24 h. Several tests were carried out on LD and SM	↑ Tenderness in treatments 6, 7 and 8 for LD and 4, 7 and 8 for SM. Sensory tenderness for LD was significant for treatments 4, 6, 7 and 8	Treatments had no effect on L^* and a^*. b^* was decreased by NaCl	Treatments 7 and 8 had off-flavour due to the high salts	Microbiological status: no effects in the most effective tenderizing treatments. The juiciness of LD was parallel to the tenderness and only significantly higher for treatments 4, 7 and 8. Higher pH_U in LD in treatments 2, 5, 6, 7 and 8	Individually, the examined compounds had no effect on improving the tenderness. Combining the compounds created synergistic effects. The increase in pH_U by the mixtures of compounds improved cooking losses and tenderness	Murphy and Zerby (2004)
Cattle	0.23% dextrose, 0.21% glycerine, 0.14% phosphate blend and 0.1% maltose (tenderizing blend)	SS, LTL, ST Dairy cows after artery infusion in 10% live weight, carcasses were kept in a holding cooler at 2–4°C	Tenderness and protein extractability were improved after infusion	–	–	Infusion had no effect on water-holding capacity	Tenderness was improved and dressed carcass yield was increased	Farouk *et al.* (1992a)

MPSC (98.52% water, 0.97% saccharides, 0.23% NaCl and 0.28% phosphate blend), MPSC + C (MPSC + 500 ppm VitC)	LL and ST were obtained After artery infusion in 10% live weight, carcases were placed in a spray-chill cooler at 0–2°C for 12 h, and then placed in a cooler at 2°C	Infusion had no effect on tenderness of LL/ST	–	Infusion had inconsistent effect on flavour profile of cooked beef	No effect on purge loss from vacuum-packaged muscles	MPSC infusion treatment had higher dressing percentage than control	Yancey *et al.* (1999)
Vitamin D	Animals (n= 20 and 47 over two trials) fed Vitamin D supplemented in diet at different levels 0, 2.5, 5.0 or 7.5 ×10^6 IU of vitamin D_3 per day) for different periods (up to 10 days) before slaughter. The consumed diet contained 91g of $CaCO_3$ in the supplement. Ca^{2+} in blood plasma and LD. WB shear force and trained panel tasting	↑ Tenderness in 7 days PM samples supplemented by Vit D but not in 14 and 21 days. Trained panel scored the control as more tender than treated samples (7 days PM)	–	No effect	Increase in Ca^{+2} in blood plasma and LD. Lower μ-calpain and calpastatin activities in Vit D supplemented LD	Increased activation of μ-calpain due to higher Ca^{2+} (about 50% more than in control)	Swanek *et al.* (1999)
Sodium chloride and Sodium pyrophosphate 0.1125 M $Na_4P_2O_7$+ 0.1125 M $Na_2H_2P_2O_7$+ 0.2 M NaCl (PPi)	Pre-rigor heifer SM and BF muscles (n = 12) 10% target weight. 8 h at 2°C	↑ Tenderness in injected meat	–	–	pH ↓ rate of decline over the 48 h PM temperature lower during the first 3 hours post-injection. ↑ % exudates in injected samples. No difference in cooking %	The rates of degradation of titin and troponin-T as well as the appearance of 95 and 30 kDa peptides were faster in the PPi-injected muscles than the controls	Lee *et al.* (2000a)

Continued

Table A11.2. Continued.

Animal	Treatment / infused compounds	Experimental parameters	Impact on meat				Outcome	Reference
			Tenderness	Colour	Sensory flavour/ acceptability	Other		
	Lactic acid (0.5M)	M. *Pectoralis profundus* from cull cows (3–4 years). Meat injected (target weight 10% w/w) at 1 or 24 h PM and aged for 2 or 14 days	↑Tenderness in lactic acid injected meat after 2 days of aging and no further improvements with prolonged tenderization for 14 days	–	–	Sharp drop of the meat pH after 4 h of injection (≈ pH 5)	Faster release of lysosomal enzymes into the cytosol in injected meat. No change in collagen solubilization due to injection indicating collageno-lytic rather than acid solubilization effects	Berge *et al.* (2001)
	MPSC + 500ppm VitC, MPSC + 500ppm VitE, MPSC + 500ppm VitC + 500ppm VitE (MPSC + C + E)	LT, PM, SM. After artery infusion in 10% live weight, carcases were spray chilled in a cooler at 1–2°C for 12h, and then placed in a cooler at 1–2°C.	–	LT from control had more uniform and cherry red colour than infusion treatment samples. Infusion had no effect on colour or display-colour stability of LT. Infusion solution with VitE improved colour stability of ground beef	–	–	–	Yancey *et al.* (2001)

98.52% water, 0.97% saccharides, 0.23% NaCl and 0.28% phosphate blend (MPSC), MPSC + 1000ppm VitC (MPSC + C)	LL and ST. Same procedure as Yancey *et al.* (2002b)	MPSC infusion had no effect on tenderness of LT & ST	–	MPSC had inconsistent effects on flavour of cooked beef. Additional VitC to MPSC increased soapy/chemical flavours	–	The differences were largely masked by the long ageing time	Yancey *et al.* (2002a)
MPSC [98.52% water, 0.97% saccharides, 0.23% NaCl and 0.28% phosphate blend], MPSC + C [MPSC + 500 ppm VitC] MPSC + E [MPSC + 500 ppm α-tocopherol] MPSC + E + C [MPSC + 500ppm α-tocopherol + 500ppm VitC]	LT and ST obtained at 48 h from carcasses infused 10% live weight, placed in a spray-chill cooler at 0–2°C for 12 h, and then placed in a cooler at 2°C The samples were VP and stored until 14 days PM at 2–4°C	No effect on tenderness	–	No effect on flavour intensity or off-flavour. Soapy flavour with Vit E but it may be due to the preparation of the infusion solution	–	No added benefit from the solution used in the study on LT and ST quality	Yancey *et al.* (2002b)
0.2 M solutions of, acetic acid citric acid lactic acid and citrus juice	PM, SM and shin beef slices at 48h PM (5 mm thick) were marinated for 20 h	Lower SF with all marinated samples. (38–210% increase in the tenderness rating)	–	–	Very low pH (≈3.1), less cooking loss, higher juiciness scores	Tenderness seemed to be due to liquid uptake and solubilization of collagen	Burke and Monahan (2003)
MPSC (0.3M $CaCl_2$, 98.52% water, 0.97% saccharides, 0.23% NaCl and 0.28% phosphate blend)	LTL, ST, QF. After artery infusion in 10% live weight, carcasses were chilled at 2°C using a 1-min spray-chill cycle every 15 min for 8 h after cooler entry followed by 16 h of air chilling	$CaCl_2$ had no effect on ST tenderness but it decreased LL tenderness due to severe muscle contraction	–	$CaCl_2$ had no effect on ST flavour, but it reduced the flavour intensity of LL steak and ground beef	–	Negative impact on tenderness	Dikeman *et al.* (2003)

Continued

Table A11.2. Continued.

Animal	Treatment / infused compounds	Experimental parameters	Impact on meat: Tenderness	Colour	Sensory flavour/ acceptability	Other	Outcome	Reference
	A solution of sodium lactate, sodium tripolyphosphate, and sodium chloride	Paired muscle samples were cut in half (20.32 × 15.24 × 15.24 cm) and assigned to one of four injection treatment groups (0, 5, 7 and 9%) All injection treatments were formulated to contain 0.25% sodium tripolyphosphate, 0.35% sodium chloride, and 2% sodium lactate Following injection, samples were stored for 24 h at 4 ± 1°C	Injected treatments were more tender ($p<0.05$) than control products, as measured by Warner–Bratzler shear force and consumer sensory panel ratings	–	The samples used for sensory evaluation were randomly assigned to a storage period (21, 28 and 35 days) Panellists found flavour at 28 days to be significantly more desirable For tenderness, panellist ratings were higher ($P < 0.05$) at 21 and 28 days when compared to 35 day samples	Injected treatments had lower ($P<0.01$) cooking and re-heating loss percentages when compared to control samples Lipid oxidation in injected treated samples was significantly reduced as compared to control meat samples	Injection of sodium tripolyphosphate, sodium chloride and sodium lactate is one method that may be used to help traditionally less tender beef cuts. Additional research is needed to determine the optimum levels of sodium tripolyphosphate, sodium chloride, and sodium lactate	McGee *et al.* (2003)

Samples were randomly assigned to four marination treatments: 1, 2% NaCl (w/v) control-CON; 2, 2% NaCl + 0.5% sodium tripolyphosphate (STP); 3, 2% NaCl + 0.5% citric acid (CA); 4, 2% NaCl + 0.5% dicalciumhydrogen phosphate (CHP)	LD at a ratio of 1:1 (meat: liquid) in plastic boxes and stored at 4°C for 24 h	All treatments significantly affected hardness, chewiness and resilience values of steaks The lowest hardness value was found in steaks marinated with CA STP and CHP treatments also improved tenderness	The highest lightness was found in steaks marinated with the CA solution The CA solution resulted in yellowness of steaks Steaks marinated with STP and CHP solutions were darker	–	Marinating with STP and CHP solutions resulted in lower cooking losses Marinade uptake and area gain were lower for control steaks and steaks marinated with the CA solution	Marination with 2% NaCl+0.5% STP or 2% NaCl + 0.5% CHP can be successfully used to enhance tenderness reduce cooking losses and improve water holding capacity in cattle meat	Önenç *et al.* (2004)
Potassium lactate (Lact) Sodium chloride (NaCl) Sodium acetate (Acet) All the formulations contained (0.3% $Na_5P_3O_{10}$ + 0.058% rosemary extract) Cont (0.3% NaCl) LC (1.5% Lact + 0.3%NaCl) LHS (1.5% Lact +0.6% NaCl) LA (1.5% + 0.3% NaCl + 0.1 Acet)	8 days post-mortem (n=4). 8.5% target weight increase Parameters were evaluated at 2, 9 and 14 days of storage in MAP (80% O_2 and 20% CO_2) post injection Sensory panel (n=6) and objective WB shear force	Objective Cont = LC ≥ LHS > LA Sensory Cont > LC = LHS		Rancid flavour Cont = LHS >> LC = LA	Juiciness Cont > LC = LHS	Sodium acetate and KL both improve sensory attributes of injection-enhanced beef	Knock *et al.* (2006)

Continued

Table A11.2. Continued.

Animal	Treatment / infused compounds	Experimental parameters	Impact on meat: Tenderness	Colour	Sensory flavour/ acceptability	Other	Outcome	Reference
	Blade tenderization (BT) Brine 1 (0.6% salt + 0.3% sugar) Brine 2 (0.6% salt + 0.4% sugar + 0.25% sodium bicarbonate)	LD, RF and VL (n = 90 each) from 4.5 years old dairy cows were commercially processed and aged for 7 days at 2°C. The muscles were assigned to one of the following treatments, Control, BT, Brine 1, Brine 2	Tenderization of the muscles were in the following order: Brine 2 > Brine 1 > BT > control	–	The brines reduced the beef flavour in fried beef and produced more intense beef flavour in boiled beef. WOF slightly increased in brine-treated samples	BT treated samples were less juicy compared with the other treatments	Tenderness of RF treated with Brine 2 was higher than the LD control	Rosenvold *et al.* (2006)
	Infusions of 2% KCl and 3% of a 1:1 mixture of sodium hexametaphos-phate and sodium pyrophosphate (PP) plus either 8% NaC1, 2% glucose (G) plus 6% NaCl, 6% G plus 2% NaC1, or 8% G	*Longissimus* muscle sections were excised from eight pork carcasses 1 h post-mortem and sectioned into six 0.5-kg roasts After infusion, the roasts were held at 7°C for 3 h At 4 h post-mortem, the roasts were vacuum-packaged and held at –20°C for 4 to 11 days until evaluated	The infused treatments were rated more ($p < 0.001$) tender by taste panellists and by WBS measurement	–	The infused groups were juicier ($P < 0.001$) than either the Cold Processed (CP) or Hot Boned (HB) controls. The infused groups were rated more salty than either the CP or HB controls	The infused groups were higher in moisture and ash but lower in protein content than either the CP or HB controls ($P < 0.05$) The fat content of the infused groups was lower than of the HB control but was not different from that of the CP control. Either 2% NaCl plus 6% G or equal amounts (4%) of NaCl and G produced the most tender and juicy product.	A hot-boned fresh pork chop was produced that is tender and juicy but lower in sodium	Wu *et al.* (1990)

				–	The substitution of 4% glucose for NaCl not only reduced the NaCl content of the infusion solution, but also improved the palatability of the meat			
Pigs	5% Salt 5% polyphosphate 3% bicarbonate & combinations	1°C after 2 h post-mortem for 24 h LTL injected 10% initial weight	Final tenderness bicar > phosph > salt > control	–	There was a slight 'soapy' note with phosphate It was difficult to detect off-flavours with bicarbonate.	Weight gain: bicar = phosph = salt > control pH: bicar > phosph > salt = control	Significant improvement	Sheard and Tali (2004)
	Commercial marinades Kip (contains the following di/triphosphate, sodium acetate, sodium citrate and sodium ascorbate), pH = 7.7 Kip vers (contains the following: glucose syrup, salt, sodium acetate, sodium citrate and sodium ascorbate), and water. pH = 10.3	14% Marinade. Injected to a target weight of 112% of the initial weight. 24 h PM LD, 3 days VP, Injected, left overnight at 1°C, VP and frozen	Final tenderness Kip = Kip vers > control = water	–	Flavour: Control = water > Kip = Kip vers Liking: control = water > Kip = Kip vers	Weight gain: Kip > Kip vers > control > water pH: Kip = Kip vers > control = water Juiciness: Kip = Kip vers > control = water	pH was positively correlated with tenderness and juiciness. Higher yield in treated samples	Sheard *et al.* (2005)

Continued

Table A11.2. Continued.

Animal	Treatment / infused compounds	Experimental parameters	Impact on meat: Tenderness	Colour	Sensory flavour/ acceptability	Other	Outcome	Reference
	Sodium citrate or acetate and post-rigor injection of phosphate plus salt	Pre-rigor injection. 40 pork carcass sides were assigned to one of four treatments: pre-rigor citrate (CIT) or acetate injection (ACE), post-rigor phosphate and salt injection (PHOS), and non-injected control (CON). Loins in 20 sides were injected at 50min post-mortem with 4% solutions of CIT or ACE to approximately 110% of projected loin weights, and 10 loins were injected at 24h post-mortem to 106.6% with a solution of 4.4% PHOS and 2.2% salt	Pre-rigor CIT injection improved tenderness	No detrimental effects on colour or flavour found with PHOS, but neither CIT nor ACE altered glycolytic metabolites or improved firmness, wetness, or fresh visual colour over CON	Although CIT increased pH ($p < 0.05$), neither CIT nor ACE altered ($p > 0.05$) glycolytic metabolite concentrations. The pH increase in muscles from the CIT treatment was probably due to its buffering ability rather than to its glycolytic inhibition			Stephens *et al.* (2006)
Bison	0.5% Sodium chloride and 0.3% sodium tripolyphosphate	SM at 7 days PM (several chilling/freezing/thawing regimes. Injected to 110% of its original weight	↑ Tenderness (40%↓ in shear force)	HunterLab a^* (redness) and b^* (yellow-ness) values did not differ ($p > 0.05$) between injection treatments, however, injected steaks had lower L^* values (darker) compared with controls	↑ Acceptability but it was dependent on the cooking temp	↑4.4% Yield ↑ Juiciness (29%)	–	Dhanda *et al.* (2002)

Elk	Sodium tripolyphosphate (STPP, 0.1, 0.2, or 0.3%) and 0.5% NaCl	Muscles, LL and SM, from each side of the carcass were divided into 2 equal portions to yield 4 sections, each from the 2 muscle groups per animal One section was kept as a non injected control, the other 3 were injected with brine solutions (2.3 ± 0.5°C). Injected to a target weight gain of 10% of the original mass	Addition of 0.5% NaCl with 0.2% or 0.3% STPP to the marinade was best at improving the cooking yield and tenderness of the resultant products All marination levels improved shear values of elk roasts	Injected samples had significantly (P <0.05) lower L^*, a^*, and b^* values		Cooking yields for the roasts obtained from the injected B (0.2% STPP) and C (0.3% STPP) muscle sections were significantly (P <0.05) greater than those from injected A (0.1% STPP) and control non-injected sections when cooked by either dry or moist-heat	Marination by injection has a great potential to improve the tenderness and juiciness of elk meat	Dhanda *et al.* (2003)

List of abbreviations in the Appendix.

BF, *Biceps femoris*; Blade eye, combination of 10 muscles and reported as blade eye; IS, *Infraspinatus*; MS, Inside round = *Semimembranosus*; LL, *Longissimus lumborum;* LTL, *Longissimus thoracis et lumborum;* PsM, *Psoas major*; QF, *Quadriceps femoris*; LT, Rib-eye = *Longissimus thoracis*; RF, *Rectus femoris*; ST, *Semitendinosus*; SS, *Supraspinatus*; TF, *Tensor fascia latae*; GM, Top sirloin = *Gluteus medius*; TB, *Triceps brachii*; VL, *Vastus lateralis*; VM, *Vastus medialis*; MPSC, Mixture of salt, saccharides and polyphosphates; PM, Post-mortem; WBSF, Warner–Bratzler shear force.

12 Processed Camel Meats

Mustafa M. Farouk[1] and Alaa El-Din Bekhit[2]
[1]*AgResearch Limited, Ruakura Research Centre, Hamilton, New Zealand;* [2]*Department of Food Science, Division of Sciences, University of Otago, Dunedin, New Zealand*

12.1 Introduction

Camels can be raised under environmental conditions that are not suitable for other domestic livestock. The camel is therefore an important source of food and income in arid and semi-arid regions of the world (Yousif and Babiker, 1989). The world production of camel meat was estimated at 361,000 t in 2009 (FAOstat, 2011). How much of the meat was further processed into finished products during the same period has not been established. Nevertheless, it is safe to assume that some of the meat is processed for the purpose of extending shelf life and improving availability and utilization of the meat. In this chapter, processed camel meat is defined as the camel meat that has gone through physical or chemical treatments beyond the simple process of meat fabrication into cuts and trimmings. The treatments that processed camel meat undergoes include salting, curing, drying, smoking, cooking, fermenting or a combination of those that are done in order to improve palatability and/or shelf life.

Because of the high ambient temperatures and the lack of refrigeration capacity in many of the areas where camel meat is produced, most of the traditional processing involves the use of drying technology on its own or in combination with fermentation. Some of these dried products are now becoming novelty items to many non-traditional camel meat consumers around the world, particularly tourists.

This chapter discusses the different types of processed camel meats available around the world, categorizes the products on the basis of the principal technology involved in their preservation, and provides the generalized (as opposed to detailed) aspects of their traditional as well as modern processing methods.

12.2 Categories and Importance

12.2.1 Categories of processed camel meats

Available literature indicates that camel meat can be successfully processed into many products that are currently manufactured from beef and other red meats (Dawood, 1995; Warfield and Tume, 2000; Ulmer *et al.*, 2004; Abdallah, 2008; El Malti and Amarouch, 2009). There is, therefore, a potential for a wide range of processed meat products that may be manufactured from camel meat. However, because of the complexities in the manufacturing, methods of preservation and even packaging of the wide range of processed meats available to consumers today, it is extremely difficult to group processed camel meat products into representative categories. There is, therefore, no single classification system that would completely satisfactorily categorize these products. Earlier, Long *et al.* (1982) categorized processed meats into cured meats; sausages; luncheon meats, meat loaves and spreads; and miscellaneous canned meat products. Pearson and Gillett (1999) simplified the grouping when they categorized processed meats into cured-not-smoked meats; cured-smoked/dried or cooked meats; and convenience meats. Heinz and Hautzinger (2007) categorized processed meat products according to the processing technology applied into six categories: fresh; cured; raw cooked; precooked-cooked; raw (dry) fermented sausages; and dried meat. For this chapter, processed camel meats will be grouped into the following categories based on the preservation technologies involved:

- Non-cured processed camel meats.
- Cured/smoked processed camel meats.
- Non-fermented semi-dry/dried processed camel meats.
- Fermented semi-dry/dried processed camel meats.
- Canned/pouched processed camel meats.
- Speciality processed camel meats.

12.2.2 Importance of processed camel meats

Data on the global production of processed camel meat are not available. The importance of processed meats in general, and that of processed camel meat, is therefore gauged by the volume and value of processed meat products in the major camel-producing and camel-consuming regions compared with the world total amounts of these products imported during the same period (Table 12.1). It is clear from the tabulated data that the value of the two broad categories of processed meats imported in the camel producing regions constitutes only about 4–7% of the world total imports, which may be indicative of the heavy reliance on locally traditionally processed meats in these regions.

The total Western European processed meat market alone, which includes delicatessen, frozen convenience meat, canned meat, cured meat and bacon and ham, was worth Euro 116.5 billion (FFT, 2007).

Processed meats are also used as ingredients in ready-to-eat (RTE) meals. The value of the global RTE meals market was US$71.6 billion in 2009 and is expected to increase to US$83.4 billion between 2009 and 2013 (BI, 2010). About US$29 billion of the total RTE meal market value in 2009 was contributed by the meat/poultry-based RTE meals sub-category (BI, 2010). In a 2011 ranking of the top 125 processors of meat and poultry in USA, the top 20 of the 125 ranked processors all manufacture processed meats (Clyma, 2011).

12.3 Non-cured Processed Camel Meats

These are processed meats prepared with no nitrates or nitrites used to improve colour or

Table 12.1. Estimated value (US$1million) of sausages and prepared and preserved meats imported into regions of high camel population compared with the total world import of processed meats in 2010 (source: International Trade Centre, 2011).

Region	Sausages and similar products	Prepared and preserved meats
Africa	133	127
Gulf Cooperation Countries (GCC)	57	143
Magreb (four north African countries)	6	8
Middle East	91	220
World	4,000	12,000

taste in their manufacture. Products in this category can be prepared from whole tissue where the muscle/meat structure is more defined or remained intact such as in balangu and shawarma, or the meat is diced into cubes as in tsire or ground coarsely as in camel burgers (camburgers) and camel meatballs (Dawood, 1994; Igene, 2008; Kilic, 2009). Some of these products could be classified as RTE meats because they can be eaten without any additional preparation.

The general flow diagrams for the manufacture of non-cured camel meat products are shown in Figs 12.1 and 12.2. There may be variations to the flow charts in manufacturing those products that are being used around the regions where the products are consumed. The important steps are, however, essentially the same with the type of ingredients used contributing to most of the differences existing between regions or processors.

12.3.1 Balangu

Balangu is a ready-to-eat boneless whole-tissue beef, mutton, chevon or camel meat that is slowly roasted over an open hearth. It is a very popular street meat in west and central Africa (Igene, 2008). Raw meat preparation for balangu (Fig. 12.1) involves slicing chunk of meat/cuts to about 1-cm thickness often by continuous butterflying of the cut until the chunk is reduced to a long (30–40 cm) spread of meat depending on the thickness of the initial chunk/cut. The width of the raw meat spread depends on the initial length of the chunk/cut. Thin meats are generally not butterflied. The raw meat spread is then salted, spiced and placed on wire mesh on a hearth of burning charcoal and cooked slowly with frequent turning until the meat is well done. Vegetable oil is sprinkled on the spread as it cooks when very lean meat is used. Some consumers prefer balangu that has not been spiced,

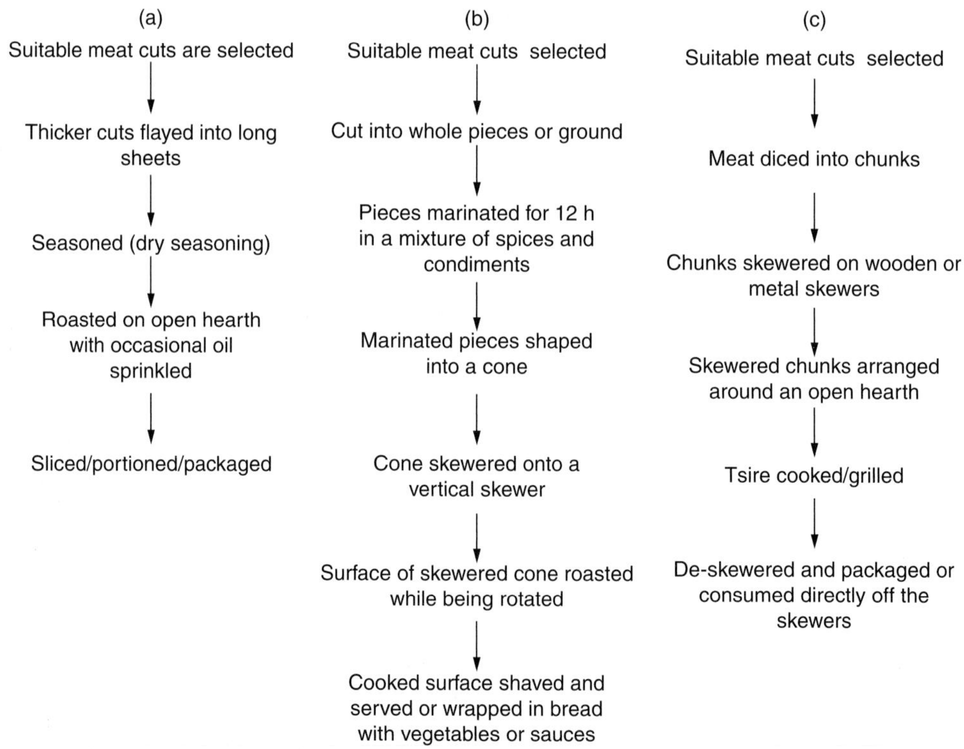

Fig. 12.1. Generalized flow charts for the manufacture of (a) balangu; (b) shawarma; and (c) tsire.

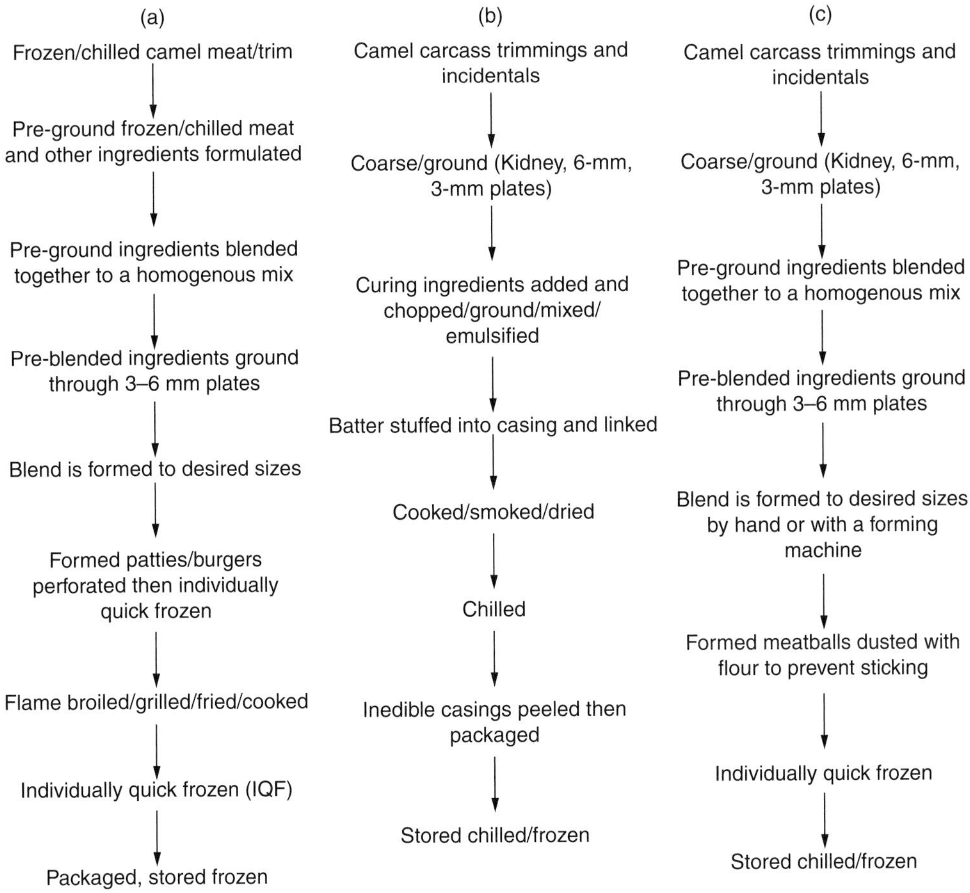

Fig. 12.2. Flow chart for the manufacture of (a) pre-cooked camel meat burger (camburger); (b) pre-cooked camel sausages; and (c) camel meatballs (camballs).

thus many balangu processors would only use salt before the raw meat is spread prior to roasting to give the customer the choice of eating balangu with or without spices. If a customer opts for spices, then the spices are either sprinkled on slices of balangu or are packaged in a polyethylene bag or wrapped in a newspaper cutting for the consumer to sprinkle on or dip the slices in at consumption. Balangu is packaged after purchase in paper or plastic.

12.3.2 Camel meat burgers (camburgers)

Hamburgers or patties are mostly produced from beef, but other meats including camel meat can be used (Dawood, 1995; Warfield and Tume, 2000). Processing methods for hamburgers can vary considerably, but the common general steps (Fig. 12.2) include: (i) suitable meat is selected and thawed if frozen; (ii) meat is coarsely ground or flaked; (iii) the fat content of the hamburgers is determined and where necessary the meat re-blended to achieve the desired fat content; (iv) non-meat ingredients are blended in; (v) the mix is reground through a finer plate; (vi) hamburgers are formed into different sizes, weights and shapes; and (vii) formed burgers may be stored chilled or frozen, and may be raw or pre-cooked. If the burgers are to be stored frozen they are individually quick frozen (IQF) to prevent sticking and to improve free-flow before packing for storage. These steps were essentially followed by Dawood

(1995) to produce camburgers from camel chuck at 0, 5 and 10% fat levels using a basic formulation that included salt, ground black pepper, onion, and cardamom and ginger powders. The individual burgers weighed 200 g, and they were 1.5 cm thick and 8 cm in diameter and were roasted at 163°C to an internal temperature of 80°C.

Hoogenkamp (2001) presented several patty formulations from a variety of meats from around the world with particular emphasis on the use of soy proteins in the formulations. Ibrahim and Nour (2010) prepared burgers from beef replaced at five levels with camel meat (0, 25, 50, 75 and 100%). The camel meat, beef and composite burgers were prepared as follows. The meat and fat were ground through 8-mm and 6-mm plates, respectively, the rest of the meat and the other ingredients were thoroughly mixed by hand and the mixture was reground through a 5-mm plate, and finally 100-g burgers, 4 inches diameter and 5 mm thickness were formed, frozen and packed in plastic bags. The cooked burgers were evaluated by a sensory panel and the results indicated the tenderness, flavour, juiciness and colour of the burgers increased significantly with the increase in the level of camel meat in the burger. Al-Khalifa and Atia (1997) also prepared camburgers of different fat content and with different levels of soybean hull and reported that sensory panellists preferred camburgers containing 10% fat and 6% soybean hulls.

In 2009, a fast-food restaurant in Saudi Arabia had young camel meat burgers (Hashi Burger) on offer on their menu, and, in 2010, the Local House Restaurant started offering quarter-pound camel burgers in Dubai, UAE (Shyoukhi, 2009; Anon, 2010a). Camel burger is also on offer in a Somali restaurant in Minnesota, USA (Anon, 2011).

12.3.3 Camel meatballs (camballs)

Meatballs can be prepared from camel meat in a similar way to the way they are prepared from other meats. Camel meatballs form a common dish in Morocco where they are usually served grilled with bread (Whitesel, 2011). Camel meatballs were sold in barbecue packs in Australia in 2000 and the pre-cooked flame grilled version was identified as a potential retail product in supermarket and food service outlets in Australia (Warfield and Tume, 2000).

There are many ways to prepare, flavour and consume meatballs, but the basic preparation steps are all the same (Fig. 12.2). The steps include grinding the meat, mixing the meat with mostly dry ingredients, adding a binder such as eggs, forming the blend into balls of different sizes, and finally cooking the formed balls using dry or wet heat or canning meatballs. Ulmer *et al.* (2004) reported a camel meatball formulation and described how it is processed. The traditional Turkish Çiğ kőfte prepared by kneading ground meat with bulgur, red pepper, onions and other condiments for two hours and consumed raw without cooking is an example of the other meatballs variants found in many parts of the world (Kilic, 2009).

12.3.4 Shawarma

Shawarma is a sandwich-like wrap of grilled shaved meat that has become a Middle Eastern fast-food staple across the world (http://en.wikipedia.org/wiki/Shawarma). Shawarma can be prepared from many types of meats including camel meat (Kadim *et al.*, 2008; Anon, 2010b). Kilic (2009) described the production of shawarma, also known as Doner kebab or just Doner, as follows (Fig. 12.1). Meat is cut into whole-tissue pieces or ground; the meat pieces are marinated for about 12 h in a mixture of spices, condiments, oils, fruit juices, milk products including yoghurt, vegetables and binders to meet the taste or preference of the consumers. The marinated pieces are then shaped into a cone, the cone impaled onto a vertical skewer, the skewered cone of raw shawarma is then rotated to cook the surface using an open gas or electric cooker, and the cooked meat surface shaved off and served on bread/wraps with vegetables and sauces.

12.3.5 Tsire

Tsire is a RTE roasted spicy skewered boneless whole-tissue beef, mutton, chevon or camel meat. Tsire processing has been described in detail by Farouk (1983), Igene and Abulu (1984), and Igene and Ekanem (1985). The steps in the traditional processing of tsire (Fig. 12.1) include cutting meat into chunks, skewering the chunks onto wooden or iron skewers, dusting the skewered chunks of meat with a blend of dry ingredients including spices and condiments, arranging the spiced skewered meat around a glowing fire, and roasting with occasional turning until well done. Roasted tsire meat chunks are often removed from the skewers and packaged at the point of sale in paper or plastic bags. The skewers are then available for re-use. Finished tsire could be likened to kebabs or satay. Igene (2008) reported that tsire can be prepared using domestic ovens.

12.4 Cured/Smoked Camel Meat

The major difference between uncured and cured processed meats is the nitrites or nitrates used in the cure of the latter to provide the characteristic colour and taste associated with cured meats. In addition to the role of nitrites in adding colour and flavour, nitrites also prevent botulism, increase the shelf life of products and were recently shown to be effective in reducing listeriosis (Booren, 2010; Honikel, 2010). Examples of cured products include hams, pastrami, bologna, hot dogs and other emulsion sausages.

12.4.1 Banda/kundi

Banda (hausa) and kundi (yoruba) are non-cured cooked-smoked whole-tissue meat popular in west and central Africa. Banda/kundi is mostly prepared from beef and camel meat but can be prepared from all types of meat including game meat (Farouk, 1983; Igene and Tukura, 1986; Fakolade and Omojola, 2008). Banda is produced by cutting meat into chunks, partially sun-drying the pieces, cooking/smoking the partially dried pieces on an open kiln, and then packaging in sacks or drums (Igene, 2008). The fresh or partially dried meat chunks can be boiled before smoking or cook-smoked directly without pre-boiling.

12.4.2 Camel meat sausages

There are too many types of sausages available for consumers, each with its special appeal to some part of the population, to classify sausages using one or multiple criteria (Pearson and Gillett, 1999). Sausages can be cured/uncured (fresh), cooked/uncooked, smoked/unsmoked or a combination of any of these operations. Cooked sausage is a RTE food that can be eaten cold or heated, as part of a meal or on its own (Puolanne, 2010). Camex Australia Pty, a supplier of camel meat products to the international market, had on offer plain and gourmet camel sausages in their list of products (Anon, 2012a).

The steps in processing sausages, whether at the cottage or industrial level, include some of the following (Fig. 12.2):

1. Selection of meat or a combination of meats.
2. Grinding the meat/meats to varying levels of coarseness.
3. Mixing and blending the ingredients including the ground meat.
4. Chopping the blend to different levels of fineness.
5. Emulsifying the chopped blend or batter.
6. Stuffing the batter into casings of different sizes and shapes.
7. Linking and tying the sausages to varying lengths.
8. Smoking and cooking of the raw sausages.
9. Chilling the cooked sausages.
10. Peeling of non-edible casings.
11. Packaging or directly consuming the finished products.

Heikal *et al.* (1972a) produced smoked sausages from 100% camel meat and compared them with similar sausages prepared from camel meat replaced up to 20, 30 and 40% with fresh or boiled beans and found that good quality smoked sausage could be prepared using 100% camel meat or camel meat replaced with up to 20% beans. Gheisari *et al.* (2008) produced emulsion sausages from fresh, defrosted and actinidin-tenderized camel meat and compared it with similarly prepared sausages from beef. They found that the properties of both sausages types were quite similar, except that camel meat sausages had a better flavour than beef sausages and sausages produced from tenderized meat samples were more acceptable than those made from fresh and defrosted meats. The authors concluded that camel meat is suitable for replacing beef in the production of emulsion product types.

12.5 Non-fermented Semi-dry/Dried Camel Meat

Processed meat products in this category are dried at some point in the manufacturing process using a range of technologies. The drying lowers the water activity (A_w) of the meat. The extent of the drying or the A_w achieved determines the characteristics of the product and its shelf life in terms of texture and chemical and microbiological stability. Semi-dry/dried processed meats can be manufactured from meats of different levels of comminution from whole-tissue to fine emulsion. Most dried products are RTE that are eaten as snacks, in home-cooked meals with or without prior reconstitution, and added to flavour some traditional dishes. Lowering the A_w to ≤0.8 will ensure that all spoilage and pathogenic microorganisms are controlled (Zukál and Incze, 2010). Lonnecker (2010) recently surveyed a number of jerky processors in the USA and reported an A_w of 0.74 as the average for all the beef jerky sampled in the survey. Farouk and Swan (1999) described a two-step process of manufacturing softer jerky with A_w (0.72–0.75) similar to other dried beef products manufactured using traditional methods. Other processes of manufacturing semi-dried and dried jerky and other related RTE meats with or without nitrites or starter cultures from whole-tissue or ground meat have been reported recently with basic steps essentially similar to Figs 12.3 and 12.4 (Choi *et al.*, 2008; Liao *et al.*, 2009; Lonnecker *et al.*, 2010).

12.5.1 Biltong

Biltong is a South African cured and air-dried meat that is now found in many countries around the world. Biltong can be produced from all types of meats of domesticated or game animals. Springbok Butchery and Wagonwheel Butchery (Anon, 2012b) had camel biltong on offer on their list of authentic South African delicacies in their small family business in UAE.

Biltong processing is simple. Igene (2008) and Naidoo and Lindsay (2010) outlined both the traditional and modern methods of biltong production; the traditional method (Fig. 12.3) includes the selection of meat cuts, soaking cuts into vinegar, marinating the pieces in spices (consisting of salt, nitrates, black pepper, coriander, salt and brown sugar) for up to 12 h, hanging the marinated pieces in biltong drying chambers to dry until the marinated meat loses 75% of its original weight, and then the dried biltong is packaged in polyethylene or cellulose bags. The only difference between the traditional and modern methods is that in the modern method the soaking in vinegar is skipped (Naidoo and Lindsay, 2010).

12.5.2 Dambun nama/meat floss/shredded meat

Dambun nama (a Hausa word for meat floss) is a RTE meat delicacy that is produced from red meats and poultry including camel meat. Dambun nama is produced from camel meat in the northern part of Nigeria where it is considered a delicacy and a treat served at special occasions, such as welcoming a guest, naming a child and wedding ceremonies, and as a provision during

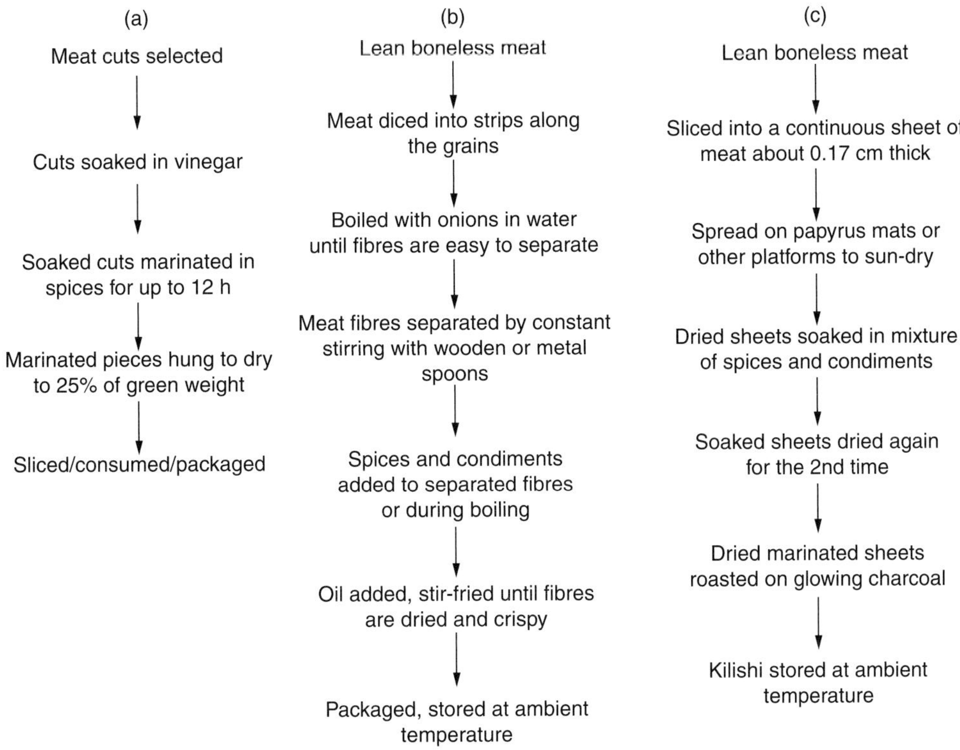

Fig. 12.3. Flow chart for the manufacture of (a) biltong; (b) dambun nama/camel meat floss; and (c) kilishi.

long-distance travel such as the annual pilgrimage trip to Mecca.

The manufacturing of dambun nama involves:

- Selection of meat cuts, preferably thick hindquarter cuts.
- Cuts are diced into strips, preferably along the grains.
- Diced cuts are boiled with salt, spices/flavourings and onions or onions only until the meat fibres begin to separate.
- Excess water in the boiled strips is poured off and boiled strips are further heated and continuously stirred until the fibres are separated.
- Oil and condiments are added to the shredded fibres and stir-fried until crispy golden.
- End product is packaged or stored in lidded dishes (Fig. 12.3).

The process of manufacturing dambun nama is not too dissimilar to the one described by Liao *et al.* (2009) for pork floss. Properly processed and packaged dambun nama has been known to be shelf-stable for at least 1 year under ambient conditions.

12.5.3 Kadid

Kadid is a sun-dried product commonly found in north Africa and the Middle East. Kadid has been consumed in Arabia since pre-Islamic times (Anon, 2005). There are two forms of kadid. The first type is found in north Africa and is prepared from strips or pieces of meat mixed with salt, garlic and oil and sun-dried for 7 days (Essid *et al.*, 2007). The product is sold in dried form or covered with oil that solidifies with time. About 17 strains of *Lactobacillus plantarum* have been isolated from Tunisian kadid. All the isolated strains had antimicrobial and acidifying activities as well as proteolytic activities. However, these strains did not show any lipolytic activity. The second type of kadid is the one found in Egypt, which is prepared from meat pieces fried in tallow till low moisture content is achieved and the finished product is preserved for extended periods of time covered with tallow fat to help protect the product from contamination and oxidation.

12.5.4 Kilishi

Kilishi is a sun/heat-dried shelf-stable RTE meat product that is widely produced and distributed in the Sahelian regions of Africa (Igene, 2008). The process of manufacturing kilishi has been described by Farouk (1983), Igene *et al.* (1990), Igene (2008) and Mgbemere *et al.* (2011). The traditional process used at the cottage level (Fig. 12.3) involves thick cuts of meat being sliced into thin sheets; the sliced sheets spread on papyrus or raffia mats to dry under the sun; the dried sheets infused with a mixture of spices and condiments; the infused sheets dried again for the second time; and the dried infused sheets roasted briefly on smokeless burning charcoal (Farouk, 1983; Igene *et al.*, 1990; Kalilou and Zakhia, 1999). Igene (2008) reported an improved factory-based kilishi manufacture, which includes the following unit operations:

- Fresh meat trimmed of excess connective tissue and cleaned.
- Trimmed meat sliced to 0.17–0.2 cm thick.
- Sliced meat first-stage dried to 20–25% moisture.
- Dried meats infused (1:4 ratio of meat to ingredients).
- Infused slices of meat second-stage dried to 15–20% moisture content.
- Dried infused sheets are then third-stage dried in an oven for 30–60 min until a moisture content of 10–15% is reached.
- The final product is then packaged.

Kilishi could be related to jerky, the snack product popular around the world. Some methods of making jerky are similar to kilishi except for most jerky processes the meat pieces are only dried once in contrast to the two-stage drying in kilishi manufacturing. A survey of Midwestern small to medium jerky processors in the USA revealed a wide range of processing methods and technology used for jerky manufacture (Lonnecker *et al.*, 2010). Heinz and Hautzinger (2007) described a simple process of manufacturing jerky. The process involves trimming fat and adhering connective tissues, cutting the meat into strips 1–2 cm wide and 15–20 cm long, marinating the strips in a mixture of condiments with or without nitrite, and the marinated strips are then dried under the sun, in solar driers or hot-air ovens. Farouk and Swan (1999) adopted the two-step process of manufacturing kilishi to produce a soft beef jerky suitable for the Asian plate with A_w (0.72–0.75) similar to other dried beef products manufactured using traditional and modern methods.

Camel jerky is considered a novelty in some parts of the world where camel meat is not traditionally consumed. Warfield and Tume (2000) recommended soft camel jerky as potential products for the Asian market and for retailing in airports and snack outlets. Territory Jerky in Australia offers a 250 g pack of camel jerky among the other jerky products it sells (http://www.territoryjerky.com.au/); the camel jerky has been reviewed by many Internet bloggers.

12.5.6 Odka/muqumad

Odka/muqumad is a traditional Somali intermediate moisture meat product that is popular with Somali people both at home and abroad. The process of manufacturing odka has been described by Ismail and Swan (2000) and Igene (2008). To prepare odka, muscles/meat are cut into long strips and salted; the strips are dried, the dried strips cut into small pieces, then fried in ghee flavoured with herbs, and the end product is stored, with or without extra ghee, at ambient temperature (Fig. 12.4). Odka is stored in ghee in traditional containers known as tebed or gumbe for up to 2 years (Igene, 2008). Vacuum-packed laboratory-scale produced odka was kept at 10°C for 10 months without any marked deterioration in flavour (Ismail and Swan, 2000).

12.5.7 Qwanta

Qwanta is a traditional Ethiopian dried RTE meat product. Qwanta is now part of the menu in many Ethiopian restaurants in Western countries such as those in the list of menu items in the Hebasha Market and Carryout in Washington, DC (http://www.habeshamarket.com/menu.html) and the Queen of Sheba Restaurant in London (http://flavors.me/queenofshebalondon). The process of preparing qwanta is similar to the Somali odka with few variations. Igene (2008) reported the unit operations for qwanta manufacture, which includes: (i) lean muscles are sliced into strips 1 cm thick and 20–40 cm long; (ii) strips are rubbed or marinated with hot pepper sauce and salt; (iii) marinated pieces are sun dried; (iv) dried pieces are further exposed to wood smoke; (v) pieces are then diced and fried in fat; and (vi) the final product is stored in glass or metal containers (Fig. 12.4).

12.5.8 Sharmoot

Sharmoot is a sun-dried meat product that is commonly consumed in Sudan. A study found that sharmoot contributes about 16% of the diet of communities displaced from southern and western parts of Sudan (Osman, 1999). To prepare sharmoot (Fig. 12.4), meat is cut into thin strips, then sun-dried for 3–5 days depending on the meat thickness and weather conditions. The dried meat is then ground into a fine powder (Gailani and Fung,

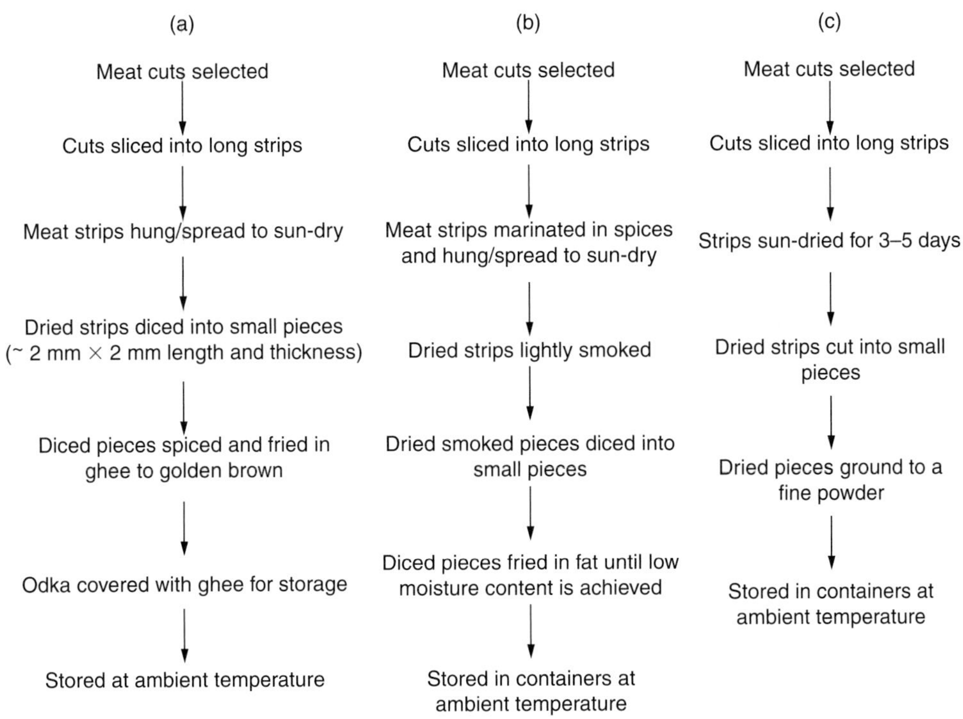

Fig. 12.4. Flow chart for the manufacture of (a) odka; (b) qwanta; and (c) sharmoot.

1986) or left as strips similar to jerky (Anon, 2010c). Under traditional processing conditions, the product is exposed to a high level of contamination. A modified method that involves a pre-cooking step, force air drying in a chamber at 65.5°C for 3.5 h and the use of water activity and antimicrobial regulators, can improve the product safety (Gailani, 1986). The nature of microorganisms in the traditional products and their contribution to the sensory properties of the final product are yet to be determined. Sharmoot is consumed as a stew/sauce called 'mullah' in Sudanese dialect (Osman, 1999; Anon., 2010c).

12.6 Fermented Semi-dry/Dried Camel Meat

Semi-dry and dry fermented sausages differ in their A_w and final pH. The final pH of semi-dry fermented sausages ranges between 4.7 and 5.4 with $A_w > 0.90$–0.91, whereas, the pH of dry fermented sausages is 5.2–5.8 with A_w in the range 0.85–0.91 (Vignolo *et al.*, 2010). Semi-dry sausages are usually fully cooked, whereas dry sausages are only lightly smoked or not at all smoked. This results in products that differ in hardness and other textural attributes (Pearson and Gillett, 1999).

12.6.1 Ndariko/jirge

Ndariko (Fulfulde/Peul) and jirge (Shuwa or Baggara Arabs) is a dried meat product found in the north-eastern part of Nigeria popular among the Fulani and Shuwa Arabs. It is prepared from the meat of all ruminant animals including camel meat. According to Farouk (1983), ndariko and jirge are prepared by cutting meat into long strips about 2 cm thick, the strips are then hung or spread on papyrus mats to dry, which usually takes about 6–7 days. The strips may or may not be salted or spiced (Fig. 12.5). The major difference between ndariko and jirge is that the latter is made from thicker cuts that have been left to start fermenting and undergo proteolysis to develop the mixture of sour and umami-like tastes unique to these products. The dried finished product is stored in earthen pots, metal containers or sacks made of natural fibres. A similar product to ndariko made from camel meat, which was cut into strips, salted and dried at ambient temperature for about 1 month, is consumed in Ethiopia (Zegeye, 1999).

12.6.2 Pastirma

Pastirma or basturma is a traditional intermediate-moisture meat product commonly consumed in Turkey, Egypt, Armenia, Greece and many other Mediterranean and Middle Eastern countries and in countries outside these regions including the West where immigrants from these regions live.

Any part of a carcass can be used in making pastirma. However, the quality of the finished product depends on the cut used. Normally elongated cuts or muscles such as the eye of the round or muscles in the shank are used. Thicker cuts are at times cut into strips up to 60 cm long and 5 cm in diameter and used in pastirma processing.

The traditional method of processing pastirma is a long process lasting several weeks (Aktaş *et al.*, 2005). The traditional pastirma process includes the following steps:

- Meat cut selection and curing for 2–5 days. The muscle or meat strips are rubbed and covered with salt and nitrate, piled up and kept for up to 5 days. During this time, the salted meat may be turned or re-salted again.
- Rinsing for as long as necessary. Salted meat is rinsed with water in order to remove excess salt.
- Drying for 2–3 days. Rinsed pieces are air dried at room temperature for 2–3 days in summer and up to 15–20 days in winter.
- Pressing (for 1 day) and then drying for 7–13 days. The dried meat blocks are piled up and pressed with heavy weights or a mechanical device for up

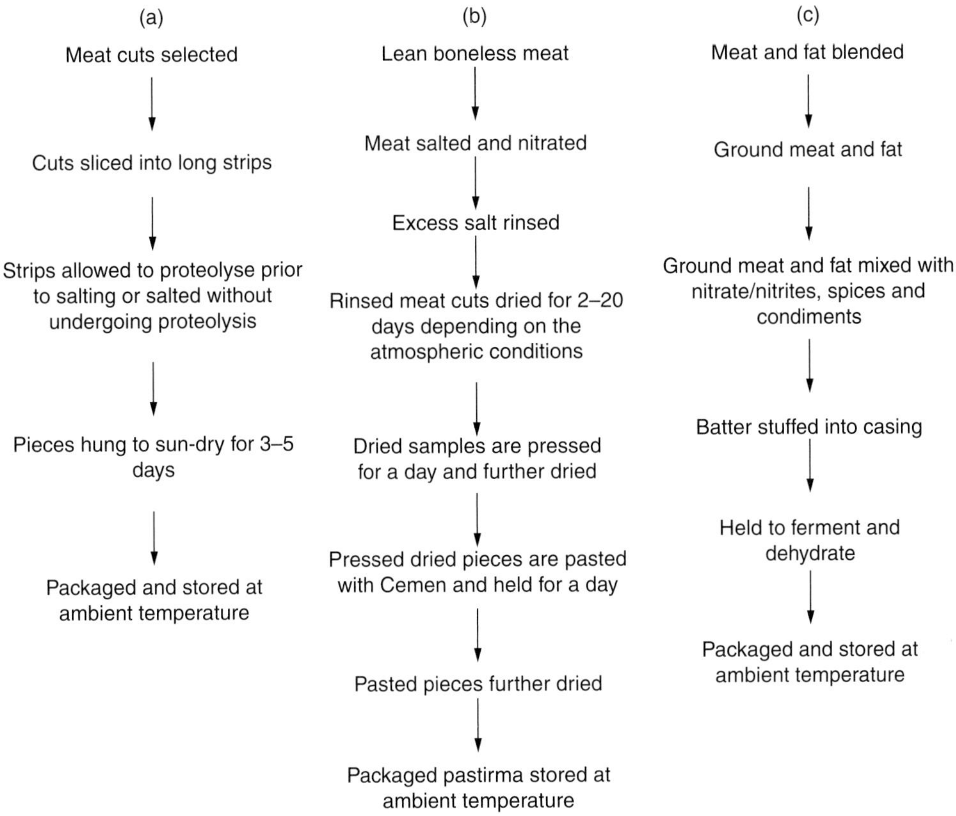

Fig. 12.5. Flow chart for the manufacture of (a) ndariko/jirge, (b) pastirma and (c) sucuk.

to 12 h at a time, then dried again for 2–3 days and pressed for another 12 h and then dried again for another 5–10 days.

- Pasting (for 1 day): the entire surface of the meat is covered with a paste called Cemen. Cemen is made up of freshly ground garlic, fenugreek to act as a binder, paprika, mustard and water. Other spices such as cumin and coriander or high-quality flours may be added to the Cemen mixture. The pasted meat pieces are stored for a day.
- Drying of pasted meat at room temperature for another 5–12 days for the pastirma to be ready.
- Finished pastirma is packaged for storage (Fig. 12.5). The finished product should have a water activity of about 0.88 and salt content of around 4.5–6%.

Various aspects of camel meat pastirma have been studied including the effects of: (i) pre- and post-rigor condition of muscles/meat on the quality of the finished product (Heikal *et al.*, 1972b); (ii) curing on palatability (Abdallah *et al.*, 1978; Yetim and Cankaya, 2001); and (iii) the use of pepsin proteolytic enzyme on microbiological and lipid stability (Goma *et al.*, 1978). The outcomes of these studies indicate that pastirma is best made from post-rigor meat rather than pre-rigor and with aged meat rather than un-aged meat. Other aspects of beef and buffalo pastirma production and quality have been extensively studied and could be applicable to camel meat pastirma (Kilic, 2009).

The traditional pastirma manufacturing method described previously was developed in regions where the climate is hot and at the time when modern technology such as injectors, tumblers and dryers were not available. The availability of such technologies enables the traditional process of making pastirma to be modified. Any modification of the traditional method must not, however, alter the quality of the product significantly from that of the 'real McCoy'. Farouk (2012) modified the traditional method of making pastirma and shortened it to 24 h (unpublished data) and compared it with pastirma produced by two traditional pastirma processors using different cuts of beef and subjected the products to sensory evaluation by a panel composed of a mixture of traditional and non-traditional pastirma consumers. The results of the sensory evaluation showed that the 24-h pastirma was equally acceptable to consumers as the traditionally produced pastirma.

12.6.3 Sucuk/fermented sausage

Sucuk is a dry uncooked, cured fermented sausage and one of the most important and widely consumed traditional Turkish meat products around the world (Kilic, 2009). Although sucuk is traditionally made from a mixture of beef and buffalo meat, camel meat is also used in Turkey and other African and Asian countries to make the product (Kalalou *et al.*, 2004; Öksüztepe *et al.*, 2006; Özbey *et al.*, 2007; El Malti and Amarouch, 2008).

Kilic (2009) and Kabak *et al.* (2011) reviewed many aspects of sucuk traditional and modern manufacturing methods, which are summarized as follows (Fig. 12.5). Sucuk is prepared by mixing meat with fat, sugar, salt, nitrite/nitrate, garlic, and spices including black or red pepper, paprika and cumin. The sucuk mix is then stuffed into natural casings, often the small intestine of sheep, and allowed to ferment by either microorganisms naturally present or added starter cultures and allowed to dry for several weeks at ambient temperature (22–25°C) and humidity (80–90%). The composition of the final product varies widely with moisture content, fat and pH in the range of 21.5–50, 24–51, and 4.2–6.3%, respectively.

El Malti and Amarouch (2008) manufactured fermented camel sausages by natural fermentation as follows:

- Camel meat was trimmed of visible fat and blended (90% lean meat and 10% fat).
- The blend was mixed with ingredients including salt, garlic, glucose, black pepper, cardamom, mace, sodium nitrate and sodium nitrite.
- The blend was minced through a 12-mm plate and then stuffed into collagen casings and hung in a climate chamber at 13°C with a relative humidity of 90% for 7 days; the humidity was reduced to 80% at 18°C and the sausages left to ferment and dry for 28 days.

Erkoşun and Özkal (2011) prepared sucuk using a starter culture by mincing a mixture of the meat and spices, salt, sugar, garlic, spices and nitrite through a 1.3 cm plate, and mixing starter culture composed of *Staphylococcus carnosus* and *Lactobacillus plantarum* with the sucuk dough. The mixture was held at 4°C for 12 h and then re-minced through a 3-mm plate while frozen fat was slowly added. The dough was stuffed into 38-mm fibrous casing, held at 20°C and 90% relative humidity, which was gradually reduced by 3% per day until the relative humidity was lowered to 75% at the end of the 5th day, and held like this until the end of the fermentation.

12.7 Canned/Pouched Camel Meat

Canned meat products include stews, luncheon meats and pastes that are either fully cooked before being placed into the can or pouch or are filled into the cans/pouches and then cooked (Legarreta, 2010). Canned products can be fully sterilized or pasteurized or aseptically assembled. The heat treatment that these products are subjected to determines the temperature at which the products should be stored (ambient or chilled) and their shelf life (Pearson and Gillett, 1999). The generalized flow chart for the manufacturing of canned luncheon meat summarized from Long *et al.*

(1982), Pearson and Gillett (1999), Salvage (1999) and Hoogenkamp (2001) include the following steps representative of typical processing steps that many canned or pouched products pass through during manufacturing: (i) meat is ground through a 3–6 mm plate; (ii) the ground meat is then mixed with other ingredients including nitrites under vacuum for 5–8 min; (iii) the mixture is filled into appropriately sized cans or pouches; (iv) the filled cans/pouches are then sterilized or pasteurized using the right temperature and time; (v) the sterilized cans are cooled in water; and (vi) they are stored at ambient or chilled for sterilized and pasteurized cans, respectively. Ulmer *et al.* (2004) provided recipes for canned camel meat chilli con carne, goulash, meatballs, meat in gravy and stews. Canned camel meat is being manufactured in Brunei and sold in some countries (Yahya, 2011).

12.8 Speciality Processed Camel Meat

The examples of speciality processed meat are the different types of meat loaves that are sold in many food outlets around the world. Other examples of products in this category are meat extracts, broth, concentrate, powders and other ingredients that are rarely consumed on their own but are used as ingredients in other products.

12.8.1 Camel meat loaves

The basic steps involved in the manufacturing of meat loaves include: (i) camel meat and other ingredients are ground through a 3-mm plate; (ii) remaining ingredients are added and coarsely or finely emulsified; (iii) the mixture or emulsion is filled in moulds and cooked and/or smoked to 72°C internal temperature; (iv) the cooked products are cooled in the moulds before the moulds are emptied; (v) the cooled products are sliced, portioned and shaved; and (vi) packaged and stored chilled or frozen.

12.8.2 Camel meat broth

A meat broth is a liquid resulting from cooking meat or fish in water or water containing vegetables and other condiments or additives. The term is used synonymously with bouillon. The strained liquid from broth is referred to as stock. Broth and stock could be dehydrated to produce broth or stock powder or it can be concentrated and marketed as frozen stock. Broth and stock could be prepared from high-quality cuts or from carcass by-products (Anon, 1992). They are used to add flavour to many foods (Blackmer *et al.*, 1997; Ottinger and Hofmann, 2003). The manufacturing of broth can be a simple or complex process. The simplest method involves heating or boiling cuts of meat in water or extracting solution for a period of time and straining the meat and other particulates from the extract. The extract forms the broth. The equipment needed for broth manufacture varies depending on the quantity of broth being produced and the quality or purity of the end product.

12.8.3 Camel meat extracts

Meat extract is an aqueous, dark brown aromatic meat essence concentrated to a paste and contains the water-soluble meat ingredients (Stute and Seuss, 1993). Extract is used commercially as a seasoning material and is usually produced by concentrating the solution obtained through the extraction of meat in hot water (Kuroda and Harada, 2002). There are many methods of extraction and precipitation employed in the manufacture of meat extract ranging from boiling in water to the use of enzyme or acid hydrolysis (Trojak and Tolic, 1977; Stamenkovic *et al.*, 1978; Remon *et al.*, 1985; Stute and Seuss, 1993; Kim and Yoo, 1995). Trojak and Tolic (1977) manufactured a meat extract by heating the meat in water. The broth was centrifuged to remove fat and coagulated protein, and the extract was dried in a thin film at 60–80°C under vacuum to 18–20% dry matter (DM), then followed by further evaporation to 50% DM and a final concentration in a condenser to

80% DM. The extract contained 15–20% moisture, 26–29% ash, 5% sodium chloride, 0.6–2.0% fat and ≤7% total creatinine.

12.8.4 Camel meat concentrate/ powders/flours

Camel meat or other meat powders and flours could be used in many food preparations including sausages, mayonnaise, dietetic foods or special foods for children and old people (Schmitz, 1974). Sharmoot, a popular Sudanese meat product, is essentially a meat powder in one form of its preparation. In a French patent Charier-Vadrot (1969) described the manufacture of meat powder to include the following steps:

- Meat is minced using appropriate equipment.
- The minced meat is then cooked at 100°C in 10–30% water: the water contains acetic or citric acid, antioxidant such as ascorbic acid, sucrose stearate to act as surface active agent, gum arabic, monosodium glutamate, potassium or sodium nitrate and glucose.
- The cooked meat and soup-like product is then crushed at 50°C to produce a pasty liquid phase with the meat in suspension.
- The liquid phase is fed into a column under pressure, and then vaporized.
- The fine droplets are projected into an air current at 150–220°C to precipitate powdered solids.
- The powder is collected in a separator.

12.9 Future Trends in Processed Camel Meats

Camel meat is a versatile raw material that would be suitable for producing products similar to those produced from beef and other red meats. The low fat content in camel might hinder the use of camel meat in certain products that require high fat content (e.g. salami), but this may represent a useful opportunity to utilize the hump fat in processed camel meat products.

Consumers all over the world including in the camel meat producing regions are becoming increasingly conscious of what they eat and are concerned about how that would impact on their health, happiness, overall quality of life and their environmental footprint. Future processed camel meats would therefore have to reflect these consumer outlooks in order to transition from the current cottage industry stage and its novelty-meat status to become a serious competitor in the meat protein supply space.

The current underlying trends driving new ready-to-eat product development are convenience, health, indulgence and ethical considerations (BI, 2010). Within these key trends, the following are some of the consumer demands that might have implications in the future for processed camel meats:

1. Natural: the demand for natural foods in developed economies is growing and manufacturers of processed meats are responding by avoiding the use of artificial colours, flavours and preservatives (Messenger, 2007; BI, 2010).
2. Reduced/no salt: sodium reduction in meats is a trend that started many years ago and will continue.
3. Organic/country of origin: processed meats produced from free-range and locally based producers have been growing in popularity for some time with more consumers demanding products that carry the 'organic' labels that are free from hormones, antibiotics, artificial additives and are considered 'authentic', 'real' and free of negative ingredients (Crews, 2007; Sloan, 2009).
4. Less processed: consumers are seeking less-processed foods with fewer ingredients (Browne, 2011). Many of the traditional camel meats could be considered less processed owing to the use of very few ingredients in their manufacture.
5. Food safety: the major challenge in terms of food safety is to produce processed camel meats as well as meats from other species that meet the consumer demands in all these identified trends and are safe.

The current camel husbandry practices and the leanness of camel meat compared with meat from other farmed animals mean that camel meat can be considered 'healthy'. When these attributes are coupled with the consumer perception of camel meat as novel and exotic, the future of processed camel meat could be promising. As many of the traditional meat products such as biltong and jerky continue to grow in popularity worldwide because of their unique flavour and high protein contents, there is the opportunity to promote these products from camel meat among the sports and hiking enthusiast as a 'low-carb' food source.

12.10 Conclusion

Camel meat is a very versatile material that has been successfully used in the production of various ready-to-eat and shelf-stable products. The emerging properties of camel meat as a healthy meat provide the foundation for the expansion and marketing of camel processed products. With the increasing demand for high-protein and low-fat shelf-stable products for recreational activities (hiking, cross-country and other sports), the potential of some traditional products (such jerky, qwanta and pastirma) to be marketed internationally is very high. Several obstacles can, however, face processors that can prevent access to international markets. For example, the cost associated in establishing hygienic production, processing systems, and accreditation and quality control may be prohibitive for processors in developing countries. The main challenge for potential processors who already have the required processing and quality management in place such as in Australia would be the ability to replicate the typical flavours and other eating quality attributes associated with the traditional products as currently produced artisanally in the products' country of origin and adapt these unique attributes to the Western consumer palate. This scenario warrants further research and development on processed camel meat products.

References

Abdallah, D.A.M. (2008) The effects of cooking methods on cooking loss and sensory evaluation in camels (*Camelus dromaderius*) and cattle. *Research Journal of Biological Sciences* 3, 612–615.

Abdallah, N.M., Amin, A.I. and Youssef, M.K.E. (1978) Effect of curing on organoleptic, chemical, physical, microbiological and microscopic characteristics of camel meat. *Annals of Agricultural Science, Moshtohor* 9, 125–137.

Aktaş, N., Aksu, M.I. and Kaya, M. (2005) Changes in myofibrillar proteins during processing of pastirma (Turkish dry meat product) produced with commercial starter cultures. *Food Chemistry* 90, 649–654.

Al-Khalifa, A. and Atia, M. (1997) Effect of soy hull and fat on camel meat patties. *Alexandria Science Exchange* 18, 303–311.

Anon (1992) Frozen natural stocks for the food processor. *International Food Ingredients* 2, 63–64.

Anon (2005) Food and feeding. Growth and the good health. In: Singh, N.K. and Kirmani, M.Z. (eds) *Encyclopaedia of Islamic Science and Scientist.* Vision Publishing House, New Delhi, India, p. 248.

Anon (2010a) Camel burgers: the newest healthy option in Dubai. Available at: http://www.foxnews.com/story/0,2933,584918,00.html#ixzz1UDA0Bhkq (accessed 6 August 2011).

Anon (2010b) Camel is king at one Dubai restaurant that offers a unique spin on Gulf delicacy. JakartaGlobe. Available at: http://www.thejakartaglobe.com/food/camel-is-king-at-one-dubai-restaurant-that-offers-a-unique-spin-on-gulf-delicacy/376582 (accessed 6 August 2011).

Anon (2010c) Silver lining: building a shared Sudanese identity through food. *Gastronomica: The Journal of Food and Culture* 10, 110–116.

Anon (2011) Minneapolis: Somali cuisine moves to the mainstream. Available at: http://www.bartamaha.com/?p=44931 (accessed 13 August 2011).

Anon (2012a) Halal camel meat and live camel export. Available at: http://www.alibaba.com/product/au112653984-111695548-0/HALAL_CAMEL_MEAT_and_LIVE_CAMEL_EXPORT.html (accessed 13 August 2011).

Anon (2012b) Springbok Butchery and Wagonwheel Butchery. Available at: www.sabutchery.com/ (accessed 7 August 2011)

BI (2010) Innovation in ready meals. Market drivers, NPD and alternative sales channels. Business Insights Ltd, pp. 1–107. Available at: http://www.business-insights.com/ (accessed 10 June 2011).

Blackmer, D.S., Mandigo, R.W., Eilert, S.J., Calkins, C.R. and Osburn, W.N. (1997) Effect of spray dried beef broth on the sensory, textural and cooking characteristics of grilled or broiled low-fat ground beef patties. *Journal of Muscle Foods* 8, 465–479.

Booren, B. (2010) Demystifying nitrites. *Meat & Poultry*. December, p. 55. Available at: http://www.meatpoultry.com

Browne, D. (2011) Covenience meals and processed meats. *Prepared Foods*, pp. 51–58. Available at: http://www.preparedfoods.com

Charier-Vadrot, P. (1969) Soluble meat powder. French Patent Number1568051.

Choi, J., Jeong, J., Han, D., Choi, Y., Kim, H., Lee, M., Lee, E., Paik, H. and Kim, C. (2008) Effects of pork/beef levels and various casings on quality properties of semi-dried jerky. *Meat Science* 80, 278–286.

Clyma, K. (2011) Winner's circle. Ranking the industry's top processors. *Meat & Poultry*, March, pp. 36–60.

Crews, J. (2011) Unveiling ideas. New food products highlight quality, convenience and flexibility. *Meat & Poultry*. April, pp. 105–107.

Dawood, A.A. (1995) Physical and sensory characteristics of Najdi-Camel meat. *Meat Science* 39, 59–69.

El Malti, J. and Amarouch, H. (2008) Microbiological and physicochemical characterization of natural fermented camel meat sausage. *Journal of Food Processing and Preservation* 32, 159–177.

Erkoşun, H. and Özkal, S.G. (2011) Kinetics of traditional Turkish sausage quality aspects during fermentation. *Food Control* 22, 165–172.

Essid, I., Ben Ismail, H., Ahmed S. B-H., Ghedamsi, R. and Hassouna, M. (2007) Characterization and technological properties of Staphylococcus xylosus strains isolated from a Tunisian traditional salted meat. *Meat Science* 77, 204–212.

FAOstat (2011) World production of camel meat in 2009. Available at: http://faostat.fao.org/site (accessed 27 July 2011).

Farouk, M.M. (1983) Production of Kilishi. BSc Thesis. University of Maiduguri, Maiduguri, Nigeria.

Farouk, M.M. and Swan, J.E. (1999) Boning and storage temperature effects on the attributes of soft jerky and frozen free-flow mince. *Journal of Food Science* 64, 465–468.

Fakolade, P.O. and Omojola, A.B. (2008) Proximate composition, pH value and microbiological evaluation of 'Kundi' (dried meat) product from beef and camel meat. Conference on International Research on Food Security, 7–9 October Natural Resource Management and Rural Development. University of Hohenheim, Stuttgart, Germany.

FFT (2007) Food for thought: The processed meat markets in western Europe. Available at: http://www.fft.com (accessed 10 June 2011).

Gailani, M.B. and Fung, D.Y.C. (1986) Critical review of water activities and microbiology of drying of meats. *CRC Critical Reviews of Food Science and Nutrition* 25, 159–183.

Gailani, M.B. (1986) Water activity in relation to microbiology during processing and storage of Sudanese dried beef (Sharmoot). *Dissertation Abstracts International, B*46, 2513–2514.

Gheisari, H.R., Shekarforoush, S.S. and Aminlari, M. (2008) Application of fresh, defrosted and actinidin-tenderized camel and cattle meat in the production of emulsion type sausages.*Advances in Food Sciences* 30, 207–212.

Goma, M., Zain, G.N., Dessouki, T.M., Thabet, F.M. and Bakr, A. (1978) Influence of pepsin proteolytic enzyme on the microbiological changes and lipid oxidation during manufacturing of pastirma from camel meat. Research Bulletin, Faculty of Agriculture, Ain Shams University (No. 902).

Heikal, H.A., El-Dashlouty, M.S. and Saied, S.Z. (1972a) Biochemical, histological and technological changes occurring during the production of sausage from camel meat and beans. *Agricultural Research Review* 50, 243–252.

Heikal, H.A., Saied, S.Z. and El-Dashlouty, M.S. (1972b) Histological changes in camel meat tissues during pastirma production. *Research Review* 50, 253–266.

Heinz, G. and Hautzinger, P. (2007) Meat Processing Technology for Small-to Medium-Scale Producers. RAP Publication 2007/20. FAO Regional Office for Asia and the Pacific, Bangkok, Thailand.

Honikel, K.O. (2010) Curing. In: Toldra, F. (ed) *Handbook of Meat Processing*, Wiley-Blackwell, Ames, Iowa, pp. 125–141.

Hoogenkamp, H. (2001) *Soy Protein & Meat Formulations*. Protein Technologies International – A DuPont Business. St Louis, Missouri.

Ibrahim, G.A. and Nour, I.A. (2010) Physical and chemical properties of camel meat burgers. *Journal of Camelid Science* 3, 39–43.

Igene, J.O. (2008) Traditional African Meat Products for Food Security. Traditional African Meat Products for Food Security and Agro-Industrialization: Development Challenges. Lambert Academic Publishing, Theodor-Heuss-Ring, Köln, Germany.

Igene, J.O. and Abulu, E. (1984) Chemical and bacteriological characteristics of tsire-type suya meat products. *Journal of the Science of Food and Agriculture* 37, 818–824.

Igene, J.O. and Ekanem, O. (1985) Effect of processing method on the nutritional quality of tsire (suya). *Nigerian Journal of Applied Science* 3, 1–20.

Igene, J.O. and Tukura, D.H. (1986) Effect of processing methods on product characteristics, lipid, fatty acid composition and oxidative stability of smoke-dried Beef. *Journal of the Science of Food and Agriculture* 37, 818–824.

Igene, J.O., Farouk, M.M. and Akanbi, C.T. (1990) Preliminary study on the traditional processing of Kilishi. *Journal of the Science of Food and Agriculture* 50, 89–98.

International Trade Centre (2011) Market Analysis and Research. International Trade Centre (ITC), Geneva 10. Available at: http://www.trademap.org/tradestat/Country_SelProduct_TS.aspx (accessed 27 July 2011).

Ismail, A. and Swan, J.E. (2000) Muqumad – a Traditional Somali Meat Product. Abstract of a paper presented at NZIFST/MIRINZ Joint Conference 2000, Auckland.

Kabak, B. and Dobson, A.D.W. (2011) An introduction to the traditional fermented foods and beverages of Turkey.*Critical Reviews in Food Science and Nutrition* 51, 248–260.

Kadim, I.T., Mahgoub, O. and Purchas, R.W. (2008) A review of the growth and of the carcass and meat quality characteristics of the one-humped camel (*Camelus dromedarius*). *Meat Science* 80, 555–569.

Kalalou, I., Faid, M., and Ahami, A.T. (2004) Improving the quality of fermented camel sausage by controlling undesirable microorganisms with selected lactic acid bacteria. *International Journal of Agriculture and Biology* 6, 447–451.

Kalilou, S. and Zakhia, N. (1999) Traditional methods of processing meat in Niger. *Tropical Science* 39, 18–22.

Kilic, B. (2009) Current trends in traditional Turkish meat products and cuisine. *Food Science and Technology* 42, 1581–1589.

Kim, Y.B. and Yoo, I.J. (1995) A study on the extraction conditions of goat meat extract. *Korean Journal of Animal Science* 37, 179–185.

Kuroda, M. and Harada, T. (2002) Distribution of GAMMA-glutamyl-BETA-alanylhistidine isopeptide in the macromolecular fractions of commercial meat extracts and correlation with the color of the macromolecular fractions. *Journal of Agricultural and Food Chemistry* 50, 2088–2093.

Legarreta, I.G. (2010) Canned products and Pâté. In: Toldra, F. (ed.) *Handbook of Meat Processing.* Wiley-Blackwell, Ames Iowa, pp. 337–349.

Liao G., Xu, X. and Zhou, G. (2009) Effects of cooked temperatures and addition of antioxidants on formation of heterocyclic aromatic amines in pork floss. *Journal of Food Processing and Preservation* 33, 159–175.

Long, L., Komarik, S.L. and Tressler, D.K. (1982) *Food Product Formulary. vol 1. 2nd edn. Meats, Poultry, Fish, Shellfish.* The Avi Publishing Company, Inc., Westport, Connecticut.

Lonnecker, S.M., Boyle, E.A.E., Getty, J.K., Buege, S.R., Ingham, S.C., Searl, G. and Harper, N.M. (2010) Production methods and product characteristics of jerky produced by small and very small meat processing businesses. *Journal of Muscle Foods* 21, 826–833.

Messenger, J. (2007) Deli-cious trends. IDDBA's Carol Christison talks about deli trends with M and P. *Meat & Poultry* May, pp. 28–30.

Mgbemere, V.N., Akpapunam, M.A. and Igene, J.O. (2011) Effect of groundnut flour substitution on yield, quality and storage stability of Kilishi – A Nigerian indigenous dried meat product. *African Journal of Food, Agriculture, Nutrition and Development* 11, 4718–4738.

Naidoo, K. and Lindsay, D. (2010) Survival of Listeria monocytogenes, and enterotoxin-producing Staphylococcus aureus and Staphylococcus pasteuri, during two types of biltong-manufacturing processes. *Food Control* 21, 1042–1050.

Özbey, G., Kok, F. and Muz, A. (2007) Isolation of Salmonella spp. in camel sausages from retail markets in Aydýn, Turkey, and Polymerase Chain Reaction (PCR) confirmation. *Turkish Journal of Veterinary and Animal Sciences* 31, 67–71.

Öksüztepe, K.G., Ilhak, O.I. and Patir, B. (2006) Chemical and microbiological quality of fermented sausages made from camel meat. *Medycyna Weterynaryjna* 62, 893–896.

Osman, F.M. (1999) Iron deficiency anaemia in preschool children (1-6 years) in Al Shigia area (Khartoum State): Prevalence and Aetiology. PhD Thesis, University of Khartoum, Sudan.

Ottinger, H. and Hofmann, T. (2003) Identification of the taste enhancer alapyridaine in beef broth and evaluation of its sensory impact by taste reconstitution experiments. *Journal of Agricultural and Food Chemistry* 51, 6791–6796.

Pearson, A.M. and Gillett, T.A. (1999) *Processed Meats*, 3rd edn. Aspen Publishers Inc., Gaithersburg, Maryland.

Puolanne, E. (2010) Cooked sausages. In: Toldra, F. (ed.) *Handbook of Meat Processing.* Wiley-Blackwell, Ames, Iowa, pp. 313–325.

Remon, J.P., Aerde, P.V. and Severen, R.V. (1985) Effect of fluidized bed granulation on characteristics of meat extract tablets. *Voedins middelentechnologie* 18, 24–25.

Salvage, B. (2010) Poultry on the menu. The affordability and versatility of chicken make it a foodservice mainstay. *Meat & Poultry*, December, pp. 20–34.

Schmitz, W. (1974) Method of producing meat powder of high protein content. German Federal Republic Patent Number 2319640.

Shyoukhi, N. (2009) Saudi fast food restaurant serving camel burgers. Available at: http://www.reuters.com/article/2009/07/07/us-saudi-camels-idUSTRE56628U20090707 (accessed 6 August 2011).

Sloan, A.E. (2009) 10 top food trends. *Food Technology*, April, pp. 22–40.

Stamenkovic, T., Ristic, D., Stanimirovic, D. and Saponja, M. (1978) Rendering of extracts from pork and beef bones to be used as raw materials for table-ready foods. *Technologija Mesa* 19, 102–107.

Stute, R. and Seuss, I. (1993) Process for manufacture of meat broth or meat extracts. German Federal Republic Patent Number DE4216078C1.

Trojak, V. and Tolic, M. (1977) Manufacture and application of a meat extract. *Technologija Mesa* 18, 349–358.

Ulmer, K., Herrmann, K. and Fischer, A. (2004) Meat products from camel meat. In: Farah, Z. and Fischer, A. (eds) *Milk and Meat from the Camel. Handbook on Products and Processing.* Vdf Hochschulsverlag AG an der ETH Zurich, Zurich/Singen, pp. 137–226.

Vignolo, G., Fontana, C. and Fadda, S. (2010) Semidry and dry fermented sausages. In: Toldra, F. (ed.) *Handbook of Meat Processing.* Wiley-Blackwell, Ames, Iowa, pp. 379-398.

Warfield, B. and Tume, L. (2000) Marketing analysis and plan for the camel industry. A report for the Rural Research and Development Corporation (RIRDC). RIRDC Publicatiom No 00/9. RIRDC, Barton, Australia.

Whitesel, T. (2011) Moroccan finger foods. Food eHow. Available at: http://www.ehow.com/list_6087503_moroccan-finger-foods.html (accessed 4 August 2011).

Yahya, R.A. (2011) Canned camel meat promotion. Available at: http://old.brudirect.com/public_html/DailyInfo/News/Archive/July03/290703/bb03.htm (accessed 7August 2011).

Yetim, H. and Cankaya, H. (2001) The effects of $CaCl_2$ and curing technique on the tenderness of pastirma, a Turkish dry meat product. *Gida* 26, 203–207.

Yousif, O.K. and Babiker, S.A. (1989) The desert camel as a meat animal. *Meat Science* 26, 245–254.

Zegaye, A. (1999) A note on the influence of heat treatment, salting and smoking on the acceptability of camel meat products. *Meat Science* 53, 217–219.

Zukál, E. and Incze, K. (2010) Drying. In: Toldra, F. (ed.) *Handbook of Meat Processing.* Wiley-Blackwell, Ames, Iowa, pp. 219–229.

13 Nutritive and Health Value of Camel Meat

Alaa El-Din Bekhit and Mustafa M. Farouk
[1]*Department of Food Science, Division of Sciences, University of Otago, Dunedin, New Zealand;* [2]*AgResearch Limited, Ruakura Research Centre, Hamilton, New Zealand*

13.1 Introduction

Camel meat is a significant source of animal protein in many African and Asian countries. The culinary and cooking practices, as well as the palate for meat, in several African and Arabian countries have evolved to prefer camel meat over other meat animal species because of beliefs in medicinal benefits, its availability and/or affordable price.

This chapter examines the nutritional value and the availability of muscle bioactive compounds in camel meat. A comparison of the nutritional properties of camel meat with those of other meats (e.g. beef, lamb) will be limited to the studies where meat samples from the different species were analysed in the same study. This is to eliminate the effects of confounding factors such as environment, diet and practices normally used in cattle and sheep farming in Western countries, which vary greatly from those normally found in camel farming systems.

13.2 Proximate Composition

The proximate composition of camel meat is similar to meats from other species where an inverse relationship exists between the moisture and protein contents and the fat content (Table 13.1). In addition to its importance in determining the nutritional value of meat, proximate composition is also an important indicator of meat functionality. For instance, the moisture content of camel meat plays an important role in the keeping and eating qualities of meat (Kadim *et al.*, 2006), whereas protein and fat contents dictate the manufacturing quality of meat.

13.2.1 Moisture content

A wide range of moisture content has been reported for camel meat (67.84–78.85%; Table 13.1). Different cuts from the same animal seem to have similar moisture contents (Babiker and Yousif, 1990; El-Faer *et al.*, 1991; Al-Shabib and Abu-Tarboush, 2004; Alfawaz, 2004). However, in animal trials where the diet was manipulated, the *Biceps femoris* muscle had a higher moisture content (74.2%) than the *Longissimus dorsi* muscle (69.2%) because of the higher fat content in the *Longissimus dorsi* (Shehata, 2005). The moisture content of camel meat decreases with the increase in the animal age (Dawood and Alkanhal, 1995; Gheisari *et al.*, 2009). The differences between the maximum and minimum moisture contents of camel *Longissimus thoracis* were 3.2, 6.4 and 12.3% for the 1–3, 3–5 and 6–8 year age groups,

Table 13.1. Published proximate analysis of camel meat as affected by muscle/meat cut and animal age.

Muscle/cut	Moisture	Protein	Fat	Ash	References
Longissimus thoracis	65.70	19.50	2.10	1.20	Kadim *et al.* (2011)
Longissimus dorsi	–	–	7.95	–	Gheisari and Motamedi (2010); Gheisari (2011)
–	73.82	23.69	3.62	–	Al-Bachir and Zeinou (2009)
Biceps femoris[a]	73.00	22.8	1.05	0.75	Gheisari *et al.* (2009)
Triceps brachii[a]	72.00	21.2	1.35	0.81	
Longissimus dorsi[a]	68.30	21.5	1.60	0.69	
Biceps femoris[b]	71.40	22.2	1.56	0.98	
Triceps brachii[b]	70.50	20.25	2.44	1.06	
Longissimus dorsi[b]	67.84	20.52	2.54	0.95	
Longissimus thoracis	74.80	21.10	2.76	1.34	Kadim *et al.* (2009)
Longissimus thoracis	71.70	22.70	4.40	1.10	Kadim *et al.* (2006)
	71.00	20.90	7.00	1.10	
	70.30	20.50	8.30	1.10	
–	–	–	2.28	–	Sallam and Morshedy (2008)
Leg	78.71	–	0.84[f]	–	Alfawaz (2004)
Shoulder	78.11	–	0.80[f]	–	
Leg	75.19	21.34	2.26	1.10	Al-Sheddy *et al.* (1999)
Loin	77.50	18.20	2.90	1.20	Al-Shabib and Abu-Tarboush (2004)
Leg	77.20	18.40	2.90	1.20	
Combined leg and loin	77.2	19.30	2.60	0.90	Elgasim and Alkanhal (1992)
Chuck + ribeye + leg[c]	75.99	20.55	4.14	1.10	Dawood and Alkanhal (1995)
Chuck + ribeye + leg[d]	71.56	20.39	7.19	1.03	
Chuck + ribeye + leg[e]	68.79	18.95	9.79	1.11	
Chuck	72.82	19.36	6.28	1.12	
Ribeye	69.55	20.26	10.58	1.01	
Leg	74.57	20.27	4.27	1.11	
Shoulder	78.25	19.52	1.24	1.09	El-Faer *et al.* (1991)
Thigh	78.40	18.88	1.40	1.13	
Ribs	78.85	18.71	1.85	1.04	
Neck	78.52	19.23	1.60	1.08	
Longissimus dorsi	75.89	21.63	1.43	1.05	Babiker and Yousif (1990)
Semitendinosus	75.81	21.41	1.40	1.38	
Triceps brachii	75.23	22.13	1.42	1.22	

[a]From 1-year-old animals; [b]from 5-year-old animals; [c]from 8-month-old animals; [d]from 16-month-old animals, [e]from 26-month-old animals; [f]well trimmed.

respectively. This indicates that the variation in moisture content within the samples is greater in older animals (Kadim *et al.*, 2006). Studies that compared the moisture content of camel meat with that of other species at the same age and sex found no species effects (Gheisari *et al.*, 2009). The moisture content of trimmed Australian beef, veal, lamb and mutton varied between 72.9% and 74.8% (Williams, 2007) and the *Longissimus dorsi* from mature animal of different species are in the range 76.8–77.0% (Lawrie, 2002) suggesting that the moisture content of camel meat is similar to red meat from other species with the exception of animals >6 years old.

13.2.2 Protein content

The protein content of camel meat is in the range 18.2–23.7% (Table 13.1). Slight differences exist between different meat cuts and meat from animals from different age groups (El-Faer *et al.*, 1991; Dawood and Alkanhal, 1995; Kadim *et al.*, 2006). Genetic and diet effects might cause slight differences. Studies from Saudi Arabia (El-Faer *et al.*, 1991; Elgasim and Alkanhal, 1992; Al-Shabib and Abu-Tarboush, 2004) reported lower protein content than those from United Arab Emirates, Iran, Sudan and Syria (Babiker and Yousif, 1990; Kadim *et al.*, 2006, 2009; Al-Bachir and Zeinou, 2009; Gheisari *et al.*, 2009). Meat from young camels has similar protein content to those in young cattle, lamb and goat meats (Elgasim and Alkanhal, 1992; Kadim *et al.*, 2009). The protein contents of three skeletal muscles (*Triceps brachii* and *Longissimus thoracis*) were not different in young male and female camels and cattle. Higher protein content was found, however, in the *Biceps femoris* muscle from adult cattle than in the *Biceps femoris* muscle from the adult camel (Gheisari *et al.*, 2009). Total collagen content is higher in camel *Longissimus thoracis* muscle than in *Semitendinosus* or *Triceps brachii* muscles possibly because of morphological requirement for stabilizing the hump attached to the *Longissimus thoracis* (Babiker and Yousif, 1990).

13.2.3 Fat content

The fat content of camel meat ranged from 1.4 to 10.6% (Table 13.1). Slight differences in the fat content were reported in different cuts and muscles with significant variation in fat content between these reports (Table 13.1). Clearly the animal's age has a great effect on the fat content with camel meat from older animals' containing more fat than meat from younger animals (Table 13.2). However, other on-farm factors seem to affect the fat content in camel meat from animals within similar age groups (E1-Faer *et al.*, 1991; Elgasim and Alkanhal, 1992; Dawood and Alkanhal, 1995; Kadim *et al.*, 2006, 2009; Gheisari *et al.*, 2009). Camel meat contains less fat than beef, lamb and goat meat and slightly higher fat content than ostrich meat (Table 13.2). This makes the camel meat a healthy option and advantageous in special diets.

13.2.4 Ash content

The ash content in camel meat has been reported in the range 0.75–1.38% (Table 13.1). Some reports suggested that the ash content varies with muscles and meat cuts (Babiker and Yousif, 1990; Dawood and Alkanhal, 1995; Gheisari *et al.*, 2009) and in meat from camel carcasses of different ages (Gheisari *et al.*, 2009), whereas others found no effect for age and meat cut on ash content (El-Faer *et al.*, 1991; Al-Shabib and Abu-Tarboush, 2004; Shehata, 2005; Kadim *et al.*, 2006). Camel meat has relatively lower ash content than beef, lamb and goat meat (Elgasim and Alkanhal, 1992; Gheisari *et al.*, 2009).

13.2.5 Amino acid composition

Essential amino acids

The essential amino acid content of camel meat is not affected by the animal age (Dawood and Alkanhal; 1995). Camel meat has a comparable essential amino acid contents to beef, lamb and goat meat (Table 13.3) but camel meat has a higher lysine and methionine content than ostrich meat (Al-Shabib and Abu-Tarboush, 2004). The amount of camel meat required to supply the daily requirements of essential amino acids for an adult (70 kg body weight) is similar to that from lamb (based on methionine, which has the lowest content in meat) but is less than the amount required from beef (Fig. 13.1).

As reported for other meats, leucine (7.08–9.51% of protein) and lysine (8.33–9.85% of protein) are the highest essential amino acids in camel meat (Table 13.3). The camel meat essential amino acids contents varied slightly among meat cuts with greater differences between studies (Table 13.3). For example, the essential amino acid contents in loin and leg meats differed by >2.1% with the exception of leucine, methionine and tryptophan, which differed by 18.5%, 25.4% and 14.6%, respectively (Al-Shabib

Table 13.2. Effect of age and muscle/meat cut on the fat content of camel meat compared with the fat content in other meats reported in the same study.

Muscle/cut	Fat (%)	Fat content of other species	Age (years)	References
Longissimus thoracis	2.1	–	2–3	Kadim *et al.* (2011)
Longissimus dorsi	7.95	Beef = 5.35	(Mature)	Gheisari and Motamedi (2010); Gheisari (2011)
–	3.62	–	1	Al-Bachir and Zeinou (2009)
Biceps femoris	1.05	Beef = 2.71		
Triceps brachii	1.35	Beef = 3.14	1	
Longissimus dorsi	1.60	Beef = 3.25		
Biceps femoris	1.56	Beef = 4.26		Gheisari *et al.* (2009)
Triceps brachii	2.44	Beef = 4.50	5	
Longissimus dorsi	2.54	Beef = 4.74		
Longissimus thoracis	2.76	Beef = 7.83	Camel (2–3) Cattle (1–3)	Kadim *et al.* (2009)
–	2.28	Beef = 3.58 Lamb = 4.75	Camel (2–7) Cattle (1.5–3.5) Lamb (1–2.5)	Sallam and Morshedy (2008)
Longissimus thoracis	4.40	–	1–3	
	7.00	–	3–5	Kadim *et al.* (2006)
	8.30	–	6–8	
Longissimus dorsi	8.31	–	0.83–1	Shehata (2005)
Biceps femoris	4.32	–		
Leg	0.84	Beef = 0.22 Lamb = 2.48–2.96	Cattle = 1 Camel = 1 Lamb = 0.5–0.75	Alfawaz (2004)
Shoulder	0.80	Beef = 0.65 Lamb = 3.73–5.12	0.75	
Loin	2.90	Ostrich = 2.10	Camel = 0.8–1.0	Al-Shabib and Abu-Tarboush (2004)
Leg	2.90	Ostrich = 1.70	Ostrich = 0.5–0.8	
Leg	2.26	–	1.37	Al-Sheddy *et al.* (1999)
Average of chuck, rib-eye and leg	4.14	–	0.67	
	7.19	–	1.37	Dawood and Alkanhal (1995)
	9.79	–	2.17	
Combined leg and loin	2.60	Beef = 4.70 Lamb = 6.20 Goat = 3.30	Camel = 2 Cattle = 0.58 Lamb = 0.5 Goat = 0.42	Elgasim and Alkanhal (1992)
Shoulder	1.24	–		
Thigh	1.40	–	1–3	El-Faer *et al.* (1991)
Ribs	1.85	–		
Neck	1.60	–		

and Abu-Tarboush, 2004). Similarly, essential amino acid contents in chuck, rib-eye and leg samples differed by >4.2% with the exception of isoleucine, methionine, threonine, tryptophan and valine, which differed between 8 and 42% (Dawood and Alkanhal, 1995). On the other hand, differences in essential amino acids reported across different cuts ranged between 0.6 and 166.7% (Elgasim and Alkanhal, 1992; Dawood

Table 13.3. Reported composition of the amino acids in camel meat.

Factor	Essential amino acids									Non-essential amino acids									References
	His	Ileu	Leu	Lys	Met	Phe	Thr	Trp	Val	Ala	Arg	Asp	Cys	Glu	Gly	Pro	Ser	Tyr	
Effect of meat cut																			
Longissimus thoracis	4.4	4.7	8.29	9.35	2.9	4.3	4.5	–	5.6	6.5	6.6	9.3	–	15.9	4.3	3.9	3.6	3.5	Kadim *et al.* (2011)[a]
Loin	3.4	4.2	7.1	9.1	1.6	5.6	4.8	1.6	4.7	–	–	–	–	–	–	–	–	–	Al-Shabib and Abu-Tarboush (2004)
Leg	3.4	4.3	8.4	9.1	1.3	5.5	4.8	1.9	4.6	–	–	–	–	–	–	–	–	–	
Chuck	4.7	5.3	8.6	8.4	2.6	4.1	4.2	0.5	4.9	6.3	7.5	9.3	–	17.1	6.0	5.4	3.5	3.0	Dawood and Alkanhal (1995)
Rib-eye	4.3	5.4	8.3	8.6	2.2	4.4	4.7	0.7	5.3	6.2	7.1	9.3	–	17.3	5.9	4.9	3.8	3.4	
Leg	4.5	4.9	8.3	8.3	2.5	4.2	4.2	0.6	5.4	6.3	7.5	8.6	–	16.4	5.9	5.9	3.6	3.3	
Combined leg and loin	5.6	5.9	9.5	8.9	3.6	4.7	4.8	–	6.3	3.9	7.1	10.8	–	18.6	6.1	3.9	3.2	3.8	Elgasim and Alkanhal (1992)
Effect of age (months)																			
8	4.3	5.2	8.4	9.6	2.7	4.0	4.2	0.6	5.2	6.3	7.3	8.8		16.6	6.1	5.7	3.5	3.1	Dawood and Alkanhal (1995)
26	4.3	5.3	8.7	8.1	2.1	4.2	4.4	0.6	5.1	6.0	7.5	9.4	–	17.2	5.5	5.8	3.8	3.4	
Effect of species																			
Camel	5.6	5.9	9.5	8.9	3.5	4.7	4.8	–	6.3	3.9	7.1	10.8	–	18.6	6.1	3.9	3.2	3.8	
Beef	6.2	6.5	10.7	9.1	2.7	5.7	5.5	–	6.6	7.7	7.1	10.8	–	16.5	6.2	4.5	4.2	4.1	Elgasim and Alkanhal (1992)
Lamb	5.9	5.8	9.6	8.5	3.3	4.9	4.2	–	5.9	6.7	6.9	10.3	–	17.9	5.5	3.8	2.9	3.5	
Goat	4.7	6.0	7.9	10.9	3.9	6.5	4.4	–	6.8	4.7	7.1	10.8	–	15.6	5.2	3.8	3.6	5.9	
Camel	3.4	4.3	7.7	9.1	1.4	5.5	4.8	1.8	4.7	6.5	6.9	9.7	–	17.0	6.2	–	4.3	3.3	Al-Shabib and Abu-Tarboush (2004)
Ostrich	2.8	3.8	7.4	4.3	0.5	4.9	4.2	1.8	3.8	5.6	5.9	8.3	–	15.4	4.5	–	3.7	2.8	

[a]Calculated from mg/100 DM values using the average DM and protein contents

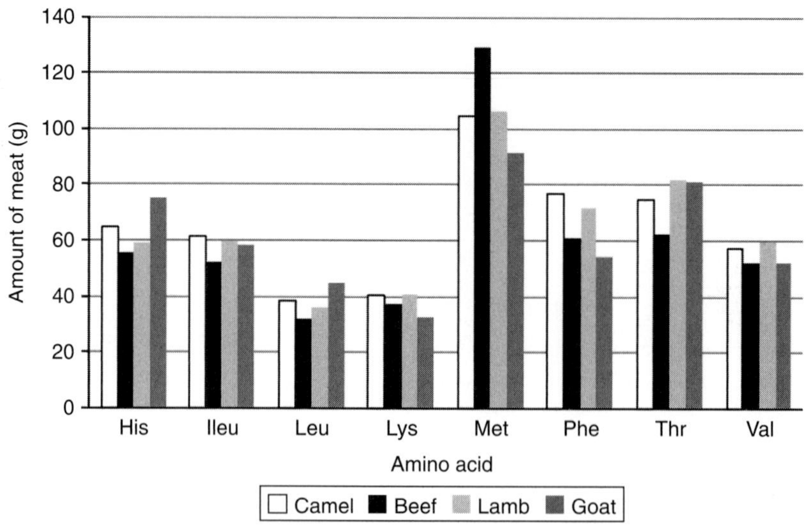

Fig. 13.1. The amounts of meat from different species required to supply the daily requirements of indispensible amino acids for adults (70 kg body weight) based on mean nitrogen requirement of 105 mg nitrogen/kg/day (0.66 g protein/kg/day) recommended by WHO (2002).

and Alkanhal, 1995; Al-Shabib and Abu-Tarboush, 2004), indicating the confounding effects of on-farm and analytical factors. Tryptophan concentration in camel meat was lower than in other meats (Dawood and Alkanhal, 1995). However, a recent study by Al-Shabib and Abu-Tarboush (2004) found tryptophan concentration was 1.76% of the total amino acids, which was higher than the 1.28% reported for beef (Kadim *et al.*, 2008). Given that these values were obtained from individual studies, the amino acid profiles in camel and other red meats need to be assessed in same study.

Non-essential amino acids

Glutamic (15.95–18.60% of protein) and aspartic (9.30–10.80% of protein) acids are the major non-essential amino acids in camel meat (Table 13.3). As with the essential amino acids, non-essential amino acids contents also vary slightly among cuts within studies and a larger variation is found between studies. Age effects were reported for aspartic acid only (Dawood and Alkanhal, 1995). Generally speaking, camel meat is similar or maybe a better source of non-essential amino acids compared with beef, lamb, goat and ostrich meats (Table 13.3). Elgasim and Alkanhal (1992) reported a very low alanine concentration in camel meat compared with other red meats. These findings are different from later studies that did not report a lower concentration of alanine in camel meats relative to other red meats (Dawood and Alkanhal, 1995; Al-Shabib and Abu-Tarboush, 2004; Kadim *et al.*, 2011).

13.2.6 Bioactive compounds

Several bioactive compounds have been investigated in meat (Arihara, 2006) that are nutritionally important and can potentially be useful in marketing red meat. Carnosine (β-alanyl-L-histidine) and its derivative anserine (β-alanyl-L-methyl-L-histidine) are important dipeptides which are found in high concentrations in the muscle and brain of mammalian and avian species (Tomonaga *et al.*, 2006). They function as antioxidants and putative neurotransmitters in the brain (Tomonaga *et al.*, 2006). High concentrations of about 365 and 400 mg/100 g have been reported in beef and lamb, respectively

(Purchas *et al.*, 2004) and in red deer, 290–329mg/100g (Purchas *et al.*, 2010). The average levels of carnosine and anserine (mmol/kg dry matter) in camel middle gluteal muscles are shown in Fig. 13.2. On a fresh weight basis, camel muscle has 181.7mg carnosine/100g fresh weight and 268.6mg anserine/100g fresh weight (Dunnett and Harris, 1997; Dunnett *et al.*, 1997). Given the wide variation in carnosine concentrations between muscles in beef and lamb (Purchas *et al.*, 2004; Purchas and Busboom, 2005), similar variations in carnosine and anserine have been found between camel muscles too. This has been attributed to differences in metabolic activity and diet (Dunnett *et al.*, 1997). Thus, the concentrations of these bioactive molecules in different camel muscles and how they compare with other animal species under similar environments require more research.

L-Carnitine (β-hydroxy-γ-trimethyl amino butyric acid) plays an important physiological role in producing energy during exercise through transporting long-chain fatty acids across the inner mitochondrial membranes. The L-carnitine concentrations in chicken, pork, beef, horse and venison have been reported as 0.69–4.95µmol/g fresh weight (Shimada *et al.*, 2004). Alhomida *et al.* (1995) reported 5.17, 2.60 and 7.77µmol/g fresh weight of free carnitine, acylcarnitine and total carnitine, respectively, in camel meat (Fig. 13.3). Although the significance of the concentration cannot be objectively determined because these results have been generated from different laboratories, it is possible that camel meat could potentially be one of the best sources of L-carnitine after goat meat (11.36µmol/g fresh weight; Shimada *et al.*, 2004).

Other bioactive compounds that are available in meat are vitamin E, coenzyme 10, choline, vitamin B groups and glutathione. There is, however, a dearth of information on the levels of these compounds in camel meat.

13.2.7 Fatty acid composition

The available data on fatty acid composition of camel meat are rather limited (Table 13.4) with most of the available literature focused more on the composition of the hump (Mirgani, 1977; Emmanuel and Nahapetian, 1980; Abu-Tarboush and Dawood, 1993; Kadim *et al.*, 2002). Extensive characterization of the fatty acids of camel meat was reported by Rawdah *et al.* (1994) who identified 22 fatty acids in camel meat. Major fatty acids in camel meat were also reported by Al-Bachir and Zeinou (2009) and Kadim *et al.* (2011). The composition of major fatty acids seems to be controversial partially because of the number of identified fatty acids that will affect the percentage of individual fatty acids (Table 13.4). For example, although Rawdah *et al.* (1994) reported 18.93% oleic (C18:1) and 12.07% linoleic acid (C18:2) of the fatty acids in the camel meat, about twice the percentage of oleic (C18:1) and less than half the percentage of linoleic acid (C18:2) were reported by Al-Bachir and Zeinou (2009) and Kadim *et al.* (2011). Linoleic acid is derived entirely

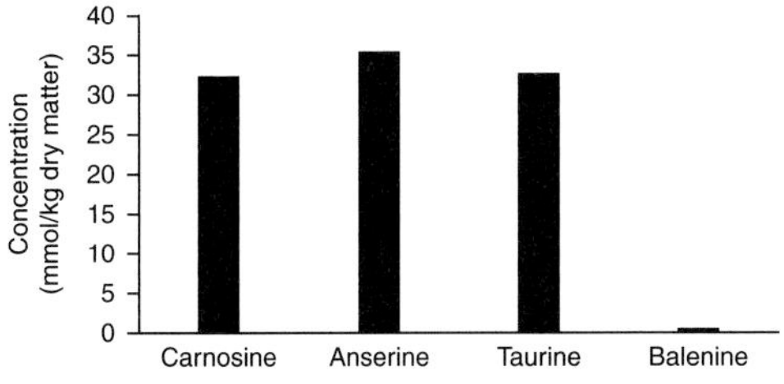

Fig. 13.2. The average concentrations of carnosine, anserine, taurine and balenine in camel middle gluteal muscle. From Dunnett *et al.* (1997) and Dunnett and Harris (1997).

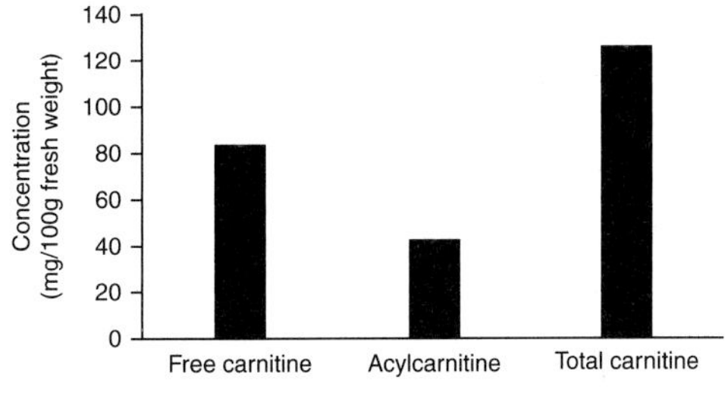

Fig. 13.3. Average concentrations of free carnitine, acylcarnitine and total carnitine in camel muscle tissue (calculated from Alhomida *et al.*, 1995).

Table 13.4. Reported composition of the fatty acids in camel meat.

Fatty acids (%)	Rawdah *et al.* (1994)	Al-Bachir and Zeinou (2009)	Kadim *et al.* (2011)
Saturated			
14:0	7.68	4.53	3.10
15:0	1.66	–	2.10
16:0	25.98	30.29	28.50
17:0	1.48	2.54	–
18:0	8.63	25.51	19.30
20:0	trace	–	–
22:0	trace	–	–
Unidentified	2.55	–	–
Monounsaturated			
14:1	1.0	–	1.60
16:1	8.06	–	6.30
17:1	0.94	–	–
18:1	18.93	32.01	33.50
20:1	trace	–	–
Unidentified	0.97	–	–
Polyunsaturated			
18:2ω6	12.07	5.13	3.20
20:2ω6	0.11	–	–
18:3ω3	0.52	–	1.20
20:3ω9	0.37	–	–
20:3ω6	0.30	–	–
20:4ω6	2.84	–	1.20
22:4ω6	0.10	–	–
20:5ω3	0.32	–	–
22:5ω3	0.48	–	–
22:6ω3	0.10	–	–
Unidentified	1.34	–	–
P/S	0.36	–	0.11
Total saturated	51.54	–	53.00
Total monounsaturated	29.90		41.40
Total polyunsaturated	18.55	–	5.60
ω3/ω6	0.092	–	–

from the diet (Wood *et al.*, 2008) and such differences are not unexpected from studies from different regions. The major saturated, monounsaturated and polyunsaturated fatty acids in camel meat are C16:0, C18:1 and C18:2, respectively (Table 13.4). Although there is an agreement on the percentage of total saturated fatty acids among the published reports (51.5–53%), different percentages for monounsaturated (29.9% and 41.4%) and polyunsaturated (18.6% and 5.6%) fatty acids have been reported (Rawdah *et al.*, 1994; Kadim *et al.*, 2011). Unfortunately, there are no reports comparing the fatty acids of meat from different species within the same study. Given that diet plays an important role in modifying the fatty acid profile (Wood *et al.*, 2003, 2008), a comparison with literature values would not be appropriate.

The camel hump is important and commonly used for cooking in African camel producing countries. On a fresh weight basis, the camel hump contributes about 64.2–84.8% fat with a very high content of saturated fatty acids of about 63.0% (Rawdah, *et al.*, 1994; Kadim *et al.*, 2002). Palmitic acid (C16:0), stearic acid (C18:0) and oleic acid (C18:1) are the most abundant fatty acids in the hump. The composition of the hump fatty acids is affected by the animal age. The highest percentage of unsaturated fatty acids and lowest percentage of saturated fatty acids are found in animals of less than 1 year, whereas the opposite trend applies in animals in the 1–3 year old age group (Kadim *et al.*, 2002).

Despite camel meat having a higher catalase and glutathione peroxidase content compared with beef and chicken, a higher lipid oxidation was reported in camel meat compared with beef and chicken (Gheisari, 2011). This contradicts the reported better lipid stability of raw, cooked and frozen camel meat compared with beef, lamb and chicken (Alfawaz, 2004).

Cholesterol

The adipose fat from the camel carcass contained a similar amount of cholesterol in the hump (139 mg/100g fresh weight), which is lower than those in lamb and beef adipose tissues (196 and 206 mg/100g fresh weight, respectively) analysed in the same study (Abu-Tarboush and Dawood, 1993), which supported the earlier reported low cholesterol content of camel meat compared to beef and lamb (Elgasim and Elhag, 1990). The cholesterol content in camel meat increases with the increase in animal age (135 mg per 100 g fresh weight for 8-month-old animals versus 150 mg/100 g fresh weight for 26-month-old animals) but this is mostly due to the increase in the fat content rather than actual increase in the synthesis of cholesterol considering the cholesterol content was 167 and 166 mg/100g fat (Abu-Tarboush and Dawood, 1993). This is particularly important to the Middle Eastern and African countries where the eating habits and cooking styles are very different from the Western ones and the use of animal fat in cooking is very common.

13.2.8 Mineral composition of camel meat

Minerals are generally classified as either essential/nutritional elements (Table 13.5) that are required for growth and optimal health or toxic elements (Table 13.6) that pose health risks to organisms. Both the deficiency and excess intake of essential elements, as well as exceeding the safe limits of toxic elements, can be detrimental to human health.

Nutritional elements

EFFECTS OF MEAT CUT. Calcium content (mg/100g fresh weight) was reported to be in the range of 1.33–11.48 (Table 13.5). The level of variation reported from the same laboratory (Kadim *et al.*, 2006, 2011) indicates that physiological factors play a major role in determining the calcium contents in camel meat. Small variations in calcium content can be found among different meat cuts within a study, with larger variation among studies (Table 13.5). For example, the variation between four to six different meat cuts was 19–27% (El-Faer *et al.*, 1991; Dawood and Alkanhal, 1995; Rashed, 2002), whereas up to 144% variation in calcium content can be observed among different meat cuts from different studies (Table 13.5).

Cobalt and chromium contents were reported in the range of 0.003–0.004 and 0.008–0.03 mg/100g fresh weight (Kadim *et al.*, 2006). Copper content in camel meat ranged from 0.04 to 0.12 mg/100g fresh weight (Table 13.5). The foreleg seems to have a higher copper content (scapula, front knuckle and front limb) compared with other meat cuts (Rashed, 2002).

Iron content in camel meat (1.16–3.39 mg/100 g fresh meat) varied among different meat cuts (Table 13.5), which is expected due to the different physiological requirements of myoglobin of different muscles. As with other red meat species, meat cut containing oxidative muscles (e.g. leg and neck) has higher iron content than glycolytic muscles (e.g. rib-eye and ribs).

Potassium is the major element in camel meat (193.4–379.1 mg/100g fresh weight) and magnesium content in camel meat ranged between 10.41 and 21.03 mg/100g fresh weight (Kadim *et al.*, 2009). Meat cuts from the limbs (e.g. shoulder, thigh and scapula) have a higher potassium and magnesium content than the loins and ribs (Table 13.5).

Table 13.5. Reported nutritional elements concentrations in camel meat (mg/100g fresh weight, except for those reported by Rashed, 2002).

Factor	Calcium	Co	Cr	Cu	Fe	K	Mg	Mn	Mo	Na	P	S	Zn	
Effect of meat cut														
Rump	–	0.004	–	0.12	2.53	–	–	–	0.04	–	–	–	–	Badiei *et al.* (2006)
Intercostal	8.50	0.29	0.42	0.13	51.00	515.00	29.50	0.19	–	300.50	–	–	74.00	
Scapula	10.00	0.35	0.32	0.21	54.50	670.00	51.00	0.22	–	225.00	–	–	58.00	
Sirloin	10.15	0.27	0.41	0.16	44.00	446.00	28.00	0.16	–	188.50	–	–	66.00	Rashed
Flank	8.40	0.32	0.33	0.12	49.00	810.50	49.50	0.19	–	223.00	–	–	69.50	(2002)[a]
Front knuckle	8.35	0.26	0.42	0.25	44.50	630.00	37.00	0.17	–	299.50	–	–	73.50	
Front limb	9.75	0.19	0.37	0.26	50.50	547.50	42.50	0.19	–	312.50	–	–	85.50	
Chuck	11.48	–	–	–	3.23	249.38	17.41	–	–	73.53	–	–	3.65	Dawood and
Rib-eye	8.10	–	–	–	2.86	230.55	16.27	–	–	67.08	–	–	3.73	Alkanhal
Leg	10.28	–	–	–	3.39	250.29	17.10	–	–	69.67	–	–	3.93	(1995)
Leg+loin	4.97	–	–	0.04	1.94	228.00	17.74	0.01	–	47.88	–	–	3.21	Elgasim and Alkanhal (1992)[b]
Shoulder	5.05	–	0.01	0.07	1.24	357.40	20.56	0.01	–	69.08	195.70	56.09	3.52	
Thigh	5.41	–	0.01	0.09	1.35	360.50	21.03	0.01	–	70.42	199.00	54.99	3.07	El-Faer *et al.*
Ribs	4.71	–	0.01	0.07	1.16	324.00	18.46	0.01	–	84.10	181.10	57.97	3.85	(1991)
Neck	5.61	–	0.03	0.09	1.35	338.10	18.45	0.01	–	87.30	180.70	64.38	4.80	
Effect of age (year)														
2–3			0.08	–				–				136.57	3.62	Kadim *et al.*
	1.33	–	–	0.08	–	379.08	18.10	–	–	75.82	164.27			(2011)[b]
1–3	3.88	0.003	0.01	–	–	199.15	10.41	–	0.03	40.19	105.56	–	–	Kadim *et al.*
3–5	5.39	0.003	0.01	–	–	228.29	12.01	–	0.04	47.97	126.67	–	–	(2006)
6–8	8.79	0.004	0.02	–	–	247.61	12.95	–	0.04	48.50	148.29	–	–	

0.67	10.59	–	–	–	3.40	303.20	19.77	–	–	76.23	–	–	3.34	Dawood and
1.37	8.73	–	–	–	2.84	226.70	15.50	–	–	68.23	–	–	3.57	Alkanhal
2.17	10.54	–	–	–	3.24	208.33	15.52	–	–	65.86	–	–	4.41	(1995)
Effect of species												–		
Camel	5.96	0.003	0.008	–	–	193.42	12.95	–	0.008	45.30	105.36	–	–	Kadim *et al.*
Beef	6.17	0.003	0.009	–	–	415.98	20.5	–	0.006	51.02	161.82	–	–	(2009)[b]
Camel	4.97	–	–	0.04	1.94	228.00	17.74	0.01	–	47.88	–	–	3.21	Elgasim and
Beef	6.97	–	–	0.06	2.66	277.31	24.76	0.02	–	31.23	–	–	4.07	Alkanhal (1992)[b]
Effect of environment (location/diet)											–			
Desert	9.05	0.33	0.36	0.21	45.50	576.67	46.67	0.15	–	301.67	–	–	134.00	Rashed
Town	9.19	0.28	0.38	0.19	48.92	603.17	39.58	0.18	–	258.17	–	–	71.08	(2002)[a]

[a]Average of samples from town and desert grown camels (mg/100 DM) and no moisture content was provided; [b]calculated from values on DM basis using the reported average moisture content.

Table 13.6. Reported toxic/non-essential elements concentrations (mg/100g fresh weight, except for those reported by Rashed, 2002).

Factor	Ag	Al	Au	Cd	Ni	Pb	Sr	References
Effect of meat cut								
Intercostal	0.07	–	0.11	–	0.24	–	–	
Scapula	0.06	–	0.10	–	0.38	–	–	
Sirloin	0.11	–	0.19	–	0.05	–	–	
Flank	0.09	–	0.12	–	0.13	–	–	Rashed
Front knuckle	0.12	–	0.17	–	0.19	–	–	(2002)[a]
Front limb	0.11	–	0.21	–	0.21	–	–	
Shoulder	–	0.51	–	–	–	–	0.02	El-Faer *et al.*
Thigh	–	0.15	–	–	–	–	0.03	(1991)
Ribs	–	0.12	–	–	–	–	0.02	
Neck	–	0.58	–	–	–	–	0.03	
Effect of age (year)								
2–3	–	0.58	–	–	–	–	–	Kadim *et al.*
1–3	–	–	–	0.003	0.02	0.02	–	(2011)
3–5	–	–	–	0.004	0.03	0.03	–	Kadim *et al.*
6–8	–	–	–	0.004	0.04	0.04	–	(2006)
Effect of species								
Camel	–	–	–	0.003	0.025	0.015	–	Kadim *et al.*
Beef	–	–	–	0.003	0.044	0.006	–	(2009)
Effect of environment (location/diet)								
Desert	0.12	–	0.18	–	0.17	–	–	Rashed
Town	0.09	–	0.15	–	0.20	–	–	(2002)

[a]Average of samples from town and desert grown camels (mg/100 DM) and no moisture content was provided.

Meat from camels in Saudi Arabia seems to have similar manganese content (0.01 mg/100g fresh weight) across four different meat cuts (El-Faer *et al.*, 1991; Elgasim and Alkanhal, 1992). Meat from camels in Egypt, however, seems to have higher manganese content (mg/100g dry matter) and the concentration varied among different meat cuts (Rashed, 2002).

Sodium content in camel meat is in the range 40.2–87.3 mg/100g (Table 13.5). Loins have the lowest sodium content among the different meat cuts tested (Elgasim and Alkanhal, 1992; Rashed, 2002; Kadim *et al.*, 2006).

Phosphorus is the second most abundant element in camel meat (105.6–199.0 mg/100g fresh weight) and thigh and shoulder cuts have slightly higher phosphorus than ribs and neck cuts (El-Faer *et al.*, 1991).

Sulfur content was reported to in the range 54.99–136.57 mg/100g fresh weight. Sulfur in four meat cuts varied by only 17% (El-Faer *et al.*, 1991).

Red meat is an important source of zinc. Camel meat contains about 3.07–4.80 mg/100g fresh weight (Table 13.5). The variation between different cuts has been reported as 7.6% (Dawood and Alkanhal, 1995) but a higher percentage of variation (47–56%) has been reported in other studies (El-Faer *et al.*, 1991; Rashed, 2002).

EFFECTS OF ANIMAL AGE. Calcium content in camel meat increased with increasing animal age when the animals were ≥ 1 year (Dawood and Alkanhal, 1995; Kadim *et al.*, 2006) but younger animals (10 months old) had comparable calcium contents to 26-month-old animals (Dawood and Alkanhal, 1995). A similar trend was found for cobalt but there were no effects of age for chromium, potassium, magnesium, molybdenum, sodium or phosphorus (Kadim *et al.*, 2006).

EFFECTS OF ANIMAL SPECIES. Camel meat has lower Mg, K and P, and higher Na compared with beef (Kadim *et al.*, 2009). There are no differences in calcium, cobalt, chromium, molybdenum and phosphorus contents in camel and cattle meats (Kadim *et al.*, 2009). Lower calcium, copper, iron, potassium, magnesium, manganese and zinc contents have been reported by Elgasim and Alkanhal (1992) but there were no indications of the significance of these differences.

EFFECTS OF ENVIRONMENT. Higher cobalt, copper, magnesium, sodium and zinc contents and lower calcium, iron, potassium and manganese contents have been reported in camel meat from animals in the desert compared with those grown in farms (Rashed, 2002).

Toxic elements

The concentrations of silver, gold and nickel in five camel meat cuts have been reported at 0.06–0.12, 0.10–0.21 and 0.05–0.38 mg/100g dry matter, respectively (Rashed, 2002). The concentration varied among five meat cuts by 100%, 110% and 750% (Table 13.6). A similar level of variation has been reported for aluminum (Al) in four meat cuts (El-Faer *et al.*, 1991). The concentrations of nickel, muscle beryllium and vanadium increased in the camel *Longissimus thoracis* with an increase in the animal age (Kadim *et al.*, 2006). The level of lead in camel *Longissimus thoracis* was 2.5 times the concentration in beef *Longissimus thoracis*; however, the difference was not statistically significant (Kadim *et al.*, 2009). Studies on the levels of trace and heavy elements in camel blood concluded that the camel could be less efficient than other ruminants in detoxifying its body (Al-Qarawi and Ali, 2003). Therefore, attention should be paid to monitoring the toxic levels in biological materials from camel (Faye *et al.*, 2008). Also the relationship between the deposition of elements in different organs and levels in biological fluids is very important to investigate, given that meat and offal (liver, kidney and heart) are both consumed in countries where camels are used as a source of animal protein. Indeed, monitoring the level of toxic compounds in the offal should be a priority because it is regularly consumed by low-income groups as a source of animal protein in many developing countries, sometimes raw.

Farming conditions (diet, desert versus farm, and soil composition) as well as the physiological conditions of the animals (breed, sex and age) seem to play an important role in determining the level of various elements in the meat (Tables 13.5 and 13.6) and the camel blood (Faye *et al.*, 2008). For instance, calcium content in camel meat reported from the same laboratory (Kadim *et al.*, 2006, 2011) or across different laboratories (Dawood and Alkanhal, 1995; Kadim *et al.*, 2006) supports this contention. It is worth mentioning that the biological variation of elements content even within the same herd that has a similar farming background is very high (Kadim *et al.*, 2006). A major problem in estimating the nutritional and health impact of these values is that they are either reported on dry matter or fresh wet basis which can be confusing in estimating the values consumed. A more informative system of reporting these elements would be on the basis of 100g of cooked meat which will eliminate the problems associated with variation in moisture content and cooking loss of the different meat cuts. Other elements that are important for health such as selenium, boron and iodine should be considered in future research.

Contaminants

The concentrations of organochlorine pesticides in meat, liver and kidney samples from Egyptian camel, cattle and sheep was investigated by Sallam and Morshedy (2008). Despite the longer life span of camel compared with cattle and sheep, lower levels of contaminants were found in camel meat. For example, camel meat had 77 and 70% the concentration of DDT found in cattle and sheep meat, respectively (Fig. 13.4). Similarly, lower concentrations of HCH, lindane and dieldrin were in camel meat compared with cattle and sheep meat (60 and 60%; 43 and 60%; 25 and 30%, for cattle and sheep meat, respectively). The aldrin level in camel meat was similar to that found

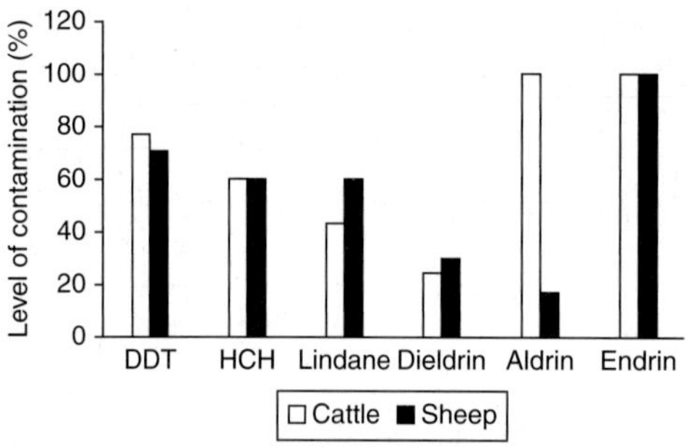

Fig. 13.4. The produced level of contamination in camel meat compared with cattle and sheep meat (calculated from Sallam and Morshedy, 2008).

in beef, but was about 17% of that found in sheep meat (Fig. 13.4). The same trend of lower contaminants in camel liver and kidney compared with beef and sheep meat was reported (Sallam and Morshedy, 2008). This might indicate that the ability of camel to detoxify its body from organochlorines is more efficient compared with cattle and sheep, but this needs to be confirmed in properly designed experiments. Such a low level of organochlorines in camel meat confirms the healthiness of camel meat as an alternative to other red meats and future research should investigate this potential claim.

13.3 Cooking Loss

The *Longissimus thoracis* and *Biceps femoris* muscles from mature camels had 37.95 and 37.07% cooking loss, respectively, which was higher than the 33.23% cooking loss in *Semitendinosus* muscle (Babiker and Yousif, 1990). Young camels (10–12 months old) had a higher cooking loss (40.80–42.96%; Shehata, 2005). *Longissimus thoracis* from 2–3-year-old camels had a significantly lower cooking loss (24.3%) than the values mentioned above (Kadim *et al.*, 2009). The cooking loss of camel *Longissimus thoracis* was not different from that of cattle *Longissimus thoracis* of the same age. However, the use of electrical stimulation increased the cooking loss in camel *Longissimus thoracis* by 33% compared with non-stimulated camel *Longissimus thoracis* (from 24.3 to 32.4%), whereas the cattle *Longissimus thoracis* was unaffected (Kadim *et al.*, 2009). Cooking loss is important because of its potential to change the level of nutrients in the meat once it is cooked. For example, although it is generally regarded that the protein content of camel meat is similar to that of other red meats (Elgasim and Alkanhal, 1992; Gheisari *et al.*, 2009), the higher cooking loss in camel meat (33–38%), especially from electrically stimulated carcasses (32.4%) compared to beef (24.6%), will generate a more nutritionally dense cooked meat (Kadim *et al.*, 2009).

13.4 Health Aspects of Camel Meat

Meat is a valuable food source rich in many essential amino acids, minerals (e.g. iron, zinc and selenium), vitamins (e.g. vitamin E and vitamin B groups), bioactive compounds (Q10, carnosine, anserine, glutathione) and some essential fatty acids such as omega 3 fatty acids (Williams, 2007; Schonfeldt and Gibson, 2008). Apart from the nutritional value of meat, it provides several eating attributes and fulfilling experiences that are normally not achieved by other protein sources. Beef, lamb, pork, poultry and fish are considered the major sources of meat protein worldwide. However, in African, Middle Eastern and some Asian countries, especially in arid and semi-arid regions, camel meat is regarded as a main source of animal protein that equals and in some cases surpasses other meats in commercial importance.

Several epidemiolgical studies linked health problems such as obesity and high saturated fat and cholesterol intake to increased consumption of animal products (Biesalski, 2005; Chao *et al.*, 2005). This has led to a concern that total dietary fat intake should be restricted by consuming smaller portions less frequently (Schonfeldt and Gibson, 2008) or replacing red meat consumption with white meat. The growing evidence of low cholesterol and fat content in camel meat could potentially support its healthiness as a better alternative to the high fat and cholesterol meats such as mutton and beef.

The low bioaccumulation of pesticides in camel meat (Sallam and Morshedy, 2008) is particularly of interest because many African countries still have major problems with organochlorine abuse in terms of the inventory of obsolete pesticides or the lack of control over their use, which consequently leads to health problems (Daba *et al.*, 2011). Camels are, however, mostly grown in arid regions where the use of pesticides is limited; it might be the lack of exposure rather than natural low bioaccumulation that is the cause of the low organochlorines observed by Sallam and Morshedy (2008). Further research is required to ascertain this phenomenon. Regardless of the outcome (either through lack of exposure or low bioaccumulation mechanisms), however, the potential of the lower pollutant levels in camel meat in the diet cannot be disregarded.

There are a few reports indicating a lower prevalence of different microorganisms in camel meat compared with lamb, goat and beef (Rahimi *et al.*, 2010) or the availability of natural antagonists against *Listeria monocytogenes* (El Malti and Amarouch, 2008).

13.5 Camel Meat as Medicine

Meat in general is considered a functional food for cures of many ailments and for improved performance in many cultures around the world (Migdal and Živković, 2007). Camel meat and offal such as liver are believed to have medicinal effects and are eaten raw (Fig. 13.5; Bin Saeed *et al.*, 2005). Kadim *et al.* (2008) stated Somalis and Indians particularly believe in the health benefits of consuming camel meat. Kadim (personal communication, 2010) indicated that camel meat has traditionally been used to cure the following ailments in some Middle Eastern countries: (i) seasonal fever, sciatica and shoulder pain, as well as for removing freckles (by placing hot camel meat slices on the freckled area); (ii) camel meat soup was used to cure corneal opacity and to strengthen eyesight; (iii) camel fat was used to ease haemorrhoidal pains and the hump fat was used to remove tapeworm; and (iv) dried camel lungs used to be prescribed as a cure for asthma, especially if taken with honey. Not only the meat and offal from camel, but other products from camel including milk, cheese and even urine and dung are considered medicinal in some parts of the world. For instance, camel milk was reported to cure jaundice, TB and asthma (Khanvilkar *et al.*, 2009); oral administration of rennet from camel milk was considered therapeutic for curing impotency in medieval Persia (Ghadiri and Gorji, 2004); and camel dung infusion in water is used to cure earache and to remove eye cataracts in some parts of Nigeria (Ibrahim *et al.*, 2010).

Fig. 13.5. A meal of fresh raw camel meat eaten by a group of Ethiopians in a local butchery. The meat is eaten with hot chilli paste. The picture is kindly provided by Prof. A.A. Bekhit (School of Pharmacy, University of Addis Ababa).

Kurtu (2004) reported that the majority of camel meat consumers believe it is a healthier option during the dry season in which cattle are infected with various zoonotic diseases. This belief probably originated from the historical use of animals' organs, including meat, in folklore and traditional medicine. Lev (2006) cited the use of camel meat in a remedial formulation by Al-Tabari, al-Kindi and al-Qazwini that indicates the roots of some of the current beliefs. However, the exposure of camels to wild animals (such as rats) in the desert, the common grazing ground, can be a health risk (Bin Saeed *et al.*, 2005) because human plague was linked to the consumption of raw liver from camels grazed near infected rats.

13.6 Taboos of Camel Meat

Although camel meat is enjoyed in many countries, there are some taboos surrounding its consumption in certain cultures and religions. Camel meat is not consumed in Europe and North America and eating camel is prohibited in the Torah for practising Jews (Anon, 2011). Wilson (1984) reported that there is often some resistance to the consumption of camel meat in developing countries. The slaughter and eating of camel meat is forbidden for the Raikas or Rabaris of India (Kohler-Rollefson, 1996; Ramadurai, 2011) and for the Ethiopian Christians (Haidar and Pobocik, 2009). The Wodaabe of Niger, on the other hand, do not eat camel meat for Islamic reasons they have claimed (Loftsdóttir, 2001). Some Senegalese avoid camel meat for totemic reasons (Seydi and Ba, 1993).

Cook (1989) observed that camel meat was the least favoured ruminant meat in the northern part of Nigeria. The result of a later survey of consumer attitude towards camel meat in northern Nigeria confirmed that observation that camel meat and offal was less favoured than that of sheep, cattle and goats; the reasons for the bias against camel products were found by the survey to be a mixture of superstitions, local taboos, personal beliefs and psychological fears (Farouk and Audu, 1993). The taboos/bias against camel meat that the survey revealed include: (i) that the consumption of camel meat might lead to stomach pain, vomiting and miscarriage in pregnant women; (ii) that camel meat stinks, takes longer to cook, and the camel was a dirty, ugly and shapeless animal; and (iii) that the camel is a beast of burden, like the donkey, which should not be consumed (Farouk and Audu, 1993).

13.7 Conclusion

The nutritional value of camel meat is similar to other red meats but camel meat, especially from young animals (Hashi camel meat), can be considered as a healthy option because of the low fat and cholesterol contents of the meat. The natural habitat of camels and their ability to sustain very high stress suggest the availability of systems to cope with oxidative stress. Information on bioactive compounds in camel meat is scarce and this requires more research to identify compounds of interest and compare them with those of other meat species under the same experimental conditions. Similarly, the reported low levels of organochlorines and certain microorganisms in camel meat compared with other red meats need to be confirmed in future work to propagate camel meat as a healthy product compared with other meats.

References

Abu-Tarboush, H.M. and Dawood, A.A. (1993) Cholesterol and fat contents of animal adipose tissues. *Food Chemistry*, 46, 89–93.

Al-Bachir, M. and Zeinou, R. (2009) Effect of gamma irradiation on microbial load and quality characteristics of minced camel meat. *Meat Science* 82, 119–124.

Alfawaz, M.A. (2004) Lipid oxidation of raw and cooked meats during cold storage as compared to raw meats stored frozen. *Journal of King Saud University Agricultural Sciences* 16, 171–186.

Alhomida, A.S., Duhaiman, A.S., Al-Jafari, A.A. and Junaid, M.A. (1995) Determination of L-carnitine, acylcarnitine and total carnitine levels in plasma and tissues of camel *(Camelus dromedarius). Comparative Biochemistry Physiology*, 111B, 441–445.

Al-Qarawi, A.A. and Ali, B.H. (2003) Variations in the normal activity of esterases in plasma and liver of camels, cattle, sheep and goats. *Journal of Veterinary Medicine* 50, 201–203.

Al-Shabib, N.A., and Abu-Tarboush, H.M. (2004) Nutritional value and some functional properties of protein in Ostrich and Camel meat. *Arab Journal of Food Nutrition* 5, 6–20.

Al-Sheddy, I., Al-Dagal, M. and Bazaraa, W.A. (1999) Microbial and sensory quality of fresh camel meat treated with organic acid salts and/or bifido bacteria. *Journal of Food Science* 64, 336–339.

Anon (2011) Taboo food and drink. *Diabetes Daily.* Available at: http://www.diabetesdaily.com/wiki/Taboo_food_and_drink (accessed: 11 August 2011).

Arihara, K. (2006) Strategies for designing novel functional meat products. *Meat Science* 74, 219–229.

Babiker, S.A. and Yousif, O.K. (1990) Chemical composition and quality of camel meat. *Meat Science* 27, 283–287.

Badiei, K., Mostaghni, K., Pourjafar, M. and Parchami, A. (2006) Serum and tissue trace elements in Iranian camels (*Camelus dromedarius*). *Comparative Clinical Pathology* 15, 58–61.

Biesalski, H. (2005) Meat as a component of a healthy diet – Are there any risks or benefits if meat is avoided in the diet? *Meat Science* 70, 509–524.

Bin Saeed, A.A., Al-Hamdan, N.A. and Fontaine, R.E. (2005) Plague from Eating Raw Camel Liver. *Emerging Infectious Diseases*, 11. Available at: http://www.cdc.gov/ncidod/EID/vol11no09/05-0081.htm (accessed 13 August 2011).

Chao, A., Thun, M., Connell, C., McCullough, M., Jacobs, E. and Flanders, W. (2005) Meat consumption and risk of colorectal cancer. *The Journal of the American Medical Association* 293, 172–182.

Cook, A. (1989) Nigerian markets for livestock and meat: prospects for Niger. Report prepared for USAID (Niamey). Contract 625.0971.83-3-70946. Available at: http://pdf.usaid.gov/pdf_docs/PNABL354.pdf (accessed 11 August 2011).

Daba, D., Hymete, A., Bekhit, A.A., Mohamed, A.M.I. and Bekhit, A.E.D. (2011) Multi residue analysis of pesticides in Wheat and Khat collected from different regions of Ethiopia. *Bulletin of Environmental Contamination and Toxicology*. 86, 336–341.

Dawood, A., and Alkanhal, M.A. (1995) Nutrient composition of Najdi- Camel Meat. *Meat Science* 39, 71–78.

Dunnett, M. and Harris, R.C. (1997) High-performance liquid chromatographic determination of imidazole dipeptides, histidine, 1 -methylhistidine and 3- methylhistidine in equine and camel muscle and individual muscle fibres. *Journal of Chromatography B* 688, 47–55.

Dunnett, M., Harris, R.C., Soliman, M.Z. and Suwar, A.A.S. (1997) Carnosine, anserine and taurine contents in individual fibres from the middle gluteal muscle of the camel. *Research Veterinary Science* 62, 213–216.

El-Faer, M.Z., Rawdah, T.N., Attar, K.M. and Dawson, M.V. (1991) Mineral and proximate composition of the meat of the one-humped camel (*Camelus dromedarius). Food Chemistry* 42, 139–143.

El Malti, J. and Amarouch, H. (2008) Natural antagonists for control of *Listeria monocytogenes* in Morocca Dromedary camel meat. *Revue de Microbiologie Industrielle Sanitaire et Environnementale* 2, 1–16. Available at: http://www.remise.ma/Malti%20et%20al.pdf

Elgasim, E.A. and Elhag, G.A. (1992) Carcass characteristics of the Arabian camel. *Camel Newsletter* 9, 20–24.

Elgasim, E.A. and Alkanhal, M.A. (1992) Proximate composition, amino acids and inorganic mineral content of Arabian Camel meat: comparative study. *Food Chemistry* 45, 1–4.

Emmanuel, B. and Nahapetian, A. (1980) Fatty acid composition of depot fats and rumen wall of the camel *(Camelus dromedarius). Comparative Biochemistry Physiology* 67B, 701–704.

Fallah, A.A., Tajik, H., Rohani, S.M.R. and Rahnama, M. (2008) Microbial and sensory characteristics of camel meat during refrigerated storage as affected by gamma irradiation. *Pakistan Journal of Biological Science* 11, 894–899.

Farouk, M.M. and Audu, N.U. (1993) Consumer attitude towards camel meat in a semi-arid region of Nigeria. Unpublished article. Pp. 1–19. Department of Food Science and Technology, University of Maiduguri, Nigeria.

Faye, B., Seboussi, R. and Askar, M. (2008) Trace elements and heavy metals in Arabian camel. In: Faye, B. and Sinyavskiy, Y. (eds) *Impact of Pollution on Animal Products*, Springer Science + Business Media B.V., pp. 97–106.

Ghadiri, M.K. and Gorji, A. (2004) Review of impotence. Natural remedies for impotence in medieval Persia. *International Journal of Impotence Research* 16, 80–83.

Gheisari, H.R. (2011) Correlation between acid, TBA, peroxide and iodine values, catalase and glutathione peroxidase activities of chicken, cattle and camel meat during refrigerated storage. *Veterinary World* 4, 153–157.

Gheisari, H.R. and Motamedi, H. (2010) Chloride salt type/ionic strength and refrigeration effects on antioxidant enzymes and lipid oxidation in cattle, camel and chicken meat. *Meat Science* 86, 377–383.

Gheisari, H.R., Aminlari, M. and Shekarforoush, S.S. (2009) A comparative study of the biochemical and functional properties of camel and cattle meat during frozen storage. *Veterinarski Arhiv* 79, 51–68.

Haidar, J. and Pobocik, R.S. (2009) Iron deficiency anemia is not a rare problem among women of reproductive ages in Ethiopia: a community based cross sectional study. *BMC Blood Disorders*, 9:7 doi:10.1186/1471-2326-9–7.

Ibrahim, J.A., Muazzam, I., Jegede, I.A. and Kunle, O. F. (2010) Medicinal plants and animals sold by the ‘Yan-Shimfidas’ of Sabo Wuse in Niger

State, Nigeria. *African Journal of Pharmacology* 4, 386–394.

Kadim, I.T., Mahgoub, O., Al-Maqbaly, R.S., Annamalai, K. and Al-Ajmi, D.S. (2002) Effects of age on fatty acid composition of the hump and abdomen depot fats of the Arabian camel (*Camelus dromedarius*). *Meat Science* 62, 245–251.

Kadim, I.T., Mahgoub, O., Al-Marzooqi, W., Al-Zadgali, S., Annamali, K. and Mansour, M.H. (2006) Effects of age on composition and quality of muscle *Longissimus thoracis* of the Omani Arabian Camel (*Camelus dromedarius*). *Meat Science* 73, 619–625.

Kadim, I.T., Mahgoub, O. and Purchas, R.W. (2008) A review of the growth, and of the carcass and meat quality characteristics of the one-humped camel (*Camelus dromedarius*). *Meat Science* 73, 619–625.

Kadim, I.T., Mahgoub, O., Al-Marzooqi, W., Khalaf, S.K., Mansour, M.H., Al-Sinani, S.S.H. and Al-Amri, I.S. (2009) Effects of electrical stimulation on histochemical muscle fiber staining, quality, and composition of camel and cattle *Longissimus thoracis* muscles. *Journal of Food Science* 74, S44–S52.

Kadim, I.T., Al-Ani, M.R., Al-Maqbaly, R.S., Mansour, M.H., Mahgoub, O. and Johnson, E.H. (2011) Proximate, amino acid, fatty acid and mineral composition of raw and cooked camel (*Camelus dromedarius*) meat. *British Food Journal* 113, 482–493.

Khanvilkar, A.V., Kulkarni, M.D., Yadav, G.B., Samant, S.R., Thorat, V.J. and Shisode, M.G. (2009) Desert friendly animal – the camel. *Veterinary World* 2, 240–241.

Kohler-Rollefson, I. (1996) The one-humped camel in Asia. Origin, utilization and mechanism of dispersal. In: Harris, D.A. (ed) *The Origin and Spread of Agriculture and Pastoralism in Eurasia. Crops, Fields, Flocks and Herds.* Routledge and the Taylor and Francis Group, Abingdon, UK, pp. 282–296.

Kurtu, M.Y. (2004) An assessment of the productivity for meat and carcass yield of camel (*Camelus dromedarius*) and the consumption of camel meat in the Eastern region of Ethiopia. *Tropical Animal Health and Production* 36, 65–76.

Lawrie, R.A. (2002) *Meat Science*, 6th edn. Pergamon Press, Oxford, UK.

Lev, E. (2006) Healing with animals in the Levant from the 10th to the 18th century. *Journal of Ethnopharmacology and Ethnomedicine*, 2. doi:10.1186/1746-4269-2-11.

Loftsdóttir, K. (2001) Birds of the bush: Wodaabe distinctions of society and nature. *Nordic Journal of African Study* 10, 280–298.

Migdal, W. and Žirković, B. (2007) Meat – from functional food to disease of modern civilization. *Biotechnology of Animal Husbandry* 23, 19–31.

Mirgani, T. (1977) Fatty acid composition of hump triglycerides of the camel *Camelus dromedarius. Comparative Biochemistry and Physiology* 588, 211–213.

Purchas, R.W. and Busboom, J.R. (2005) The effect of production system and age on levels of iron, taurine, carnosine, coenzyme Q10, and creatine in beef muscles and liver. *Meat Science* 70, 589–596.

Purchas, R.W., Rutherfurd, S.M., Pearce, P.D., Vather, R. and Wilkinson, B.H.P. (2004) Concentrations in beef and lamb of taurine, carnosine, coenzyme Q10, and creatine. *Meat Science* 66, 629–637.

Purchas, R.W., Triumf, E.C. and Egelandsdal, B. (2010) Quality characteristics and composition of the *Longissimus* muscle in the short-loin from male and female farmed red deer in New Zealand. *Meat Science* 86, 505–510.

Rahimi, E., Ameri, M. and Kazemeini, H.R. (2010) Prevalence and antimicrobial resistance of Campylobacter species isolated from raw camel, beef, lamb, and goat meat in Iran. *Foodborne Pathology Diseases* 7, 443–447.

Ramadurai, L. (2011) Rituals, beliefs and taboos. Camel Wiki. Available at: http://wiki.encyclopaediaindica.com/Kencyclo3/wiki/index.php?title=CAMEL (accessed 11 August 2011).

Rashed, M.N. (2002) Trace elements in camel tissues from a semi-arid region. *The Environmentalist* 22, 111–118.

Rawdah, T.N., El-Faer, M.Z., and Koreish, S.A. (1994) Fatty acid composition of the meat and fat of the one-humped camel (*Camelus dromedarius*). *Meat Science* 37, 149–155.

Sallam, K.I. and Morshedy, A.M.A. (2008) Organochlorine pesticide residues in camel, cattle and sheep carcasses slaughtered in Sharkia Province, Egypt. *Food Chemistry* 108, 154–164.

Schönfeldt, H. and Gibson, N. (2008) Changes in the nutrient quality of meat in an obesity context. *Meat Science* 80, 20–27.

Seydi, M. and Ba, Y.M. (1993) Non-religious prohibitions of meat consumption in Senegal. *Dakar Med.* 38, 33–38.

Shehata, M. F. (2005) Carcass traits and meat quality of one-humped camels fed different halophytic forages: 2-Physical, chemical, and sensory characteristics of camel meat. *Journal of Agriculture Sciences (Mansoura University)* 30, 1943–1952.

Shimada, K., Sakuma, Y., Wakamatsu, J., Fukushima, M., Sekikawa, M., Kuchida, K. and Mikami, M. (2004) Species and muscle differences in

L-carnitine levels in skeletal muscles based on a new simple assay. *Meat Science* 68, 357–362.

Tomonaga, S., Kaneko, K., Kaji, Y., Kido, Y., Denbow, D.M. and Furuse, M. (2006) Dietary β-alanine enhances brain, but not muscle, carnosine and anserine concentrations in broilers. *Animal Science Journal* 77, 79–86.

WHO (2002) Amino acid requirements of adults. In: Protein and amino acid requirements in human nutrition. Report of a Joint WHO/FAO/UNU Expert Consultation. WHO Technical Report Series 935. WHO, Geneva.

Williams, P. (2007) Nutritional composition of red meat. *Nutrition and Dietitian* 64, S113–S119.

Wilson, R.T. (1984) *The Camel*. Longman, London.

Wood, J.D., Enser, M., Fisher, A.V., Nute, G.R., Sheard, P.R., Richardson, R.I., Hughes, S.I. and Whittington, F.M. (2008) Fat deposition, fatty acid composition and meat quality: a review. *Meat Science* 78, 343–358.

Wood, J.D., Richardson, R.I., Nute, G.R., Fisher, A.V., Campo, M.M., Kasapidou, E., Sheard, P.R. and Enser, M. (2003) Effects of fatty acids on meat quality: a review. *Meat Science* 66, 21–32.

14 The Economic Potential of Camel Meat

Msafiri Mbaga
Department of Natural Resource Economics, College of Agricultural and Marine Sciences, Sultan Qaboos University, Muscat, Sultanate of Oman

14.1 Introduction

This chapter considers the economic potential of camel meat. Current and projected global meat demand trends are analysed to establish the extent to which satisfying the projected meat demand would impact on current resources notably water and land. On the basis of these expected challenges, camel meat is presented as a potential substitute for beef. To put things in the right perspective, a SWOT analysis is conducted to identify the camel meat commodity strengths, weaknesses, opportunities and threats. From the SWOT analysis, camel meat product strengths include: a 1.619 billion strong consumer base in the Muslim world; low production costs; ecologically friendly production systems; being a healthy meat product; benefiting from the already established beef specification system; and the presence of many established uses for camel meat and recipes. The weaknesses, opportunities and threats facing camel meat products are examined. Finally actions are proposed to take advantage of the opportunities, in order to counter threats and improve on the weaknesses.

14.2 Current and Projected Global Meat Demand

Animal meat production has significant impact on nearly all aspects of the environment, including air and climate change, land and soil, water and biodiversity. The impact can be direct through grazing, or indirect through the expansion of feed production.

Among all animal meats, it is beef that is the most popular and widely produced in the world. Unfortunately it is also the most inefficient animal meat to produce in terms of the amount of input needed to produce it. Grain-fed beef production, for example, takes 100,000 l of water for every kilogram of food. In terms of energy, beef cattle require an energy input to protein output ratio of 54:1 (Pimentel and Pimentel, 2003).[1]

In recent times the world has witnessed a significant shift in food consumption patterns towards more animal products such as beef, mainly due to growing economies and rising individual incomes (Bruinsma, 2003). Growing populations and incomes, along with changing food preferences, have therefore been the driving force behind the rapidly increasing demand for animal meat products. FAO projects the global meat production to more than double from 229 million t in 1999 to 465 million t in 2050 (FAO, 2006). This projection was also acknowledged by Elam (2006).

The production of more meat to meet the above projections, means more feed and forage will need to be produced, and that also means more land and water will be needed. The concept of 'water footprint' provides another perspective on the link between beef production and the use of global water resources. The water footprint is defined as the total volume of freshwater that is used to produce the goods and services consumed by an individual or community (Chapagain and Hoekstra, 2004). Beef has the highest water footprint at 15,400 m^3/t (Mekonnen and Hoekstra, 2010). This footprint is much larger than that of meat from sheep (10,400 m^3/t), pig (6000 m^3/t), goat (5500 m^3/t) or chicken (4300 m^3/t). To meet the projected demand for meat by the year 2050, it would therefore require the utilization of more land and water, consequently putting significant pressure on currently available land and water resources. In this backdrop, what are the potential alternatives to beef, given that more beef production to meet future demands might not be sustainable? Camel meat production seems to be the best alternative because, among other things, camels require fewer resources in terms of land and water.

14.3 Camel Meat as a Substitute for Beef

There are two types of camels known today: the one-humped or dromedary, and the two-humped, shorter-legged Bactrian camel. While the one-humped camel is found in the hot arid areas of the Middle East and Africa, the two-humped camel is found in Asia. The dromedary is the one that is most abundant and more commonly used for meat than the Bactrian camel and represents almost 90% of the genus *Camelus* (Kadim *et al.*, 2008).

Camels have great tolerance to harsh conditions of high temperatures, water scarcity and poor vegetation (Shalah, 1983; Kadim *et al.*, 2008). In these harsh environments, camels feed on low-quality feeds and fodder that are generally not utilized by other domestic species (Tandon *et al.*, 1988; Kadim *et al.*, 2008). As a result, camels can be raised to produce meat at a comparatively lower cost than other domestic animals such as goats, sheep and cattle.

Young camels, less than 3 years of age, produce high-quality low-fat meat (Kadim *et al.*, 2006), which is also a good source of minerals. Age is therefore an important factor in determining camel meat quality and composition (Kadim *et al.*, 2006). Old camels will generally produce tough, low-quality meat. This helps to clarify the commonly held view that camel meat is tough and coarse compared with meats from other animals.

Healthwise, camel meat has less fat as well as low levels of cholesterol compared with other animal meats (Elgasim *et al.*, 1987; Dawood and Alkanhal, 1995; Al-Ani, 2004; Kadim *et al.*, 2006). Quality-wise, meat from young camels is comparable to beef (Khatami, 1970; Knoess, 1977; Elgasim *et al.*, 1987; Kadim *et al.*, 2006). Furthermore a recent finding (Ibrahim and Nour, 2010) indicates that camel meat can be added up to 75% in burgers without compromising acceptability by consumers. It is therefore confirmed that camel meat can be used as a substitute for beef.

14.4 Global Camel Population and Camel Meat Supply

To evaluate the potential of camel meat the global camel population and camel meat production need to be examined. Table 14.1 presents the world total camel population and camel meat production for a 10-year period from 2000 to 2009.

In 2000, the world camel population was around 21.9 million and reached 25.9 million in 2009, which means a 15% increase over the 10 years. Camel meat production during the same period increased from 293,000 t in the year 2000 to 360,600 t in 2009, i.e. a 19% increase. Regional camel population numbers are shown in Table 14.2.

Table 14.1. World camel population and camel meat production 2000 to 2009 (source: FAOstat, 2011).

Year	Camel population	Camel population index 2009 = 100	Camel meat production (t)	Camel meat production index 2009 = 100
2000	21,935,635	85	293,382	81
2001	22,193,070	86	325,021	90
2002	22,518,689	87	321,041	89
2003	22,977,073	89	322,440	89
2004	23,752,517	92	332,558	92
2005	23,919,490	92	333,026	92
2006	24,468,834	94	349,941	97
2007	25,246,645	98	339,974	94
2008	25,787,448	100	361,165	100
2009	25,893,855	100	360,622	100

Table 14.2. World camel population (millions) 1961 to 2009 (source: FAOstat, 2011).

	Average 1961–2000	2001	2002	2003	2004	2005	2006	2007	2008	2009
Bactrian										
China	0.47	0.33	0.28	0.26	0.27	0.26	0.27	0.27	0.24	0.24
Mongolia	0.57	0.32	0.29	0.25	0.26	0.25	0.25	0.26	0.27	0.28
Russia	0.20	0.01	0.01	0.01	0.01	0.01	0.01	0.01	0.01	0.01
Total	1.24	0.66	0.57	0.52	0.53	0.52	0.53	0.54	0.51	0.52
Dromedary										
North-east Africa	10.22	13.73	14.05	14.38	14.92	14.90	15.01	15.43	15.71	15.66
West Africa	2.93	4.40	4.58	4.76	4.95	5.16	5.24	5.57	5.77	5.89
North Africa	0.75	0.70	0.69	0.71	0.73	0.71	0.77	0.70	0.73	0.74
Other countries										
Afghanistan	0.27	0.22	0.18	0.18	0.19	0.19	0.17	0.19	0.18	0.19
Bahrain	0.001	0.001	0.001	0.001	0.001	0.001	0.001	0.001	0.001	0.001
India	1.02	0.71	0.67	0.63	0.63	0.63	0.63	0.63	0.63	0.63
Iran	0.15	0.15	0.15	0.15	0.15	0.15	0.15	0.15	0.15	0.15
Iraq	0.11	0.01	0.01	0.01	0.01	0.03	0.05	0.05	0.06	0.06
Israel	0.01	0.01	0.01	0.01	0.01	0.01	0.01	0.01	0.01	0.01
Jordan	0.01	0.01	0.01	0.01	0.01	0.01	0.01	0.01	0.01	0.01
Kuwait	0.01	0.00	0.00	0.01	0.00	0.01	0.01	0.01	0.01	0.01
Oman	0.05	0.12	0.12	0.12	0.12	0.12	0.12	0.12	0.12	0.13
Pakistan	0.84	0.77	0.76	0.75	0.74	0.74	0.92	0.93	0.95	0.96
Qatar	0.02	0.03	0.03	0.03	0.03	0.01	0.02	0.03	0.03	0.03
Saudi Arabia	0.27	0.26	0.25	0.26	0.27	0.27	0.26	0.26	0.26	0.26
Syria	0.01	0.01	0.01	0.02	0.02	0.02	0.03	0.03	0.03	0.03
Turkey	0.02	0.001	0.001	0.001	0.001	0.001	0.001	0.001	0.001	0.001
UAE	0.11	0.23	0.25	0.26	0.25	0.25	0.36	0.38	0.38	0.38
Total	2.89	2.53	2.44	2.42	2.43	2.42	2.73	2.80	2.81	2.84
Total dromedary population	16.79	21.36	21.76	22.27	23.03	23.19	23.74	24.49	25.03	25.13
Total dromedary and bactrian population	18.03	22.02	22.33	22.79	23.56	23.71	24.27	25.03	25.54	25.65

Note: North-east Africa includes: Djibouti, Ethiopia, Eritrea, Kenya, Somalia, Sudan and Yemen; West Africa includes: Burkina Faso, Chad, Mali, Mauritania, Morocco, Niger, Nigeria, Senegal and western Sahara; North Africa includes: Algeria, Egypt, Libya, Morocco and Tunisia.

Table 14.2 shows that the world population of Bactrian camels slightly decreased from 0.66 million in 2001 to 0.52 million in 2009. The dromedary camel population, on the other hand, increased from 21.66 million in 2001 to 25.13 million in 2009. Northeast Africa is the leading region in terms of the dromedary camel population. In 2009, there were approximately 15.99 million dromedaries in this region alone, which is 64% of the entire dromedary population. Table 14.3 presents camel meat production numbers for selected 'leading' countries for the period 2000 to 2009.

In 2009, the 12 leading camel-meat-producing countries (Table 14.3) produced 324,298 t (90%) of the 2009 world total camel meat production of 360,622 t. Mauritania had the highest percentage increase in camel meat production between 2001 and 2009, followed by Sudan, UAE, Saudi Arabia, Kenya, Niger, Egypt and Somalia. The following conclusions can be drawn from the statistics in Tables 14.1 to 14.3:

1. There is a sufficient supply of camels to sustain the current demand for camel meat.
2. Assuming the observed camel population growth rate (Table 14.1) stays constant at 15%, the camel population is likely to reach more than 30 million by 2019.
3. The 2009 camel meat production of 360,622 t amounts to a slaughter rate of 1.4 million camels per year, assuming a carcass weight of 250 kg (Warfield and Tume, 2000). This slaughter rate is sustainable by the observed and projected camel population.

14.5 SWOT Analysis

One of the approaches to investigate the economic potential of camel meat as a commodity is to undertake a SWOT analysis. SWOT is an acronym that stands for strengths, weaknesses, opportunities and threats. SWOT analysis is a tool that provides direction and serves as a basis for the development of marketing strategies for a commodity. SWOT analysis is used here to assess the camel meat's strengths (positive attributes) and weaknesses (negative attributes) in addition to opportunities (potentially favourable attributes) and threats (potentially unfavourable attributes).

Camel meat strengths are those attributes that give it an edge over other rival meats in the market and weaknesses are the inherent disadvantages it has over other meats in the market. Together, strengths and weaknesses of camel meat therefore form the internal issues of the commodity. Opportunities are those favourable external factors that contribute or would favourably contribute towards the expansion of camel meat demand. Threats on the other hand, include those changes in the external environment that are unfavourable or might be unfavorable to the success of the camel meat as a commodity in the market. Opportunities and threats are the external issues of the commodity. A discussion of the strengths, weaknesses, opportunities and threats of camel meat is given below.

14.6 Strengths

14.6.1 Healthiness of camel meat

Camel meat is lean and has been described as raspberry red to dark brown in colour (Fig. 14.1). It has been scientifically proven to be much healthier than many other animal meats. It is a low-fat meat, low in cholesterol and high in protein. This makes it an ideal meat for those with dietary problems such as diabetes and high cholesterol. The meat has been used since the late 16th century in traditional Chinese medicine to improve resistance to disease, to strengthen the muscles and bones, to moisten the skin and to relieve internal pain (Zeng and McGregor, n.d.).

Studies conducted in various parts of the world (Kadim *et al.*, 2008) have shown that camel meat has lower fat and higher moisture content than beef and mutton. Fat in camel meat amounts to 1.2–1.8%, whereas beef may contain 4–8% fat. The figure for water is 20%. These percentages mean that camel meat is richer than beef in protein and minerals. The consumption of camel meat could not only lower the percentage of fat in the body, but it will also

Table 14.3. Camel meat production (t/year) in leading camel-meat-producing countries 2000–2009 (source: FAOstat, 2011).

Rank	Country	2000	2001	2002	2003	2004	2005	2006	2007	2008	2009	Percentage change 2000–2009
1	China	14,740	18,700	15,180	14,300	14,432	14,388	14,960	15,400	16,060	16,060	9.0
2	Egypt	39,650	52,000	46,000	38,800	39,000	40,000	43,800	37,370	45,250	45,000	13.5
3	Ethiopia	13,822	13,992	14,162	14,502	14,773	14,696	17,732	20,066	18,257	14,418	4.3
4	Kenya	19,800	22,046	24,940	26,623	28,500	25,500	27,000	27,870	23,110	24,801	25.3
5	Mali	6,344	7,096	7,968	8,716	10,002	11,632	13,176	15,147	17,600	18,400	190.0
6	Mauritania	19,940	20,930	22,000	23,000	23,500	24,000	22,500	21,113	23,699	21,462	7.6
7	Niger	10,950	12,000	12,150	12,300	12,450	12,600	12,820	12,960	13,155	13,200	20.5
8	Oman	6,237	6,342	6,447	6,510	6,552	6,552	6,750	6,800	6,720	6,720	7.7
9	Saudi Arabia	39,840	39,920	40,500	41,250	41,960	41,070	41,070	41,000	48,051	50,302	26.3
10	Somalia	39,100	41,650	40,800	44,200	44,540	44,710	44,200	43,968	44,200	44,200	13.0
11	Sudan	29,925	40,050	41,625	44,000	44,319	53,000	48,000	45,000	48,262	49,882	66.7
12	UAE	13,069	13,851	14,617	15,386	15,390	8,407	21,510	19,853	19,853	19,853	51.9
		253,417	288,577	286,389	289,587	295,418	296,555	313,518	306,547	324,217	324,298	

reduce the intake of saturated fats connected with cardiovascular diseases.

As far as consumers are concerned, there are differences between a commodity and a product. Meat as a commodity, in the eyes of a consumer, is the same no matter which animal produces it. A product on the other hand includes a number of distinguishing attributes for which consumers may or may not be willing to pay a premium price. When dealing with consumers, it is therefore the difference between a commodity and a product that is crucial. A producer of a commodity can only be successful if he or she is a low-cost producer. For a product, the success depends on the uniqueness of its attributes and the extent to which such attributes are valued by the consumer. Consumer tastes and preferences are continuously changing. Lately, for health reasons, there have been changes in consumer preferences towards more lean beef or other red meat. It is these changes in consumer preference that form a strong case for camel meat as a differentiated product. Camel meat therefore has great potential in the market if promoted as a differentiated meat product scientifically proven to have attributes that are good for consumer's health.

Promoting camel meat would require significant financial investment in research and advertisement. The question is: will the product still be competitive? To answer this question we need to understand the relationship between product characteristics and income. As the level of income increases (as the level of affluence increases) the demand for high-quality food products increases. High-quality products include those with attributes that are desired by consumers such as 'camel meat improves health because it contains less fat as well as having some medicinal values'. Figure 14.2 depicts a pyramid based on Maslow's hierarchy of needs. The level of affluence increases up the pyramid.

As affluence increases, consumer demands become more refined and generally affluent consumers tend to demand high-quality products bearing the attributes they care for. In this context, therefore, there is a niche market for camel meat if the meat is properly promoted, in particular focusing on its unique attributes. The target should be those with high income given their marginal propensity to pay.

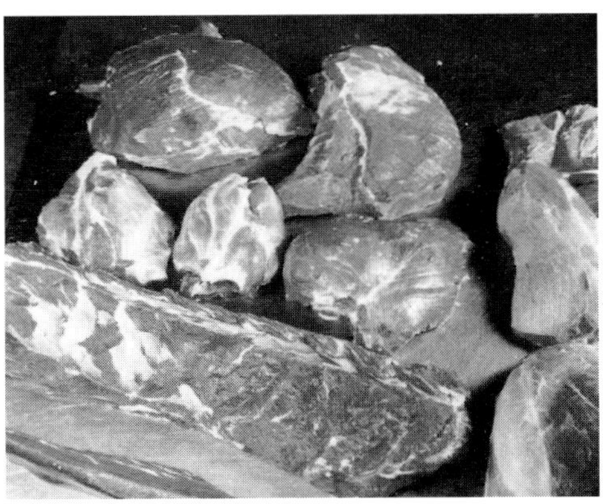

Fig. 14.1. Different cuts depicting the leanness of camel meat.

14.6.2 Association of camel meat with the Middle East and the Muslim world

Camel meat is already a popular meat product in the Muslim world, Australia and in China. According to Warfield and Tume (2000), a survey of wholesalers and retailers in Australia revealed that Muslims from the Middle East, Indonesia, Malaysia, Pakistan, India and Turkey were the most likely to buy camel meat.

The global Muslim population trends (Table 14.4) show that there were 1.619 billion Muslims in the world in 2010.

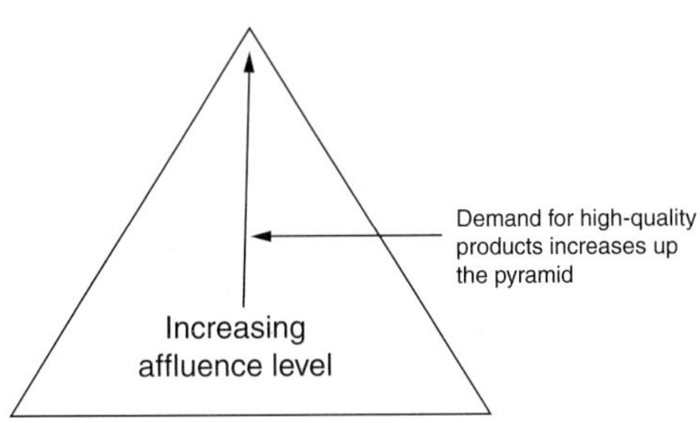

Fig. 14.2. The relationship between affluence and demand for high-quality products.

Table 14.4. Global Muslim population (source: PewResearchCenter, 2011).

	Year			
	2010		2030	
Region	Estimated Muslim population	Estimated global share of total Muslim population	Projected Muslim population	Projected global share of total Muslim population
Asia–Pacific	1,005,507,000	62.1	1,295,625,000	59.2
Middle East and North Africa	321,869,000	19.9	439,453,000	20.1
Sub-Saharan Africa	242,544,000	15.0	385,939,000	17.6
Europe	44,138,000	2.7	58,209,000	2.7
Americas	5,256,000	0.3	10,927,000	0.5
Total	1,619,314,000	100	2,190,154,000	100

By 2030, according to the Pew Research Centre, the world's total Muslim population is expected to increase by 35% over its 2010 level, to 2.2 billion people. In total, Muslims will make up about 26% of the world's population by 2030. This huge increase in Muslim population, coupled with the recent increase in the popularity of camel meat in the Muslim world, Australia and China, creates an unprecedented potential for camel meat. If camel meat is promoted accordingly, it should reach a point where it will be the first meat of choice for the 2.2 billion Muslims, as well as other consumers across the globe.

14.6.3 Low production cost of camel meat

Camels are usually reared by nomads in arid regions. In Australia they are mostly found in the wild. The animals feed mainly on annual grass, acacias, euphorbias and dwarf bushes. This type of pasture permits only extensive types of animal production systems that are not costly. Even in cases where camels are raised in commercial facilities, the production costs are lower than those for other meats. The production costs for camel ranchers in Riyadh, Saudi Arabia reported by Al-Khamis and Young (2006) range from 3279 Riyal/head/year (US$874.5 at 1 Saudi Riyal = US$0.2667) for medium herds to 2318 Riyal/head/year (US$618.5) for large herds.

Camel meat can therefore be produced cheaply compared with other competing meats. It has a competitive advantage in having low production costs, and hence low wholesale and retail prices. Table 14.5 presents a list of some cuts and their respective prices (Warfield and Tume, 2000).

14.6.4 Ecological harmlessness of camel meat production

Ecologists stress that camel grazing has very little, if any, damaging effect on desert vegetation and does not contribute to desertification. Camel foraging habits are optimally suited to areas with a low carrying capacity (Köhler-Rollefson, 1994). Camel herds disperse over huge areas instead of clustering together like sheep. Camels tend to take only one or two bites from one bush or tree before moving on to the next one, unlike goats that often ravage a whole shrub in one extended feeding session. In addition, their flat pad-like foot produces a gentle impact on the soil surface and does not carve it up like the sharp cloven hooves of small ruminants (Gauthier-Pilters and Dagg, 1981). Camels also have a very efficient feed conversion rate (Stiles, 1983). Camel meat is therefore an ecologically friendly product. Nowadays consumers, especially in affluent

Table 14.5. Camel cuts and their respective prices in Australia (source: Warfield and Tume, 2000).

Number	Camel cuts	Price (AUS$/kg)
1	Rump	8.00
2	Striploin	8.75
3	Knuckle	4.70
4	Outside	4.70
5	Bolar blade	4.70
6	Tenderloin	23.00
7	Cube roll	9.75
8	Topside	4.70
9	Chuck eye roll	3.90
10	Sausage trim	1.20

societies, tend to favour commodities that are environmentally friendly; as a result, this is a very important attribute that needs to be promoted in favour of camel meat.

14.6.5 Identified uses for camel meat as well as camel meat recipes

There are many identified uses for camel meat and recipes on how to prepare mouthwatering camel dishes. In Australia, for example, rump, striploin, tenderloin, cube roll and bolar blade are used in food service applications for a range of cooking methods depending on the muscle cuts. Topside is being made into prosciutto for the food-service sector and into stir-fry strips. Outside (silverside) and knuckle (round) are occasionally manufactured into jerky. Sausage trim is used in manufacturing sausages, patties, steakettes, formed kebabs, meatballs, and tray-packed premium mince. Tasty and easy to prepare camel meat recipes are now available making it easier for potential consumers to try camel meat. Some of these recipes can be found online in the following locations (all websites accessed 17 September 2012):

- http://www.helium.com/items/1323763-camel-meat
- http://www.abc.net.au/tv/cookandchef/txt/s2054722.htm
- http://www.abc.net.au/tv/cookandchef/txt/s2054621.htm
- http://www.celtnet.org.uk/recipes/miscellaneous/fetch-recipe2.php?rid=misc-camel-chubbagin

14.6.6 Camel meat benefits from using the well-established beef terminology and specification

Establishing meat specifications and terminologies to represent the various specifications is important for meat brokers, exporters, butchers, executive chefs, food and beverage managers, and supermarket meat buyers. It also assists the consumer, providing an awareness of what is available and what to ask for. Camel meat specifications use the same terminology for beef, e.g. rump, topside, fillet, etc.

The Camel Australia Export (CAE), which is the registered business name of the Central Australian Camel Industry Association Inc. (CACIA), presents camel meat specifications adopted from beef specification at its website (accessed 17 September 2012) at:

- http://www.ourestore.com.au/stores/cacia/shopdisplayproducts.asp?id=8&cat=Hindquarter
- http://www.ourestore.com.au/stores/cacia/definition.asp
- http://www.ourestore.com.au/stores/cacia/shopdisplayproducts.asp?id=7&cat=Carcase
- http://www.ourestore.com.au/stores/cacia/shopdisplayproducts.asp?id=9&cat=Forequarter
- http://www.camelsaust.com.au/liveintro.htm

The adoption of these specifications by the camel meat industry greatly facilitates the camel meat trade.

14.7 Weaknesses

14.7.1 Lack of consumer awareness regarding camel meat

Consumer awareness is the ability of consumers to look into factors, such as prices and product attributes before making a decision to purchase. Informed consumers look for the most value they can get from within competing products. Consumer awareness can be increased through commercials, advertisements and word of mouth (a comment from someone you know about a product or service).

Generally there is lack of consumer awareness with regard to camel meat aside from the Muslim world where camel meat is traditionally consumed. Elsewhere few people are aware of the nutritional and health benefits from consuming camel meat. According to Warfield and June (2000), 52% of restaurants surveyed in 1999 indicated lack of customer interest as the reason for not using more camel meat. Furthermore, 64% of the restaurants indicated a lack of customer awareness of camel meat in general as the reason for low customer demand for camel meat.

To overcome this drawback the camel meat industry needs to invest in creating consumer awareness of the product. This could be done, among other ways, through setting up websites. An Internet website is a cheap and easy way to reach many potential consumers worldwide. Consumer awareness has heightened in the advent of the Internet because of the proliferation of websites where customers can provide and read consumer reviews, voice their complaints and opinions about goods and services.

14.7.2 Consumers tend to relate camel meat with the animal

Consumers have been observed to link camel meat with the camel itself, which gives rise to concerns about hygiene and cleanliness and to negative perceptions that the meat is smelly and tough (Warfield and Tume, 2000). This problem could be addressed through consumer education. As suggested by Warfield and Tume (2000), this could also be lessened by renaming camel meat and avoiding the use of a camel's image on all promotion and communication material.

14.7.3 Chewiness and toughness of camel meat

Camel meat has been described by consumers as being chewy and tough even though it is not different from beef in terms of flavour. Being tough and chewy discourages potential consumers from buying camel meat. Tough meat must be tenderized/marinated before cooking and also takes more time to cook. Camels are usually slaughtered when they reach unproductive age and often this happens when they are old. Apparently the older the camel, the tougher its meat gets. Age is therefore an important factor in determining camel meat quality and composition (Kadim *et al.*, 2006). An old camel will generally produce tough, low-quality meat. Recent findings (Kadim *et al.*, 2006) indicate that young camels under three years of age produce high-quality low-fat meat that is rich in minerals. To maintain meat quality and composition young camels less than three years of age should be slaughtered.

14.7.4 Camel meat demand in high-value export markets

Consumers in high-value export markets are increasingly demanding meat that is ready to use. This is because home cooking is becoming increasingly difficult to do. In today's fast-paced life, the need for convenience food has grown and so has the market for the ready to use/cook, convenience food products. After a busy day at work even the most enthusiastic and talented cook may not be able to face the practicalities of having to create a meal from scratch. Therefore processed food products that require minimal or no preparation before cooking have a competitive advantage. The camel meat industry is at a disadvantage because it doesn't have access to the high-technology meat processing and packaging equipment needed to produce denuded cuts.

14.7.5 Lack of Halal certification

Halal is a term describing any object or an action that is permissible to use or engage in, according to the Islamic law. The term is used to designate food seen as permissible to consume according to Islamic law. Islam has laws regarding the proper method of slaughtering an animal for consumption. It is forbidden in Islam to eat meat from animals that are improperly slaughtered. Halal certification is recognition that the products are permissible under Islamic law. These products are thus edible, drinkable or usable by Muslims. The Muslim world is currently the largest and most important market for camel meat. As indicated in Table 14.4, the market (the global Muslim population) currently stands at 1.62 billion people and is expected to reach 2.2 billion people by 2030.

Halal certification is therefore very important for meat slaughtered in any slaughtering facility across the globe to be accepted by Muslims. The lack of Halal certification for many of the camel-slaughtering facilities outside the Muslim world automatically excludes their products from entering the global Muslim market. This is a serious weakness that needs to be addressed because the Muslim world is a significant market for camel meat.

14.7.6 Lack of knowledge about camel-meat cuts

There is a lack of confirmed knowledge about camel-meat-cut characteristics in relation to eating quality, and consequently the suitability of these cuts for different recipes and market segments. This type of information is very important to consumers. The beef industry has very well-documented information on various grade cuts and the suggested uses, as indicated in Table 14.6 for some key grades.

The camel meat industry is already benefiting from using the established beef grades for marketing camel meat (see, for example, http://www.exoticmeatmarket.com/camelmeat.html, accessed 17 September 2012). What is needed now is to document the characteristics of these camel meat cuts in relation to eating quality, and consequently appreciate the suitability of these cuts for different recipes and market segments. This should also include guidelines for cooking and preparing camel meat to ensure that it meets consumer expectations.

Table 14.6. Key beef grades, characteristics and suggested use (source: Purdue University Animal Sciences – Meat Quality and Safety, n.d.).

	Grade	Characteristics	Suggested use
1	Prime	Has abundant marbling and is generally sold in restaurants and hotels	Prime roasts and steaks are excellent for roasting, broiling and grilling (dry heat methods)
2	Choice	Has less marbling than prime grades, but is still high quality	May be cooked with dry heat. Be careful not to overcook roasts from rump, round, and blade chuck. A meat thermometer can be helpful in cooking to a safe temperature
3	Select	Leaner than the higher grades. Fairly tender but might lack some of the juiciness and flavour of higher grades	Only the loin, ribs and sirloin should be cooked with dry heat Other cuts should be marinated before cooking or cooked with moisture
4	Standard	Has no marbling. Will lack the juiciness and flavour of higher grades	May be sold as ungraded or 'store brand' meat
5	Commercial	May have marbling, but comes from a more mature animal and will lack tenderness	May be sold as ungraded or 'store brand' meat
6	Utility, Cutter, Canner	Meat from mature animals that lacks marbling	Usually only sold as ground beef or processed meat

14.8 Opportunities

14.8.1 Data collection and dissemination

Currently there is a lack of reliable data on the camel population, camel-meat production and potential opportunities available in the sector. These data are important, especially to potential investors in the camel-meat sector. Potential investors need access to accurate data to make investment decisions. Even the data available through FAO statistics (FAOstat) are not complete; some data points are missing. For example, the data on camel stocks and slaughter are available only up to 2009. Some data points are only estimates and in some cases the data are based on FAOstat's own estimates. There are opportunities for camel-producing countries to set up their own databases or work jointly with other countries to set up regional bodies to collect and disseminate accurate data. For the camel industry to attract investors and develop, the availability of accurate data is of supreme importance. Australia is one country that is trying to achieve this through the Central Australian Camel Industry Association (CACIA) and Camels Australia Export (http://www.camelsaust.com.au/, accessed 17 September 2012), which is the registered business name of the CACIA.

14.8.2 Game meat consumers

The game meat market is expanding worldwide. Research done in Australia (Warfield and Tume, 2000) found that consumers of game meats are likely to consume camel meat as well, and the biggest users of game meats are 5-star hotels and premium dining restaurants. There are therefore opportunities to promote camel meat by targeting game meat consumers, 5-star hotels and premium dining restaurants.

14.8.3 Lack of consistent good-quality product supply

In a survey of the food-services sector in Australia, McKinna (1999) found that all respondents rated quality, consistency, tenderness and taste/flavour as very important factors influencing the design of menus. This is very important for the camel-meat industry because the above factors need to be met if camel meat is to be included in the menu. The camel industry therefore needs to develop and implement strict quality-assurance programmes to guarantee a consistent product that meets consumer expectations of tenderness, taste, leanness, etc. Without addressing the issue of consistency and quality very little can be achieved by the camel-meat industry.

14.8.4 Camel-meat brand development

Branded meats are becoming increasingly popular within the food-service sector. Diners (especially in the affluent societies/ high-value markets) and meat shoppers for home cooking are increasingly interested in the origin of the food and the production system. There is an opportunity here for camel meat. The camel-meat industry should consider developing brands if quality-assurance systems and consistent tenderness can be delivered.

14.8.5 Adoption of extended shelf-life technologies

Generally camel meat products are slow moving in the chain from production to consumption. To maintain the freshness of the product over a longer period of time it is necessary for the camel meat industry to adopt extended shelf-life technologies. These technologies include vacuum packaging and modified atmosphere packaging (MAP).

14.9 Threats

14.9.1 Competition with other meats

Camel meat has to compete with other meats, especially beef and game meats. It is therefore important for the industry to position camel meat as a meat of choice. This could be done by, among other things, highlighting important attributes such as leanness, low cholesterol and vitamin and mineral richness.

14.9.2 Traceability requirements in the export market

Several livestock- and meat-related crises in recent years have given rise to increased worldwide consumer concern over meat safety and an increased desire for information about the meat products they purchase. During the past several years, a series of food safety and animal disease crises has occurred in the European Union (EU), including dioxin contamination of livestock feed, the announcement of the possible link between Bovine Spongiform Encephalopathy (BSE) and new-variant Creutzfeldt–Jakob disease (nvCJD), and outbreaks of foot-and-mouth disease and classical swine fever. Consumers in the export markets, especially in the EU, have lost confidence in the safety of meat products (especially beef) and in the ability of regulatory agencies to protect the food supply. The EU now leads most other countries in the development and mandatory implementation of traceability protocols for livestock and meat products.

Traceability is the comprehensive concept of tracking the movement of identifiable products along the marketing chain. Meat traceability is the ability to follow products forward from their source animal (i.e. birth or ancestry), through growth and feeding, slaughter, processing and distribution, to the point of sale or consumption (or backward from the consumer to the origin of the animal). Traceability can be used to convey information about a product, such as what it contains, how it was produced and every place to which it has been. For camels and camel meat at the moment, no total trace-back system is possible. It will be challenging to setup a traceability system for camels and camel meat because of the structure of its supply chain. It is important, however, to have the traceability system developed because without it the camel meat industry will not be able to export camel meat to EU and other high-value markets.

14.10 Discussion and Recommendations

One of the weaknesses identified in the SWOT analysis is the lack of awareness of consumers regarding camel meat. Consumers are mostly not fully aware that camel meat is a healthy meat product. Creating consumer awareness is therefore very important, because consumers won't buy a product if they have never heard of it. There are many different ways to communicate the message and create consumer's awareness that camel meat is a product that meets their needs. This can be done through commercial advertising in the newspapers, radio and television and through the Internet. Lessons could also be drawn from the approaches taken in promoting other meat products such as kangaroo in Australia and game meat in South Africa and Kenya. Camel meat needs to be promoted accordingly to raise consumer awareness by highlighting its unique attributes that distinguishes it from other animal meats. Consumers should be aware that camel meat is much healthier than many other animal meats, due to its low-fat, low-cholesterol and high-protein attributes.

Consumers were found to relate camel meat with the animal itself and, as a result, they tend to dislike the meat just for that reason. It is therefore necessary to avoid using labels that show the picture of the camel itself. Because camel meat is known to be a healthy product, a picture of a healthy person could be a possible alternative.

Camel meat has been described by consumers as being chewy and tough even though consumers agree that it is not different from beef in terms of flavour. Research has established, however, that the older the camel the tougher the meat gets. Age is therefore an important factor in determining camel-meat quality and composition. To obtain high-quality low-fat meat that is rich in minerals, young camels of less than three years of age should be slaughtered.

The export market is always very competitive and, to be successful, the export commodity must be of high quality. As a new product, it will be difficult for camel meat to meet consumer demands in high-value export markets. To meet consumer demands in such markets the camel-meat industry needs to access high-tech meat processing and packaging equipment necessary to produce the needed cuts.

There is a lack of halal certification by some meat slaughtering facilities outside the Muslim world. Camel meat is popular in the Muslim world where people traditionally eat Halal meat. The Muslim world is a very important consumer base for camel meat because the Muslim population, which is currently 1.619 billion, is expected to reach 2.19 billion by 2030. It is therefore imperative to make sure that camel meat produced outside the Muslim world is slaughtered in Halal-compliant facilities so that it can be easily exported to the Muslim world market.

Information on the characteristics of camel meat cuts in relation to eating quality, and the suitability of these cuts for different recipes, is lacking. This information is very important to consumers because other competing meat products such as chicken and beef (Table 14.6) have well-documented information on various grade cuts and the suggested uses. The camel-meat industry needs to do the same or even better than the other competing meat products.

There are mainly two threats to the success of camel meat identified in the SWOT analysis. There is the competition from other meats such as beef, chicken and game meats. The industry needs to focus on promoting the strengths of the camel meat as a healthy product and therefore a meat of choice. This could be done by highlighting important attributes such as leanness, low cholesterol, and a good source of vitamins and minerals. The second threat identified is the traceability requirement in the export markets, which at the moment is difficult to develop for camel meat because of the structure of the supply chain. Efforts are needed to develop some form of a traceability system in order to take advantage of the export market opportunities.

Camel meat has a number of opportunities that need to be exploited in order to counteract threats and improve on weaknesses highlighted above. The opportunities include:

- Collection of accurate data on camel production, camel-meat production and potential opportunities available in the industry. The availability of such data would help to attract investors for the industry.
- Targeting game meat consumers for promoting camel meat because research has shown that game meat consumers have the highest likelihood to consume camel meat.
- Improving the quality, consistency and tenderness of the camel meat. A survey of the food-services sector in Australia indicated that these are important factors that influence the design of menus in hotels and restaurants.
- The need to develop the camel meat brand because branded meats are becoming increasingly popular within the food-service sector. For example, camel meat could be branded as an organically produced meat product.
- The adoption of extended-shelf-life technologies such as vacuum packaging and MAP, which are necessary to maintain the freshness of the meat products for longer periods of time.

14.11 Conclusion

Camel meat has the potential to be the meat of the future especially now that consumers are becoming increasingly concerned with the environmental footprint of the commodities they consume. Camel meat production is, in general, ecologically harmless. Ecologists (Köhler-Rollefson, 1994) stress that camel grazing has very little, if any, damaging effect on environment because its foraging habits are optimally suited to areas with a low carrying capacity.

The objective of this chapter was to investigate the potential of camel meat as a substitute for beef and other meats. The investigation was carried out by conducting a SWOT analysis to identify the strengths, weaknesses, opportunities and threats of camel meat over other meats. The most important strengths of camel meat include the fact that it is healthier than many other animal meats because it is low in fat and cholesterol. It also has a potential to benefit from the global Muslim market, which is expected to reach 2.2 billion consumers by 2030.

With regard to weaknesses, it has been observed that generally there is a lack of consumer awareness with respect to the benefits of camel meat. Consumers also tend to relate camel meat to the animal itself. Most of the observed weaknesses and threats are minor and could be very easily addressed through public awareness and marketing campaigns.

Note

[1]Pimentel is a Professor of Ecology at Cornell University's College of Agriculture and Life Sciences.

References

Al-Ani, F.K. (2004) Use and production of camels. In: Al-Ani, F.K. (ed.) *Camel Management and Diseases* 1st edn. Al-Sharq Printing Press, Qatar, pp. 91–114.

Al-Khamis, K. and Young, D.L. (2006) An analytical study of production cost for camel herds in Riyadh area, Saudi Arabia. *Journal of King Saud University Agriculture Science* 18, 73–87.

Bruinsma, J. (2003) *World Agriculture: Towards 2015/2030: an FAO Perspective*. Earth Scan, London.

Chapagain, A.K. and Hoekstra, A.Y. (2004) *Water Footprints of Nations*, Volume 1: Main Report, Volume 2: Appendices. Earth Scan, London.

Dawood, A. and Alkanhal, M.A. (1995) Nutrient composition of Najdi- Camel Meat. *Meat Science* 39, 71–78.

Elam, T.E. (2006) Projections of global meat production through 2050. Farmcon, Center for Global Food Issues. Available at: http://www.cgfi.org/2006/08/dr-thomas-e-elam-projections-of-global-meat-production-2050/ (accessed 17 September 2012).

Elgasim, E.A., El-Hag, G.A. and Elnawawi, F.A. (1987) Quality Attributes of Camel Meat. 2nd Congress report. The Scientific Council, King Fasil University, Alhash, Saudi Arabia.

FAO (2006) Livestock's Long Shadow: Environmental Issues and Options. Available at: http://www.fao.org/docrep/010/a0701e/a0701e00.htm (accessed 17 September 2012).

FAOstat (2011) FAOstat database. Available at: http://faostat.fao.org/site/567/default.aspx# ancor (accessed 17 September 2012).

Gauthier-Pilters, H. and Dagg, A.I. (1981) *The Camel. Its Evolution, Ecology, Behaviour, and Relationship to Man*. University of Chicago Press, Chicago, Illinois.

Ibrahim, G.A. and Nour, I.A. (2010) Physical and chemical properties of camel meat burgers. *Journal of Camelid Science* 3, 39–43.

Kadim, I.T., Mahgoub, O., Al-Marzooqi, W., Al-Zadjali, S., Annamalai, K. and Mansour, M.H. (2006) Effects of age on composition and quality of muscle *Longissimus thoracis* of the Omani Arabian camel (*Camelus dromedarius*). *Meat Science* 80, 555–569.

Kadim, I.T., Mahgoub, O. and Purchas, R.W. (2008) A review of the growth, and of the carcass and meat quality characteristics of the one-humped camel (*Camelus dromedarius*). *Meat Science* 73, 619–625

Khatami, K. (1970) Camel meat: A new promising approach to the solution of meat and protein in the arid and semi-arid countries of the world. Ministry of Agriculture, Tehran, Iran.

Knoess, K.H. (1977) The camel as a meat and milk camel. *World Animal Review* 22, 39–44.

Köhler-Rollefson, I. (1994) Camel pastoralism. an indigenous arid land exploitation strategy. *Journal of Camel Practice and Research* 1, 1–6.

McKinna, D. (1999) *Marketing of New Animal Products*. Rural Industries Research and Development Corporation, Canberra, Australia.

Mekonnen, M.M. and Hoekstra, A.Y. (2010) The green, blue and grey water footprint of farm animals and animal products. Volume 1: Main Report. Value of Water Research Report Series No. 48. Institute for Water Education, UNESCO.

Pew Research Center (2011) Pew Research Center's Forum on Religion and Public Life, The Future of the Global Muslim Population. Washington, DC. Available at: http://features.pewforum.org/FutureGlobalMuslimPopulation-WebPDF.pdf (accessed 17 September 2012).

Pimentel, D. and Pimentel, M. (2003) Sustainability of meat-based and plant-based diets and the environment. *American Journal of Clinical Nutrition September* 78, (Suppl) 660S–663S.

Purdue University Animal Sciences – Meat Quality and Safety (n.d.) Available at: http://ag.ansc.purdue.edu/meat_quality/ext_ed_meat_grading.html (accessed 17 September 2012).

Shalah, M.R. (1983) The role of camels in overcoming world meat shortage. *Egyptian Journal of Veterinary Science* 20, 101–110.

Stiles, D. (1983) Camel pastoralism and desertification in northern Kenya. *Desertification Control* 8, 2–15.

Tandon, S.N., Bissa, U.K. and Khanna, N.D. (1988) Camel meat: Present status and future prospects. *Annals of Arid Zone* 27, 23–28.

Warfield B. and Tume, L. (2000) Marketing analysis and plan for the camel industry. A report for the Rural Industries Research and Development Corporation. February 2000 RIRDC Publication No 00/9 RIRDC Project No DAQ-218A. Available at: http://www.austcamel.com.au/library/market.pdf (accessed 17 September 2012).

Zeng, B. and McGregor, M. (n.d.) Review of commercial options for management of feral camels. In: *Managing the Impacts of Feral Camels in Australia: a New Way of Doing Business.* Desert CRC, pp. 221–282. Available at: http://www.desertknowledgecrc.com.au/resource/DKCRC-Report-47-Ch09_Zeng-and-McGregor_Review-of-commercial-options-for-management-of-feral-camels.pdf (accessed 17 September 2012).

Index

Note: **bold** page numbers indicate figures; *italic* page numbers indicate tables.